AF477134

Application of Wood Composites

Application of Wood Composites

Editors

Ľuboš Krišťák
Roman Réh

MDPI • Basel • Beijing • Wuhan • Barcelona • Belgrade • Manchester • Tokyo • Cluj • Tianjin

Editors
Ľuboš Krišták   Roman Réh
Department of Physics, Electrical Department of Wood Technology
Engineering and Applied   Technical University in Zvolen
Mechanics   Zvolen
Technical University in Zvolen   Slovakia
Zvolen
Slovakia

Editorial Office
MDPI
St. Alban-Anlage 66
4052 Basel, Switzerland

This is a reprint of articles from the Special Issue published online in the open access journal *Applied Sciences* (ISSN 2076-3417) (available at: www.mdpi.com/journal/applsci/special_issues/ Application_Wood_Composites).

For citation purposes, cite each article independently as indicated on the article page online and as indicated below:

LastName, A.A.; LastName, B.B.; LastName, C.C. Article Title. *Journal Name* **Year**, *Volume Number*, Page Range.

ISBN 978-3-0365-1772-8 (Hbk)
ISBN 978-3-0365-1771-1 (PDF)

Contents

Editorial

Application of Wood Composites

Ľuboš Krišťák * and Roman Réh

Faculty of Wood Sciences and Technology, Technical University in Zvolen, T.G. Masaryka 24, 96001 Zvolen, Slovakia; roman.reh@tuzvo.sk
* Correspondence: kristak@tuzvo.sk

Citation: Krišťák, Ľ.; Réh, R. Application of Wood Composites. *Appl. Sci.* **2021**, *11*, 3479. https://doi.org/10.3390/app11083479

Received: 3 April 2021
Accepted: 12 April 2021
Published: 13 April 2021

Publisher's Note: MDPI stays neutral with regard to jurisdictional claims in published maps and institutional affiliations.

Wood composites are the key material for a number of structural and non-structural applications for interior and exterior purposes, such as furniture, construction, floorings, windows and doors, etc. They can be successfully produced with predetermined specific properties matching the required end uses. Wood composites ranging from fiberboard to laminated beams need to be better known, and more attention must be paid to their research. Laboratories worldwide do innovative research, and new challenges, approaches, and ideas are continuously increasing, allowing us to mirror an exciting and interesting research future [1–4].

The time when this Special Issue had been continuously compiled was mostly a stressful period for all of us, marked with the widespread outbreak of the COVID-19 pandemic, but we hope all *Applied Sciences* readers are healthy and well. We do see a glimmer of hope to return to normalcy on the horizon.

This Special Issue "Application of Wood Composites" addressed various aspects of these important wood materials' use, e.g., mechanical processing of wood composites including their cutting, milling, or sanding incorporating the current analysis of wood dust or grain size measurements and composition of particles [5–9], scientific views on the influence of various adhesives in the creation process of wood composites and the analysis of their behavior in contact with various wood elements under different conditions [10–14], the analysis of input raw materials forming wood composites, including various wood species, but also non-wood lignocellulosic raw materials and, last but not least, the analysis of bark, which in the recent years has become an important and promising raw material involved in the construction of wood composites [15–19]; the study of the development of the sliding table saw also suitably complements this Special Issue [20].

If we take a closer look at the main topic of this publication, it is clear that wood composite materials are engineered and produced with tailored physical and mechanical properties appropriate for a wide variety of applications, known or not discovered yet. Additionally, indeed, the utilization of wood composites in various areas has increased recently due to their outstanding properties, allowing them to successfully and sustainably replace solid wood and other conventional materials. We have tried to respond to this newly created situation with this publication.

One of the publications was aimed at providing the reader with new information on the recent practices in laser cutting of wood and wood composites, and determining the optimal set of cutting parameters by a method of a low-power CO_2 laser in particular. Three factors were investigated, namely the effect of the laser power, cutting speed, and number of annual rings [7]. Other new insights are emerging in another type of wood processing: Sanding. The research results indicate that the factors determining sanding efficiency are the type of wood, and, secondly, the grit size of sanding belts. Maximum sanding efficiency for the softwood surface of wood composites ranged from 1 to 2 min, while for the hardwood species composites surface, it ranged from 2 to 4.5 min at the start of sanding and then decreased [5]. The accompanying phenomenon of sanding is wood dust that poses a serious threat to the health of workers and employees as well as a significant fire and explosion hazard; it accelerates the wear of machines, worsens the quality of processing, and requires large financial outlays for its removal [21]. Therefore, the aim was to investigate the extent to which the grit size of sandpaper influences the size of the wood dust particles and the proportion of the finest particles which may constitute the respirable fraction [10]. The grain size measurements of wood dust samples from selected tropical wood species were investigated, as well as the conditions under which wood composites are processed, e.g., impact of thermal modification of wood composites surface saturated by steam and its influence on the particle size distribution of the sawing and milling process [8,9].

It is remarkable how many opportunities will arise when using modern methods and high-quality instruments and equipment for the analysis of wood composites. With a compact Time-of-Flight Secondary Ion Mass analyzer, integrated in a multifunctional focused-ion beam scanning-electron-microscope, it was possible to show that the Ga^+ ion source could be detected and visualized in 3D ion molecular clusters specific to polymeric 4,4′-diphenyl methane diisocyanate (pMDI) adhesive and wood [10]. The bonding of wood with assembly adhesives is crucial for manufacturing wood composites. Various adhesives in the context of their application to various types of wood must be analyzed for the formation of quality wood composites, e.g., polyvinyl acetate (PVAc), lignin-based formaldehyde-free adhesives (lignosulfonates), or new and improved adhesive mixtures of urea-formaldehyde (UF) resin, e.g., with soy flour [11,12]. It has been shown that the properties of wood composites can be improved by using these new and less used combinations of adhesives. The fabricated wood composites achieved close-to-zero formaldehyde content of 1.1 mg/100 g, i.e., the super E0 emission grade ($\leq$1.5 mg/100 g), which allowed their classification as eco-friendly, low-emission wood-based composites [13,14,22,23].

Without a detailed analysis of input raw materials, it is not possible to form meaningful wood composites. The analysis of input wood raw materials must be carried out no matter what kind of wood composites are produced. The surface roughness constitutes one of the most critical properties of wood veneers for their extended utilization, affecting the bonding ability of the veneers with one another in the manufacturing of wood composites, the finishing, coating and preservation processes, and the appearance and texture of the material surface. The surface roughness was examined by applying a stylus tracing method on typical wood structure areas of each wood species, as well as around the areas of wood defects (knots, decay, annual rings irregularities, etc.), to compare them and assess the impact of the defects on the surface quality of veneers [18]. The production possibilities of oriented strand boards (OSB) in the laboratory from a mixture of softwood species and hardwood species were tested [16]. In our times, it is becoming increasingly important to use secondary products in wood processing or to use less known and less used lignocellulosic materials as sustainable alternatives of wood [24–26], such as the bark of various woody plants which has not been fully utilized yet, or utilization of walnut and hazelnut shells which are agricultural by-products, available in high quantities during the harvest season. Very interesting and promising research results were achieved [17,19]. As invasive alien species are one of the main causes of the loss of biodiversity, and thus of changes in ecosystem services, it is important to find the best possible solution to their usability. Research showed that it will be possible to deal with such a problem as well and the production of wood plastic composites is a viable solution [15].

We would like to thank our Section Managing Editor Dr. Kyle Ke for his professional attitude and assistance with publishing.

It is good that such a book publication was created. The topic "Application of Wood Composites" is still relevant, new possibilities for application of wood composite materials are emerging, and therefore it is understandable that MDPI has already opened access to a new Special Issue "Application of Wood Composites II" within the journal Applied Sciences with the possibility of publishing new high-quality original research articles and reviews on the latest advancements in wood composites materials and their applications.

Funding: This research received no external funding.

Acknowledgments: This publication was supported by the Slovak Research and Development Agency under contract No. APVV-18-0378, APVV-19-0269 and VEGA1/0717/19.

Conflicts of Interest: The authors declare no conflict of interest.

References

1. Pizzi, A.; Papadopoulos, A.N.; Policardi, F. Wood Composites and Their Polymer Binders. *Polymers* **2020**, *12*, 1115. [CrossRef] [PubMed]
2. Papadopoulos, A.N. Advances in Wood Composites. *Polymers* **2020**, *12*, 48. [CrossRef] [PubMed]
3. Papadopoulos, A.N. Advances in Wood Composites II. *Polymers* **2020**, *12*, 1552. [CrossRef] [PubMed]
4. Papadopoulos, A.N. Advances in Wood Composites III. *Polymers* **2021**, *13*, 163. [CrossRef]
5. Sydor, M.; Mirski, R.; Stuper-Szablewska, K.; Rogoziński, T. Efficiency of Machine Sanding of Wood. *Appl. Sci.* **2021**, *11*, 2860. [CrossRef]
6. Pędzik, M.; Stuper-Szablewska, K.; Sydor, M.; Rogoziński, T. Influence of Grit Size and Wood Species on the Granularity of Dust Particles during Sanding. *Appl. Sci.* **2020**, *10*, 8165. [CrossRef]
7. Kubovský, I.; Krišťák, Ľ.; Suja, J.; Gajtanska, M.; Igaz, R.; Ružiak, I.; Réh, R. Optimization of Parameters for the Cutting of Wood-Based Materials by a CO_2 Laser. *Appl. Sci.* **2020**, *10*, 8113. [CrossRef]
8. Vandličková, M.; Marková, I.; Makovická Osvaldová, L.; Gašpercová, S.; Svetlík, J.; Vraniak, J. Tropical Wood Dusts—Granulometry, Morfology and Ignition Temperature. *Appl. Sci.* **2020**, *10*, 7608. [CrossRef]
9. Kminiak, R.; Orlowski, K.A.; Dzurenda, L.; Chuchala, D.; Banski, A. Effect of Thermal Treatment of Birch Wood by Saturated Water Vapor on Granulometric Composition of Chips from Sawing and Milling Processes from the Point of View of Its Processing to Composites. *Appl. Sci.* **2020**, *10*, 7545. [CrossRef]
10. Klímek, P.; Wimmer, R.; Meinlschmidt, P. TOF-SIMS Molecular Imaging and Properties of pMDI-Bonded Particleboards Made from Cup-Plant and Wood. *Appl. Sci.* **2021**, *11*, 1604. [CrossRef]
11. Iždinský, J.; Reinprecht, L.; Sedliačik, J.; Kúdela, J.; Kučerová, V. Bonding of Selected Hardwoods with PVAc Adhesive. *Appl. Sci.* **2021**, *11*, 67. [CrossRef]
12. Ahmed, S.A.; Adamopoulos, S.; Li, J.; Kovacikova, J. Prediction of Mechanical Performance of Acetylated MDF at Different Humid Conditions. *Appl. Sci.* **2020**, *10*, 8712. [CrossRef]
13. Antov, P.; Jivkov, V.; Savov, V.; Simeonova, R.; Yavorov, N. Structural Application of Eco-Friendly Composites from Recycled Wood Fibres Bonded with Magnesium Lignosulfonate. *Appl. Sci.* **2020**, *10*, 7526. [CrossRef]
14. Taghiyari, H.R.; Hosseini, S.B.; Ghahri, S.; Ghofrani, M.; Papadopoulos, A.N. Formaldehyde Emission in Micron-Sized Wollastonite-Treated Plywood Bonded with Soy Flour and Urea-Formaldehyde Resin. *Appl. Sci.* **2020**, *10*, 6709. [CrossRef]
15. Medved, S.; Tomec, D.K.; Balzano, A.; Merela, M. Alien Wood Species as a Resource for Wood-Plastic Composites. *Appl. Sci.* **2021**, *11*, 44. [CrossRef]
16. Lunguleasa, A.; Dumitrascu, A.-E.; Ciobanu, V.-D. Comparative Studies on Two Types of OSB Boards Obtained from Mixed Resinous and Fast-growing Hard Wood. *Appl. Sci.* **2020**, *10*, 6634. [CrossRef]
17. Barbu, M.C.; Sepperer, T.; Tudor, E.M.; Petutschnigg, A. Walnut and Hazelnut Shells: Untapped Industrial Resources and Their Suitability in Lignocellulosic Composites. *Appl. Sci.* **2020**, *10*, 6340. [CrossRef]
18. Kamperidou, V.; Aidinidis, E.; Barboutis, I. Impact of Structural Defects on the Surface Quality of Hardwood Species Sliced Veneers. *Appl. Sci.* **2020**, *10*, 6265. [CrossRef]
19. Jiang, W.; Adamopoulos, S.; Hosseinpourpia, R.; Žigon, J.; Petrič, M.; Šernek, M.; Medved, S. Utilization of Partially Liquefied Bark for Production of Particleboards. *Appl. Sci.* **2020**, *10*, 5253. [CrossRef]
20. Orlowski, K.A.; Dudek, P.; Chuchala, D.; Blacharski, W.; Przybylinski, T. The Design Development of the Sliding Table Saw Towards Improving Its Dynamic Properties. *Appl. Sci.* **2020**, *10*, 7386. [CrossRef]
21. Papadopoulos, A.N.; Taghiyari, H.R. Innovative Wood Surface Treatments Based on Nanotechnology. *Coatings* **2019**, *9*, 866. [CrossRef]
22. Antov, P.; Savov, V.; Krišťák, Ľ.; Réh, R.; Mantanis, G.I. Eco-Friendly, High-Density Fiberboards Bonded with Urea-Formaldehyde and Ammonium Lignosulfonate. *Polymers* **2021**, *13*, 220. [CrossRef] [PubMed]
23. Antov, P.; Mantanis, G.I.; Savov, V. Development of Wood Composites from Recycled Fibres Bonded with Magnesium Lignosulfonate. *Forests* **2020**, *11*, 613. [CrossRef]
24. Bekhta, P.; Sedliačik, J.; Kačík, F.; Noshchenko, G.; Kleinová, A. Lignocellulosic waste fibers and their application as a component of urea-formaldehyde adhesive composition in the manufacture of plywood. *Eur. J. Wood Wood Prod.* **2019**, *77*, 495–508. [CrossRef]
25. Antov, P.; Krišťák, L.; Réh, R.; Savov, V.; Papadopoulos, A.N. Eco-Friendly Fiberboard Panels from Recycled Fibers Bonded with Calcium Lignosulfonate. *Polymers* **2021**, *13*, 639. [CrossRef]
26. Ihnat, V.; Lubke, H. Size reduction downcycling of waste wood. Review. *Wood Res.* **2020**, *65*, 205–220. [CrossRef]

applied sciences

Article

Efficiency of Machine Sanding of Wood

Maciej Sydor [1,*] , **Radosław Mirski** [2], **Kinga Stuper-Szablewska** [3] and **Tomasz Rogoziński** [4]

[1] Department of Woodworking Machines and Fundamentals of Machine Design, Faculty of Forestry and Wood Technology, Poznań University of Life Sciences, ul. Wojska Polskiego 38/42, 60-637 Poznań, Poland
[2] Department of Wood-Based Materials, Faculty of Forestry and Wood Technology, Poznań University of Life Sciences, ul. Wojska Polskiego 38/42, 60-637 Poznań, Poland; radoslaw.mirski@up.poznan.pl
[3] Department of Chemistry, Faculty of Forestry and Wood Technology, Poznań University of Life Sciences, ul. Wojska Polskiego 38/42, 60-637 Poznań, Poland; kinga.stuper@up.poznan.pl
[4] Department of Furniture Design, Faculty of Forestry and Wood Technology, Poznań University of Life Sciences, ul. Wojska Polskiego 38/42, 60-637 Poznań, Poland; tomasz.rogozinski@up.poznan.pl
* Correspondence: maciej.sydor@up.poznan.pl

Abstract: We hypothesized that the type of wood, in combination with the grit size of sandpapers, would affect sanding efficiency. Fixed factors were used in the experiment (a belt sander with pressure $p = 3828$ Pa, and a belt speed of $v_s = 14.5$ m/s) as well as variable factors (three sand belts (P60, P120, P180), six hardwood species (beech, oak, ash, hornbeam, alder, walnut) and three softwood species (pine, spruce, larch)). The masses of the test samples were measured until they were completely sanded. The sanding efficiency of hardwood species is less variable than for softwood species. Maximum sanding efficiency for the softwood ranged from 1 to 2 min, while for the hardwood species, it ranged from 2 to 4.5 min at the start of sanding and then decreased. The average time for complete sanding of the softwood samples was: 87 s (P60), 150 s (P120), and 188 s (P180). For hardwood, these times were 2.4, 1.5, and 1.8 times longer. The results indicate that the factors determining sanding efficiency are the type of wood, and, secondly, the grit size of sanding belts. In the first phase of blunting with the sanding belts, the sanding processes of hardwood and softwood are significantly different. In the second phase of blunting, sanding belts with higher grit numbers (P120 and P180) behaved similarly while sanding hardwood and softwood.

Keywords: softwood; hardwood; sanding; belt sander; sandpaper; abrasion; beech; oak; ash; hornbeam; alder; walnut; pine; spruce; larch

Citation: Sydor, M.; Mirski, R.; Stuper-Szablewska, K.; Rogoziński, T. Efficiency of Machine Sanding of Wood. *Appl. Sci.* **2021**, *11*, 2860. https://doi.org/10.3390/app11062860

Academic Editor: Roman Réh

Received: 20 February 2021
Accepted: 19 March 2021
Published: 23 March 2021

Publisher's Note: MDPI stays neutral with regard to jurisdictional claims in published maps and institutional affiliations.

1. Introduction

Sanding is widely used in the furniture industry. The objectives of sanding may be to achieve the required surface smoothness to be painted, to achieve the required roughness necessary for gluing on the surface, and effective and controlled material removal to obtain the desired shape or dimensional accuracy of the workpiece. When planning a technological sanding process, several key aspects should be considered. Providing appropriate working conditions by reducing the exposure of workers to respirable wood dust in the air is the first important aspect [1–6]. Another group of problems are the economic issues of the used technology; in other words, obtaining high quantitative efficiency and productivity and the expected surface quality and/or accuracy of the shape for the workpieces. These two groups of problems are solved by properly selecting the production equipment, parameters of the abrasive tools and parameters of the sanding process [7,8].

Issues resulting from the specific effect of abrasive grains on wood have been studied both from the point of view of machine tool design [9], abrasive tools (type of sandpaper and its grit size) and the technological parameters used (in particular, the contact pressure and speed of the abrasive belt, the size of the surface to be sanded and the orientation of the wood fibers during sanding [10–16]). The influence of the properties of various species of wood on the effects of sanding were also studied [17–21]. One of the most important measures of the efficiency of the sanding process is the mass of material sanded per unit of time. Sanding efficiency decreases during the process due to the blunting of the abrasive belt. Ockajova [21], analyzing the literature, identified three phases of sanding belt blunting: initial sharpness, work sharpness and sanding belt blunting. During the initial phase, there is a very large reduction in efficiency during sanding. The limit between the first and second phase is a stabilization of this reduction in efficiency, at a level of about 45–50% in relation to the initial sanding efficiency. In the second phase, where the wear of the abrasive grit dominates, a further, somewhat slower reduction in sanding efficiency is observed (by about 10–20% in relation to the initial efficiency). Characteristic for the third phase is a rapid decrease in sanding efficiency.

Wieloch and Siklienka [22] investigated the effect of long time sanding on the variation in efficiency for beech wood. The analyzed process lasted 480 min. P40, P80, and P120 abrasive belts were used, and different contact pressures were applied: $p = 1.0, 1.5, 1.85$ and $2.0\,\text{N/cm}^2$ (10,000, 15,000, 18,500 and 20,000 Pa). At a pressure of 10,000 Pa, a rectilinear decrease in sanding efficiency was observed. At a pressure of 18,500 Pa, however, the decrease was "bi-rectilinear": first, the efficiency decreased intensively and, after a certain time, the decrease in sanding efficiency slowed markedly. At higher contact pressures, the sanding performance decreased more rapidly. In a comparative study on sanding oak and beech wood, Ockajova et al. [21] found that the contact pressure that ensures long-term operation of the abrasive belt depends on the direction of sanding and the wood species (the pressure on beech wood may be higher). The species of wood in these studies had a greater influence on sanding belt efficiency than the direction of cutting. The examples described here concern studies using manual sanding belt machines. The operating conditions of these machine tools are relatively high pressure (up to 20,000 Pa) and low belt speed ($v_s < 10\,\text{m/s}$). Industrial belt sanding machines operate at higher belt speeds ($v_s > 10\,\text{m/s}$) and lower pressure ($p < 10,000\,\text{Pa}$). An example of the description of such research is the work of Saloni et al. [23], where a comparative study of industrial sanding of pine and maple wood is described. As a result of this study, a positive effect of the contact pressure and sandpaper belt speed on sanding efficiency was found, as well as a higher sanding efficiency of pine wood.

However, there is a lack of comparative studies on the influence of wood type and tool grit size on the variability of efficiency during sanding. Taking this into account, it was decided to verify the hypothesis that the type of wood, in combination with the grit size of sandpapers, affects the sanding efficiency during sanding with parameters typical for industrial applications ($p < 10,000\,\text{Pa}$ and $v_s > 10\,\text{m/s}$).

2. Materials and Methods

Wood from six hardwood species (beech, oak, ash, hornbeam, alder, walnut) and three softwood species (pine, spruce, larch) was tested. The wood material for making test samples was dried in an industrial dryer to a moisture content of 12% and stored in a freezer to preserve its physical properties. Then, the wood specimens with dimensions of $120 \times 55 \times 20$ (length $\times$ width $\times$ height in millimeters) were obtained from it. Each specimen was measured with a caliper with an accuracy of ± 0.2 mm and weighed using a WPS 510/C/2 balance (Radwag, Radom, Poland) with an accuracy of ± 0.01 g. These measurements were used to calculate the density of the wood and to determine its initial mass. The calculated volumetric mass densities of the wood materials tested and the numbers of samples in the sample sets for each wood species tested are given in Table 1.

Table 1. Characteristics of wood specimens used in sandability tests.

Type of Wood	Density	Number of Samples in Set		
	kg/m^{-3}	Sandpaper P60	Sandpaper P120	Sandpaper P180
Beech (*Fagus sylvatica* L.)	686.6	5	5	5
Oak (*Quercus robur* L)	686.4	5	5	5
Ash (*Fraxinus excelsior* L.)	621.3	4	4	4
Hornbeam (*Carpinus betulus* L.)	753.6	3	3	3
Alder (*Alnus glutinosa* (L.) Gaertn.)	446.3	3	3	3
Walnut (*Juglans nigra* L.)	641.0	3	4	3
Pine (*Pinus sylvestris* L.)	545.7	3	4	5
Spruce (*Picea abies* (L.) H.Karst.)	453.2	3	3	3
Larch (*Larix decidua* Mill.)	420.1	3	4	4

Before sanding, the samples were glued with PVAc glue to raw chipboard spacers with dimensions of $120 \times 55 \times 16$ (length × width × height in millimeters) (the purpose of this procedure was to enable complete sanding of the tested wood). The wood samples were positioned so that they were sanded along the wood fibers. The form of the test samples is shown in Figure 1.

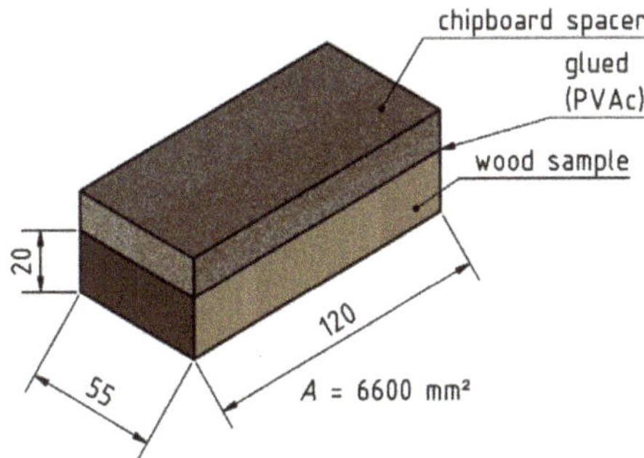

Figure 1. Form of research samples.

The samples shown in Figure 1 were conditioned for another 3 months to equalize their moisture content in the whole volume.

Abrasive belts type EKA 2000 F 2000×75 (length × width in millimeters) (manufactured by Ekamant, Poznań, Poland) with three different grit sizes (P60, P120, P180) were used for sanding; their specifications are given in Table 2.

Table 2. Sand belts specifications.

Type	EKA 1000 F (Ekamant)		
ISO/FEPA Grit designation	P 60 (medium)	P 120 (fine)	P 180 (very fine)
abrasive material	aluminum oxide, av. particle size 269 μm	aluminum oxide, av. particle size 120 μm	aluminum oxide, av. particle size 82 μm
Backing		F weight paper	
adhesive		resin	

A small industrial belt sander, Maktek S (Cormak, Siedlce, Poland), with a horizontal abrasive belt arrangement (Figure 2) was used. The gravitational clamping assembly allowed a constant pressure to be exerted by the abrasive belt on the samples (p = 3828 Pa). The speed of the sanding belt was constant and was: v_s = 14.5 m/s.

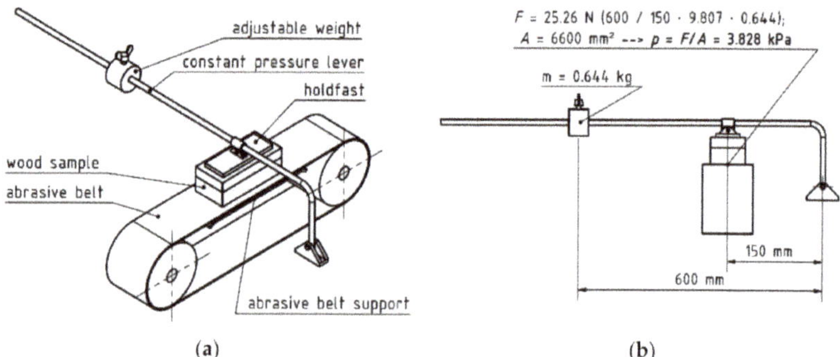

Figure 2. Construction (**a**) and kinematic diagram (**b**) of a laboratory sander.

A separate sanding belt was used for each wood species. Each sample was sanded along the wood fibers in 30 s intervals and after each interval, the sample was weighed using a WPS 510/C/2 laboratory scale (Radwag, Radom, Poland). These steps were repeated many times until the entire wood sample was sanded off from the chipboard spacer. In this way, the following time series were obtained for each tested wood species: time-varying sanding efficiency (1), time-varying wood loss (2), and time to sand each sample, which allowed the calculation of the average sanding time for each series of samples (3).

The sanding efficiency for the intervals was calculated:

$$s_e = \frac{(m_1 - m_2)}{A} / t_c \left(\frac{\mathrm{g/cm^2}}{\mathrm{min.}} \right) \tag{1}$$

where: s_e—sanding efficiency, m_1—wood sample mass at the beginning of each sanding interval (g), m_2—wood sample mass at the end of each sanding interval (g), A—sample sanded area (cm²), and t_c—sanding cycle time (min.).

The wood loss was calculated relative to the initial sample weight:

$$w_l = \frac{(m_2 - m_1)}{m_0} \; (\%) \tag{2}$$

where: w_l—weight loss, m_0—starting weight of the wood sample (g).

The mean value from the sample set measured every 30 s was taken as the s_e result, and the mean value of the mass loss measured every 30 s was taken as the w_l result until the last sample in the set was ground.

For the comparison of hardwood and softwood, the mean values of s_e and w_l were additionally calculated for all six hardwoods and three softwood species. The parameter w_l was subjected to mathematical analysis. The determination of functional equations and their similarity analysis was based on multiplicity theory for comparing functions and for narrowing functions [24].

Average time for total sanding of wood in a serie:

$$t_{AM} = \frac{1}{n} \sum_{i=1}^{n} t_i = \frac{t_1 + t_2 + \cdots + t_n}{n} \; (s) \tag{3}$$

where: t_{AM}—meantime for sanding a sample from the set, n—number of samples in a set, $t_1 + t_2 + \cdots + t_n$—sanding times of subsequent samples in a set.

3. Results

The results in the form of sanding efficiency time series are presented separately for hardwoods and softwood species for all three grades of sandpaper (Figures 3–5).

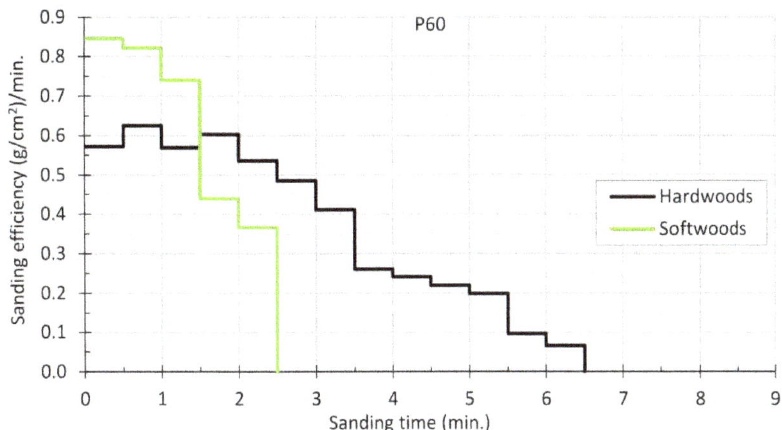

Figure 3. Mean sanding efficiencies of belts with grade P60.

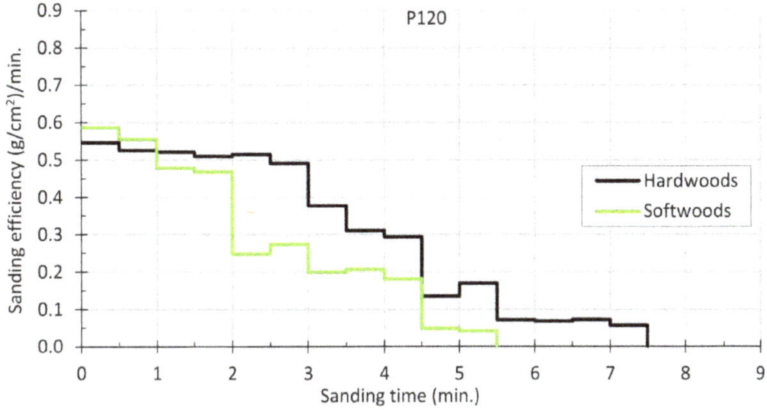

Figure 4. Mean sanding efficiencies of belts with grade P120.

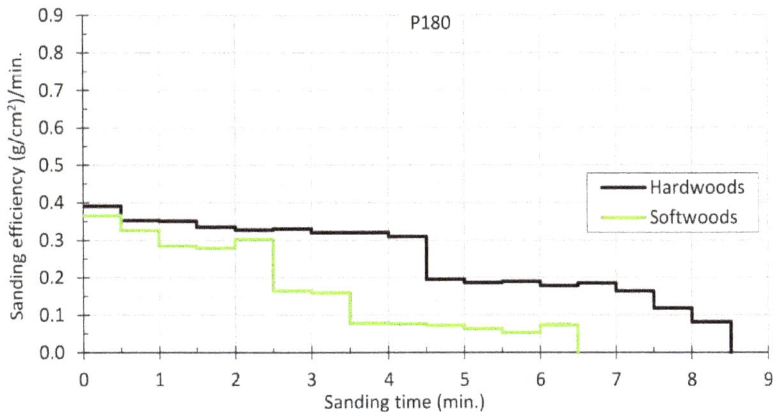

Figure 5. Mean sanding efficiencies of belts with grade P180.

Maximum sanding efficiency for hardwood species lasts from about 0.5 to 3 min, while for softwood species it lasts from 0.5 to 2 min. It is therefore apparent from Figures 3–5 that the first phase of sanding belt blunting ends quite early. In fact, for all wood species and all abrasive belt grit sizes, it is about 2–3 min after the start of machining when the sanding process moves into the second phase. In the case of softwood sanded with a P60 belt, the sample material finishes just after reaching the beginning of the second blunting phase. And in the case of pine wood sanded with the belt with the coarsest coating, it is not possible to enter the third phase of blunting of the coated abrasive before the sample wood is completely worn out.

In such a situation of the rapid progress of machine sanding at speeds higher than in the case of tests with manual sanders, it was decided to interpret the results of the experiment also by analyzing the wood removal rate for the tested wood species during the sanding. In this way, time series were obtained showing the percentage material loss during sanding (relative to a mean initial sample weight). The means were calculated separately for the sets of samples of each tested wood species. Those time series are represented by three consecutive Figures 6–8 (they show only a mean wood loss in each set of samples; without including the possible loss of a chipboard spacer).

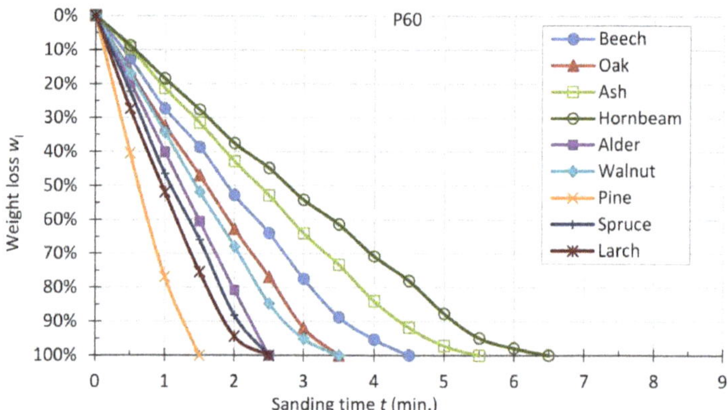

Figure 6. Average percentage weight loss of specimens sanded with abrasive belt P60.

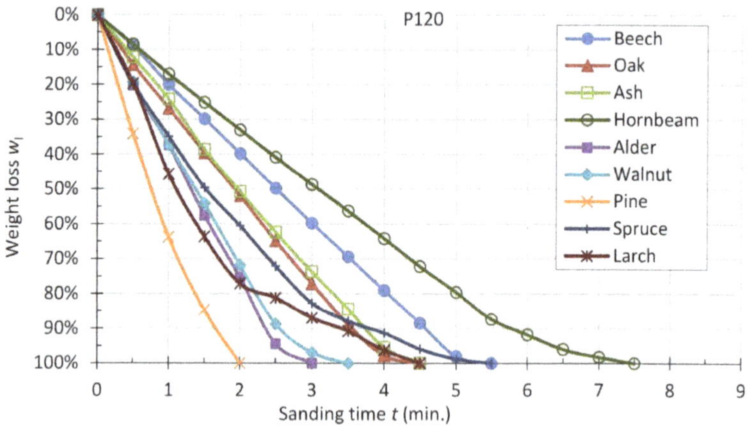

Figure 7. Average percentage weight loss of specimens sanded with abrasive belt P120.

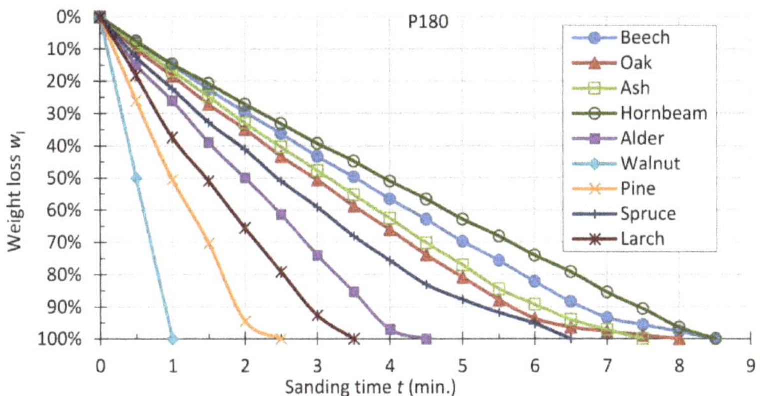

Figure 8. Average percentage weight loss of specimens sanded with abrasive belt P180.

In all the cases, the fastest sanding was performed on the pine samples and the longest on the hornbeam samples. The total loss of mass of the specimens in the case of the P60 abrasive belt occurred in 1.5 min (pine) to 6.5 min (hornbeam). For the P120 belt, it ranged from 2 min (pine) to 7.5 min (hornbeam), and for the P180 belt, it ranged from 2.5 to 8.5 min (beech and hornbeam).

4. Discussion

In Figures 3–5, different rates of decline in sanding efficiency are observed. It seems that the rapid rate of decline in the sanding efficiency is related to the high initial efficiency (the greater the initial efficiency, the more rapid its reduction). This rapid rate of decline in sanding efficiency was observed with lower density samples, especially softwoods. A possible reason for the rapidly decreasing sanding efficiency (which occurs from 0.5 min to 3 min depending on grit size and species of wood) is that the spaces between the coated abrasive become clogged more quickly by wood dust.

The weight loss of sample sets during sanding is uniform (Figures 6–8), which results from the fact that most of the experiment time takes place in the second blunting phase, for which such a course of the sanding process is characteristic. Additionally, in this way, the occurrence of a short time of the first blunting phase was emphasized. Moreover, in the case of most of the wood species, a third blunting phase occurred at the end of the experiment, when the vast majority of the sample mass had already been sanded. The rate of weight loss of the wood during sanding, as known to date, is generally greater for lower-density wood species; but the differences between the high-density species (oak) and the light softwoods (spruce, visible in the graphs) are slight. In addition, the lighter hardwood species (alder, walnut), in terms of wood removal rate, behave similarly to low-density softwoods.

The conclusions of the scientific works to date indicating the effect of wood density resulted from comparisons of a mostly small number of species. Saloni et al. [23] compared parameters of sanding hard maple (*Acer saccharum*) (hardwood) and eastern white pine (*Pinus strobus*) (softwood). They mentioned wood species as one of the factors influencing the sanding results. The wood removal rate was twice as high for pine than for maple. Ockajova et al. [21] considered only two hardwood species: European beech (*Fagus sylvatica*) and English oak (*Quercus robur*). With a slight difference in density (684 kg/m^3 for beech, 678 kg/m^3 for oak), they found significant differences between the wood removal rates of both species. Miao and Li [25] also studied two hardwood species: Manchurian ash (*Fraxinus mandshurica*), and birch (*Betula* sp.). The density of wood samples in this study was respectively 620 and 470 kg/m^3. In this case, lower values of wood removal rates in all variants of the study were for the denser and harder Ashwood. Thorpe and Brown [26] used as many as 21 species (17 hardwoods and 4 softwoods) in the study on dust production during hand sanding. They found that the quantity of wood removed during sanding varied irreversibly with wood density.

The results of these studies link the rate of the wood removed during sanding to the density and directly to the species of wood, regardless of whether it is softwood or hardwood. Therefore, to compare these two different types of wood, the total sanding times for all samples of each species were averaged. The calculated average values of these total sanding times for the six hardwood species and three softwood species separately are shown in Figures 9–11.

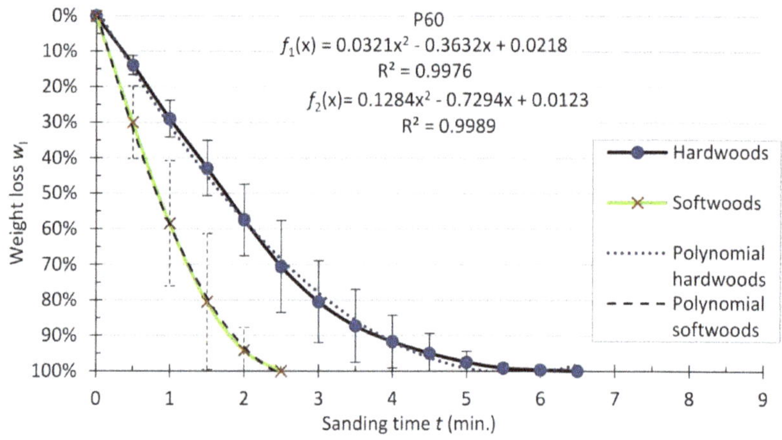

Figure 9. Comparison of average results for hardwoods (beech, oak, ash, hornbeam, alder, walnut) and softwoods (pine, spruce, larch) species; P60 sandpaper.

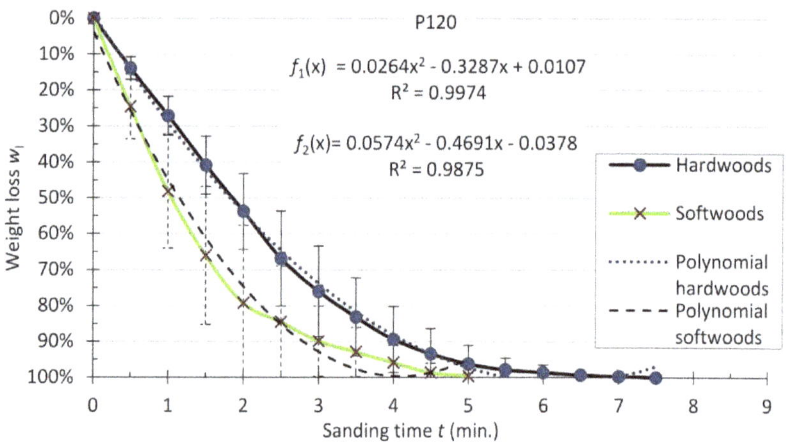

Figure 10. Comparison of average results for hardwoods (beech, oak, ash, hornbeam, alder) and softwoods (pine, spruce, larch) species; P120 sandpaper.

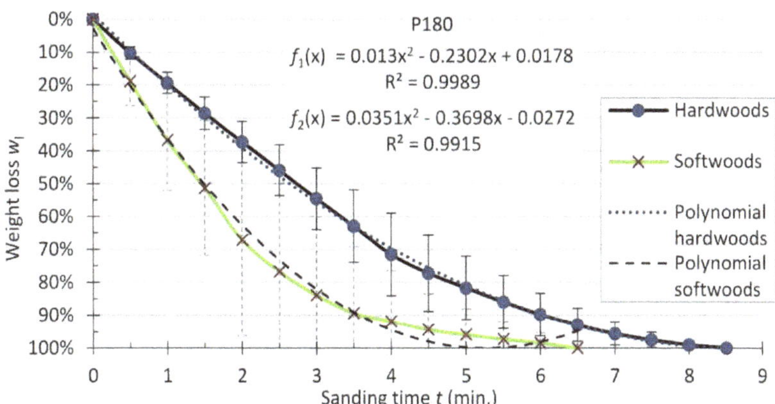

Figure 11. Comparison of average results for hardwoods (beech, oak, ash, hornbeam, alder) and softwoods (pine, spruce, larch) species; P180 sandpaper.

The graphs in Figures 9–11 show that in all tested cases, the average sanding time of the samples of softwood species was lower than that of the hardwood species. The sanding times of the softwood species were: 2.5 min (P60), 5 min (P120), and 6.5 min (P180). For hardwood species, these times were 6.5, 7.5, and 8.5 min, respectively. The spread of sanding efficiency values for softwood is much greater than in hardwood species. This is due to the difference in tribological properties of softwood and hardwood. During sanding, what is important is not only the density but also the specific physical and mechanical properties of individual wood species and their morphologies themselves (hardwood ring-porous, hardwood scattered porous, with resin content or without, chemical composition, etc.). The resin content of the softwood has a great influence on the tribological properties of wood and it may be a cause of quickly blunting of the sanding tool [27].

The mean values of the measurements were statistically analyzed. Confidence intervals were calculated using a t-distribution table ($\alpha = 0.90$). It was found that the measurement uncertainty of the means was always greater for softwood and its maximum value was independent of the sandpaper gradation. This uncertainty for hardwood was, respectively: 12.9% (P60), 13.2% (P120), 12.5% (P180). For softwood, it was: 19.1% (P60), 21.6% (P120) and 29.1% (P180).

The main hypothesis of our study is that hardwood differs from softwood in terms of the efficiency of the sanding process. For this purpose, changes in average weight loss were compared in the function of sanding time (Figures 9–11). The equations for these abrasive belts were found. Then, to adjust and compare the course of the functions, the set theory approach was used to narrow down the functions. In the first stage, for the functions to be considered equal, they must satisfy the first condition of the equality of the functions, which says that the functions $f_1(x)$ and $f_2(x)$ are equal to each other if, and only if, they have the same domains and for each point of the common domain, they assume these are the same values $f_1 = f_2 \leftrightarrow D_{f1} = D_{f2}$ and for each $x \in D_{f1} = D_{f2}$, and $f_1 = f_2$ [24].

For all three grit sizes of sanding belts P60, P120, and P180, and for both types of wood (hardwood and softwood), the domains of functions being a square function with the general formula were calculated as $y = ax^2 + bx + c$ using the procedure: (1) calculation of the root of a function Δ ($\Delta = b^2 - 4ac$); (2) determination of parameters p i q ($p = \frac{-\sqrt{\Delta}}{2a}$, $q = \frac{-\Delta}{4a}$). In the case of sanding belt P60, significant differences were found between the examined functions because their domains assumed values $D_{f1}(-5, +\infty)$ and $D_{f2}(-1, +\infty)$, thus $D_{f1} \neq D_{f2}$, so the functions are not equal. Then the second condition for the definition of function equality was checked, indicating that for each $x \in D : f_1(x) = f_2(x)$ [28]. These functions do not satisfy the equality condition and are therefore different, which indicates that hardwood differs from softwood in terms of the efficiency of the sanding process for the grit size P60.

For the sanding belt P120, it was observed that the first condition of the equality of the function was not satisfied ($D_{f1}(-1, +\infty)$ and $D_{f2}(-0, 1, +\infty)$). However, the second condition assuming that for every $x \in D$: $f_1(x) = f_2(x)$ was satisfied. Thus, a restriction was applied for functions on the selected set of points belonging to the set A $(-89.4; -100)$ where for every $x \in A$, the functions are equal : $f_1(x) = f_2(x)$ [29]. Based on the calculations related only to this set of arguments, it was found that both functions are similar to each other in the indicated range, and their domains are the same for this set of arguments. A similar situation was observed for the sanding belt P180. Using the second condition of function equality, set A was determined $(-81.7; -100)$ for which these functions are equal. Based on the mathematical analyses, it was found that the sanding belts in the second stage of blunting behave similarly. The belts of grit number (P60) dull faster than belts with higher grit numbers (P180). At the same time, differences were found between softwood and hardwood in terms of the efficiency of the sanding process.

The graph in Figure 12 shows the average sanding times with belts of different grit numbers in the individual sample sets. The graph also shows the Brinell hardness of individual wood species (the macro-hardness determination method was selected from two common hardness measurement methods [30]).

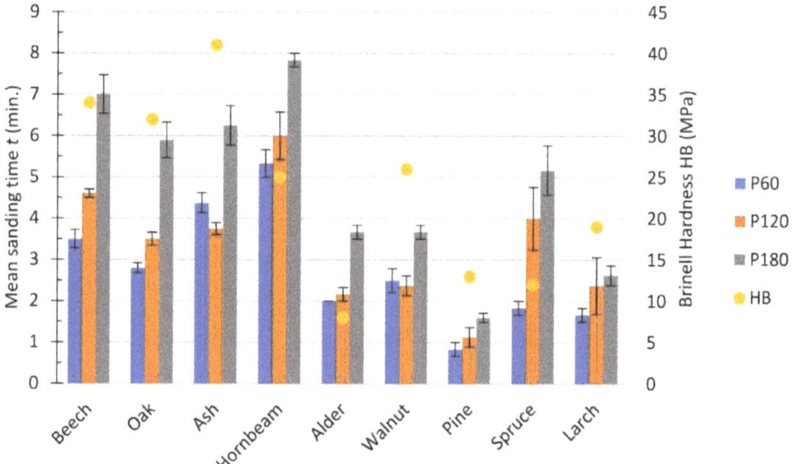

Figure 12. Average sanding time for sample sets sanded with different grit sizes.

The average sanding times for sample sets of two wood species are interesting. In the case of ash, the shortest sanding time was obtained for the P120 belt, while in the case of the other tested materials, the shortest sanding time was usually for the P60 belt (in the case of walnut, these times were more or less equal). This can be explained by the fact that it is the hardest species of wood and in this case, the optimal abrasive belt grit from the point of view of sanding efficiency fell on the belt of medium grain (according to the grain size effect described in the publication Sin et al. 1979 [31] caused by the influence of the elasticity of wood [32]). Another interesting species of wood is spruce. In the case of samples made of this material, the greatest effect of the grit number of the sanding belt on the average time for complete sanding of the sample set was observed. The sanding time with the P120 belt increased by as much as 120% compared to the sanding time with the P60 belt. For other grits, these times were either slightly shorter (by 15% for ash and 5% for walnuts) or greater (from 13 to 42%).

The times of sanding with belts of different grit numbers of the individual sets of samples do not seem to correlate with the Brinell hardness of the tested wood species. For example, walnut, which is twice as hard as alder, shows an average sanding time similar to that of alder. Among the softwood species, average sanding times of pine wood are roughly half of that of spruce, although the hardness of both species is similar. The test results show that hardness is not the only factor affecting sanding efficiency. The influence may be caused by other tribological properties, e.g., the instantaneous coefficient of friction, which depends both on the type of wood, hardness, as well as the roughness and temperature, which are time-varying during sanding and dependent on the grit of the sanding belt [33].

5. Conclusions

The results of the study on the machine sanding of different wood species with sanding belts of various grit numbers indicate that:

1. The spread of sanding efficiency values for softwood is much larger than in hardwood species. This may be due to the uneven blunting of the belts due to the specific tribological properties of the softwood (resin content).
2. The type of wood is the factor that determines the wood removal rate and thus the sanding efficiency in the initial blunting phase of the sanding belts. In the first phase of blunting of the sanding belts, the sanding processes of hardwood and softwood are significantly different. In the second phase of blunting, sanding belts with higher grit numbers (P120 and P180) behave similarly while sanding hardwood and softwood. Wood type is, therefore, another factor apart from the previously known ones (pressure force, belt speed, wood density, and the grit number of the sanding belt), which should be taken into account when designing the sanding processes of solid wood elements.

In machine sanding of wood at low pressure and high belt speed, abrasive materials with a low grit number and high sharpness affect hardwood and softwood differently. The sanding efficiency of softwood is considerably higher than hardwood in these conditions. Therefore, sanding parameters (pressure and belt speed) should be set at lower values to avoid excessive sanding or over-sanding (a situation when too much material is sanded).

Author Contributions: Conceptualization, T.R. and M.S.; methodology, T.R.; validation, M.S., R.M. and T.R.; formal analysis, K.S.-S.; investigation, T.R.; resources, T.R.; data curation, M.S.; writing—original draft preparation, T.R. and M.S.; writing—review and editing, T.R. and M.S.; visualization, M.S.; supervision, T.R.; funding acquisition, R.M. All authors have read and agreed to the published version of the manuscript.

Funding: This research was funded by the National Centre for Research and Development, BIOS-TRATEG3/344303/14/NCBR/2018.

Institutional Review Board Statement: Not applicable.

Informed Consent Statement: Not applicable.

Data Availability Statement: The data presented in this study are available on request from the corresponding author.

Acknowledgments: The authors thank Jacek Sydor for the valuable terminological comments.

Conflicts of Interest: The authors declare no conflict of interest.

References

1. Rogoziński, T.; Hlásková, L.; Wieruszewski, M.; Očkajová, A. Particle-Size Distribution of Dust Created during Sanding the Modified Ash Wood. *Ann. WULS SGGW For. Wood Tech.* **2015**, *90*, 162–166.
2. Antov, P.; Brezin, V. *a /Engineering Ecology*; Publishing house—University of Forestry: Sofia, Bulgaria, 2015; ISBN 978-954-332-135-3.
3. Očkajová, A.; Kučerka, M.; Krišťák, L.; Igaz, R. Granulometric Analysis of Sanding Dust from Selected Wood Species. *Bioresources* **2018**, *13*, 7481–7495. [CrossRef]

4. Antov, P.; Neykov, N.; Savov, V. Effect of Occupational Safety and Health Risk Management on the Rate of Work–Related Accidents in the Bulgarian Furniture Industry. *Wood Des. Technol.* **2018**, *7*, 1–9.
5. Pędzik, M.; Stuper-Szablewska, K.; Sydor, M.; Rogoziński, T. Influence of Grit Size and Wood Species on the Granularity of Dust Particles during Sanding. *Appl. Sci.* **2020**, *10*, 8165. [CrossRef]
6. Očkajová, A.; Kučerka, M.; Kminiak, R.; Krišťák, Ľ.; Igaz, R.; Réh, R. Occupational Exposure to Dust Produced When Milling Thermally Modified Wood. *Int. J. Environ. Res. Public Health* **2020**, *17*, 1478. [CrossRef] [PubMed]
7. Pahlitzsch, G. Internationaler Stand der Forschung auf dem Gebiet des Schleifens von Holz. *Holz Roh- Werkst.* **1970**, *28*, 329–343. [CrossRef]
8. Csanády, E.; Magoss, E. *Mechanics of Wood Machining*, 3rd ed.; Springer: Cham, Switzerland, 2013; ISBN 3-642-29955-5.
9. Vlasev, V.; Kovatchev, G.; Atanasov, V. Mechanism for Belt Sanding Machines with a Fixed Bearing of the Sanding Belt and Eccentric Tension. In Proceedings of the Implementation of Wood Science in Woodworking Sector, Zagreb, Croatia, 12–13 December 2019; pp. 221–224.
10. Pahlitzsch, G.; Dziobek, K. Über das Wesen der Abstumpfung von Schleifbändern beim Bandschleifen von Holz. *Holz Roh- Werkst.* **1961**, *19*, 136–149. [CrossRef]
11. Taylor, J.B.; Carrano, A.L.; Lemaster, R.L. Quantification of Process Parameters in a Wood Sanding Operation. *For. Prod. J.* **1999**, *49*, 41–46.
12. Očkajová, A.; Sikliena, M. The Influence of Chosen Factors of Wood Sanding upon the Efficiency of Sand Belt. *Drevarsky Vyskum/Wood Res.* **2000**, *45*, 33–38.
13. Carrano, A.L.; Taylor, J.B.; Lemaster, R. Parametric Characterization of Peripheral Sanding. *For. Prod. J.* **2002**, *52*, 44–50.
14. Sinn, G.; Gindl, M.; Reiterer, A.; Stanzl-Tschegg, S. Changes in the Surface Properties of Wood Due to Sanding. *Holzforschung* **2004**, *58*, 246–251. [CrossRef]
15. Gurau, L.; Mansfield-Williams, H.; Irle, M. Processing Roughness of Sanded Wood Surfaces. *Holz Roh- Werkst.* **2005**, *63*, 43–52. [CrossRef]
16. Porankiewicz, B.; Banski, A.; Wieloch, G. Specific Resistance and Specific Intensity of Belt Sanding of Wood. *Bioresources* **2010**, *5*, 1626–1660.
17. Ratnasingam, J.; Reid, H.F.; Perkins, M.C. The Abrasive Sanding of Rubberwood (Hevea Brasiliensis): An Industrial Perspective. *Holz Roh- Werkst.* **2002**, *60*, 191–196. [CrossRef]
18. Burdurlu, E.; Usta, I.; Ulupinar, M.; Aksu, B.; Erarslan, T.Ç. The Effect of the Number of Blades and the Grain Size of Abrasives in Planing and Sanding on the Surface Roughness of European Black Pine and Lombardy Poplar. *Turk. J. Agric. For.* **2005**, *29*, 315–321. [CrossRef]
19. Malkoçoğğlu, A.; Özdemir, T. The Machining Properties of Some Hardwoods and Softwoods Naturally Grown in Eastern Black Sea Region of Turkey. *J. Mater. Process. Technol.* **2006**, *173*, 315–320. [CrossRef]
20. Aslan, S.; Coşkun, H.; Kiliç, M. The Effect of the Cutting Direction, Number of Blades and Grain Size of the Abrasives on Surface Roughness of Taurus Cedar (Cedrus Libani A. Rich.) Woods. *Build. Environ.* **2008**, *43*, 696–701. [CrossRef]
21. Očkajová, A.; Kučerka, M.; Krišťák, L.; Ružiak, I.; Gaff, M. Efficiency of Sanding Belts for Beech and Oak Sanding. *Bioresources* **2016**, *11*. [CrossRef]
22. Wieloch, G.; Sikliena, M. Wpływ Wybranych Parametrów Na Wydajność Ubytkową Procesu Szlifowania Drewna/Influence of selected parameters on the loss efficiency of the wood sanding process. *Drewno* **2004**, *47*, 121–130.
23. Saloni, D.E.; Lemaster, R.L.; Jackson, S.D. Abrasive Machining Process Characterization on Material Removal Rate, Final Surface Texture, and Power Consumption for Wood. *For. Prod. J.* **2005**, *55*, 35–41.
24. Błaszczyk, A.; Turek, S. *Teoria Mnogości/Set Theory*; Wydawnictwo Naukowe PWN: Warsaw, Poland, 2007; ISBN 978-83-01-15232-1.
25. Miao, T.; Li, L. Study on Influencing Factors of Sanding Efficiency of Abrasive Belts in Wood Materials Sanding. *Wood Res.* **2014**, *59*, 835–842.
26. Thorpe, A.; Brown, R.C. Factors Influencing the Production of Dust During the Hand Sanding of Wood. *Am. Ind. Hyg.* **1995**, *56*, 236–242. [CrossRef]
27. Chand, N.; Fahim, M. *Tribology of Natural Fiber Polymer Composites*; Woodhead Publishing; CRC Press: Boca Raton, FL, USA, 2008; ISBN 978-1-84569-393-0.
28. *Proper and Improper Forcing; Perspectives in Mathematical Logic*, 2nd ed.; Springer: Berlin/Heidleberg, Germany; New York, NY, USA, 1997; Volume 5, ISBN 978-3-540-51700-9.
29. Woodin, W.H. *The Axiom of Determinacy, Forcing Axioms, and the Nonstationary Ideal*, 2nd ed.; De Gruyter: Berlin/Heidleberg, Germany; New York, NY, USA, 2010; ISBN 978-3-11-021317-1.
30. Sydor, M.; Pinkowski, G.; Jasińska, A. The Brinell Method for Determining Hardness of Wood Flooring Materials. *Forests* **2020**, *11*, 878. [CrossRef]
31. Sin, H.; Saka, N.; Suh, N.P. Abrasive Wear Mechanisms and the Grit Size Effect. *Wear* **1979**, *55*, 163–190. [CrossRef]
32. Ohtani, T.; Yakou, T.; Kitayama, S. Conditions and Origin of the Critical Grain Size Effect on the Abrasive Wear of Woods. *Mokuzai Gakkaishi/J. Jpn. Wood Res. Soc.* **1996**, *42*, 1057–1063.
33. Xu, M.; Li, L.; Wang, M.; Luo, B. Effects of Surface Roughness and Wood Grain on the Friction Coefficient of Wooden Materials for Wood–Wood Frictional Pair. *Tribol. Trans.* **2014**, *57*, 871–878. [CrossRef]

 *applied sciences*

Article

Influence of Grit Size and Wood Species on the Granularity of Dust Particles during Sanding

Marta Pędzik [1,2], Kinga Stuper-Szablewska [3], Maciej Sydor [4] and Tomasz Rogoziński [1,*]

[1] Department of Furniture Design, Faculty of Forestry and Wood Technology, Poznań University of Life Sciences, 60-627 Poznań, Poland; marta.pedzik@itd.lukasiewicz.gov.pl

[2] Wood-Based Products and Biocomposites Department, Łukasiewicz Research Network, Wood Technology Institute, 60-654 Poznań, Poland

[3] Department of Chemistry, Faculty of Forestry and Wood Technology, Poznań University of Life Sciences, 60-625 Poznań, Poland; kinga.stuper@up.poznan.pl

[4] Department of Woodworking Machinery and Machine Construction, Poznań University of Life Sciences, 60-627 Poznań, Poland; maciej.sydor@up.poznan.pl

* Correspondence: tomasz.rogozinski@up.poznan.pl

Received: 28 October 2020; Accepted: 16 November 2020; Published: 18 November 2020

Abstract: Wood dust poses a threat to the health of employees and the risk of explosion and fire, accelerates the wear of machines, worsens the quality of processing, and requires large financial outlays for its removal. The aim of this study was to investigate the extent to which the grit size of sandpaper influences the size of the wood dust particles and the proportion of the finest particles which, when dispersed in the air, may constitute the respirable fraction. Six species of hardwood (beech, oak, ash, hornbeam, alder, and walnut), and three species of softwood (larch, pine, and spruce) were used in the research. While sanding the samples under the established laboratory conditions, the following were measured for two types of sandpapers (grit sizes P60 and P180): mean arithmetic particle size of dust and finest dust particles content (<10 μm). Based on the obtained results, we found that the largest dust particle sizes were obtained for alder, pine, and spruce; the smallest size of dust particles during sanding with both sandpapers was obtained for beech, hornbeam, oak, ash, larch, and walnut. The mean arithmetic particle sizes ranged from 327.98 μm for pine to 104.23 μm for hornbeam. The mean particle size of the dust obtained with P60 granulation paper was 1.4 times larger than that of the dust obtained with P180 granulation sandpaper. The content of the finest dust particles ranged from 0.21% for pine (P60 sandpaper) to 12.58% for beech (P180 sandpaper).The type of wood (hardwood or softwood) has a significant influence on the particle size and the content of the finest dust fraction.

Keywords: wood dust; sanding; sandpaper; particle-size distribution

1. Introduction

Wood dust is a waste generated during mechanical wood processing in wood industry plants. Dust poses a threat to workers health, increases the risk of explosion and fire, accelerates the wear of machines, worsens the quality of processing, and incurs high costs for its removal.

Dust with particles smaller than 10 μm affects the respiratory system, eyes, and skin, causing health effects in the form of irritation, allergies, and diseases [1–6]. The dust toxicity is determined to a large extent by the type of wood raw material, which results from the different contents of the main chemical components, such as cellulose, hemicellulose, and lignin in coniferous and deciduous trees [7,8]. Long-term inhalation of air polluted with wood dust, including the most harmful (i.e., beech and oak wood) may contribute to cancer incidence. As a result, wood dust was classified by the International

Agency for Research on Cancer (IARC) among the most dangerous and carcinogenic materials for humans. According to Directive 2004/37/EC of the European Parliament and of the Council of 29 April 2004 on the protection of workers against the risks related to exposure to carcinogens or mutagens at work; its current limit is 3 mg/m^3. After 17 January 2023, it will be reduced to only 2 mg/m^3. To meet such high requirements, it will be necessary to use all available technical means of dust reduction.

The dustiness of the air inside production plants causes a fire and explosion hazard. Dust dispersed in the air can create an explosive mixture and settle on walls, floors, and machines, creating a risk of fire and explosion at the workplace. The inflammability of wood dust favors the spread of fire [9–13].

Extraction devices designed to remove dust from processing areas are never completely effective, especially in relation to the very small-sized dust particles. One way to reduce the amount of fine dust generated is to reduce the thickness of the furniture components and thus the diameters and depths of the holes for connectors, which reduces the volume of wood material cut. Such a procedure is effective in eliminating drilling and milling operations; however, it requires the development of new furniture fasteners [14] and does not reduce the need for sanding operations. Another way is to adjust the processing parameters to the type and properties of the material being processed. Appropriate adjustment of the treatment parameters will result in the formation of a reduced amount of fine dust dispersed in the air, could constitute a potentially dangerous inhaled fraction.

With regards to sawing and milling, the technological parameters of the processing, as well as the type and sharpness of the tools used, are particularly important. Among technological parameters, the feed per tooth, the feed rate, and the thickness of the cut layer have important influences on the amount and size of the created dust particles [15–25].

The largest amount of dust is generated during the sanding of wood; therefore, this technological operation is considered to be the source of the most serious hazards related to wood dust. The amount and size of the dust particles created when sanding wood depends on the type of sanding machine used: wide-belt, narrow-belt, or disc sanders. It is difficult to remove the dust from some special types of sanding machines, such as manual belts, discs, and oscillating sanders [26,27].

Recent studies on the size of the dust particles created during the sanding of wood concerned some of the most commonly used wood species in the industry. These tests were carried out using various types of sanding machines and sandpapers with grain sizes typical for the basic technical requirements of sanding operations. Many times, the sandpaper type studied was limited to paper tapes with the grain size of P80. Due to the method of sieve analysis used in these studies to assess the size of dust particles (the sieve with the smallest mesh of 32 μm), it was not possible to quantify the content of dust particles with the smallest sizes, which would constitute the respirable fraction after dispersion in the air [28–30]. Such particles are present in the wood sanding dust. Their presence was confirmed by spectrometric, optical, and laser methods [27,28]; however, no comprehensive and comparative studies have been performed on the content of wood dust particles that would be small enough to be dispersed in the air as a thoracic fraction [31–33].

To reduce the risk to workers' health, increase work safety, and meet future legal requirements for fine wood dust, we examined the most important factor characterizing the sanding process of wood-grit size. Therefore, the aim of the study was to investigate the extent to which the grit size of the sandpaper affects the size of the dust particles created during sanding different wood species and the proportion of the finest particles, which, when dispersed in the air, may constitute a respirable fraction.

2. Materials and Methods

2.1. Sanding and Particle Size Analysis

Three species of softwood and six species of hardwood often used in the wood industry were used in this research (Table 1). The densities of wood species were determined according to the method described in the standard ISO 13061–2:2014.

Table 1. Wood species.

Wood Species	Beech	Oak	Ash	Hornbeam	Alder	Walnut	Larch	Pine	Spruce
Density, kg/m^{-3}	687	686	621	754	446	641	546	453	420

Sanding was performed using a prototype narrow belt sanding machine designed and made in the laboratory of the Department of Furniture Design (Faculty of Forestry and Wood Technology, Poznań University of Life Sciences PULS, Poznań, Poland). EKA 1000 F sandpaper (Ekamant, Poznań, Poland) in the form of belts with dimensions 1000 × 80 mm was used (Figure 1). The grit sizes of the paper were P60 and P180. A cutting speed of 14.5 m/s and a sanding pressure of 0.65 N/cm^2 were applied.

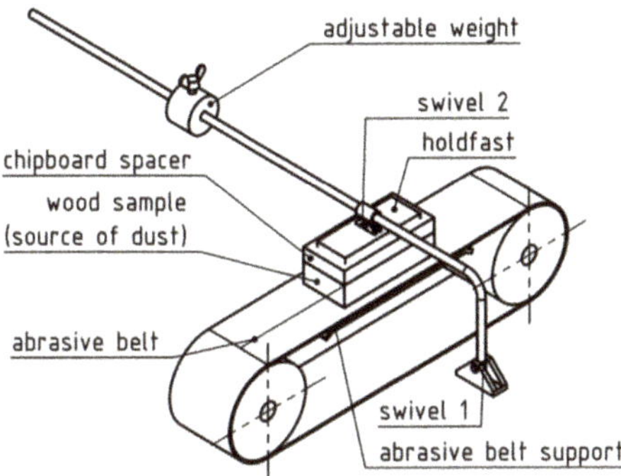

Figure 1. Schematic diagram of sanding.

Particle-size determination and calculation of the content of fine dust particles (the content of particles <10 µm) were carried out according to methods described by [20,22,34,35]. In the sieve analysis, a set of sieves with aperture sizes of 250, 125, and 63 µm was used due to the high level of wood dust fineness. Then, the content of the dust particles <10 µm in the sieve fraction <63 µm was measured using a Analysette 22 MicroTec Plus laser particle sizer (Fritsch, Idar-Oberstein, Germany). Based on the results of sieve analysis, the cumulative particle size distribution Q_3 and the particle mean arithmetic diameter $\bar{x}$ was calculated as follows:

$$Q_3 = \sum_{i=1}^{n} \bar{q}_{3,i} \Delta x_i \tag{1}$$

$$\bar{x} = \sum_{i=1}^{n} x_i \times q_{3i} \tag{2}$$

where q_3 is the particle size distribution by mass, x is the mean value of particle size class, and n is the number of particle size classes.

2.2. Statistical Analysis

The results recorded in the course of conducted tests were subjected to statistical analysis with the use of STATISTICA ver. 13.1 (StatSoft, Inc., Tulsa, OK, USA) and Microsoft$^®$ Excel 2020, Microsoft 365 (Addinsoft, Inc., Brooklyn, NY, USA) software packages. In order to compare grit sizes of the P60 and P180 sandpaper, multivariate comparison procedure was used, with identical letters denoting a lack of differences at the significance level of $P = 0.05$, lowercase letters denote significant differences between the grit sizes of the sandpaper, and uppercase letters denote significant differences between wood species. Correlation coefficients between the wood density and the mean arithmetic particle size of dust were calculated. The multiplicity factors for the mean arithmetic particle size and for the content of the finest particles obtained from paper with the grit sizes P60 to P180 were calculated. Moreover, a step linear discriminatory analysis (SLDA) and the principal component analysis (PCA) were used to separate groups of analyzed dust in the entire population.

3. Results and Discussion

The basic results of the particle size analysis came from the sieve analysis. The cumulative distributions (Figure 2) showed that, in general, the dust from sanding operations performed with the use of P180 sandpaper is finer than dust created in sanding with P60 sandpaper.

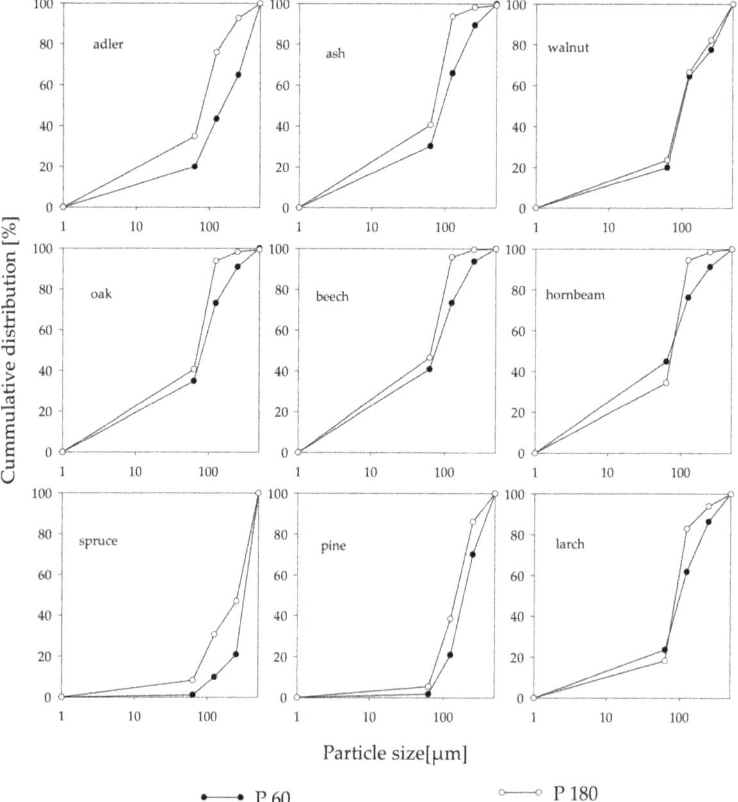

Figure 2. Particle size distributions of wood dust.

To compare the mean arithmetic particle sizes of dust created in sanding with P60 and P180 sandpapers within the entire population, that is, between wood species and within wood species, multivariance analysis was used (Figure 3). On the basis of the obtained results, the smallest size of dust particles created during sanding with both sandpapers was found for beech, hornbeam, oak, ash, larch, and walnut. Significantly larger sizes of dust were obtained for alder, pine, and spruce. By analyzing the ratio of mean arithmetic particle sizes of dust obtained from paper with grit sizes P60 and P180, the multiplicity factors were calculated and we found that the average for the entire population was 1.4, which means that the mean particle size of dust obtained from the paper with grit size P60 was 1.4 times higher than that of dust obtained from paper with the grit size P180. When analyzing the dust from hardwood and softwood species separately, we found that the multiplicity factors were 1.4 and 1.3, respectively. The lowest multiplication factor was found for walnut wood, at 1.1, and the highest for alder wood, at 1.8.

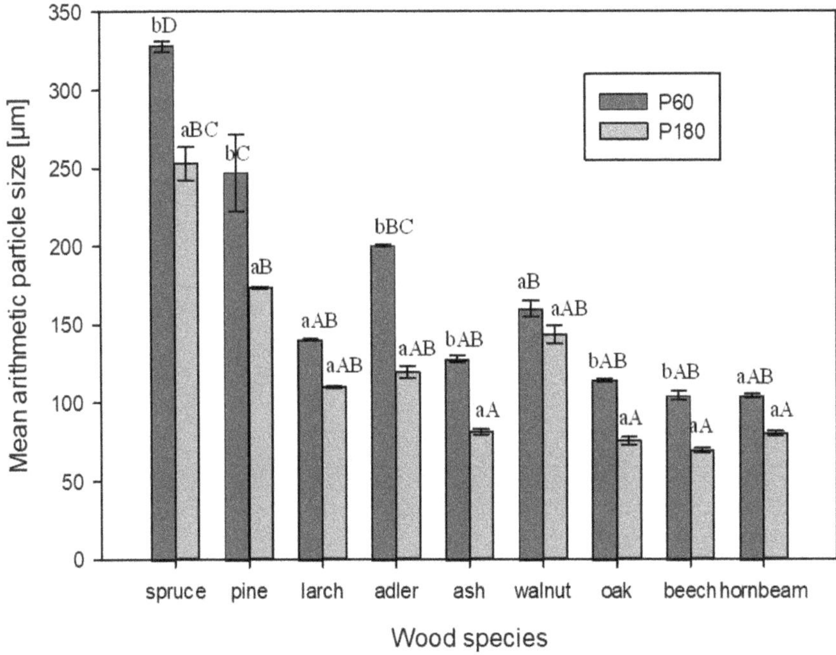

Figure 3. Analysis of the mean arithmetic particle sizes of dust; a, b: the same letters indicate no significant differences at the significance level of 0.05 between sandpaper grit size; A, B: the same letters indicate no significant differences at a significance level of 0.05 between wood species.

While analyzing the content of the finest particles (<10 μm), a multivariate analysis was also performed to compare this content for wood dust created during sanding with papers of grit sizes P60 and P180 for different species of wood and to compare the content of particles <10 μm within species (Figure 4). There were no significant differences between the content of the finest dust particles obtained from sanding with both sandpapers for pine and larch. When analyzing the differences for the species, significantly higher contents of the smallest particles were found for alder, walnut, oak, beech, and hornbeam. By analyzing the ratio of dust particles <10 μm obtained from sandpaper of grit size P60 to the paper of grit size P180, the multiplicity factors were also calculated and an analogous tendency was found for the mean arithmetic particle size. The average multiplicity factor P60/P180 was 1.4 for the entire population of wood species tested.

Then, we analyzed the main components on the basis of the value of the mean arithmetic particles sizes and the content of the finest particles for both grit sizes of sandpaper, and the second predictor, next to the wood species (Factor 1), was indicated as the type (hardwood or softwood) of wood (Factor 2). Supplementing the grouping factors with this factor showed that the obtained results could be divided into two groups, as shown in Figure 5. One region of the loop comprises hardwood and the other, softwood. The full separation was obtained by a factor of two. This clear division was also confirmed by the PCA result showing the projection of the variables on the plane (Figure 6).

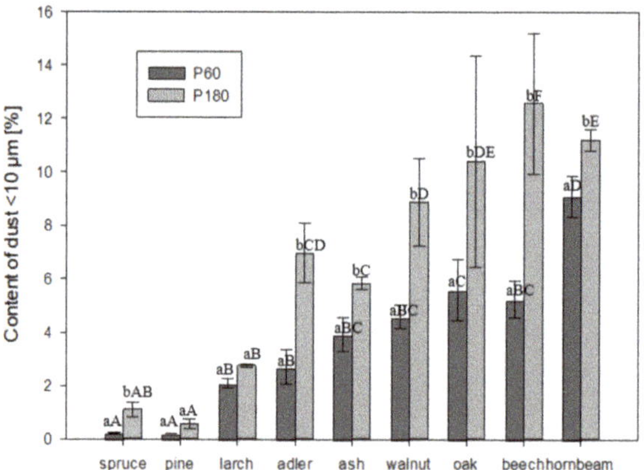

Figure 4. Analysis of the content of the finest dust particles; a, b: the same letters indicate no significant differences at the significance level of 0.05 between sandpaper grit sizes; A, B: the same letters indicate no significant differences at a significance level of 0.05 between wood species.

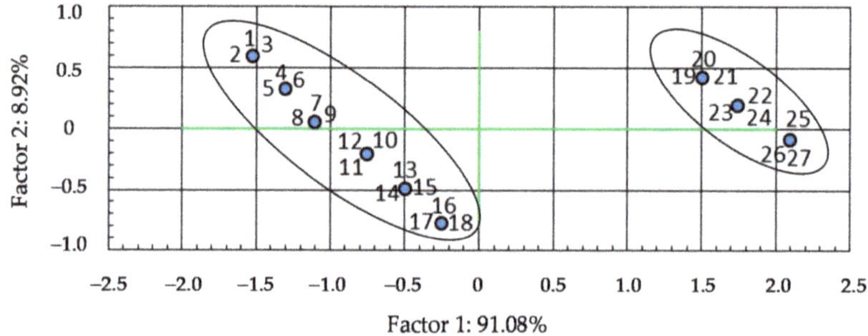

Figure 5. The spread of the mean arithmetic particle size of dust and the content of the finest particles. (1, 2, 3 are beech; 4, 5, 6 are oak; 7, 8, 9 are hornbeam; 10, 11, 12 are ash; 13, 14, 15 are alder; 16, 17, 18 are walnut; 19, 20, 21 are larch; 22, 23, 24 are pine; and 25, 26, 27 are spruce) based on the multiplicity factor of the mean arithmetic particle size of the dust created in sanding with sandpapers of grit sizes P60 and P180 and on the content of the finest particles. The clusters of results that form separate populations are marked in the loops.

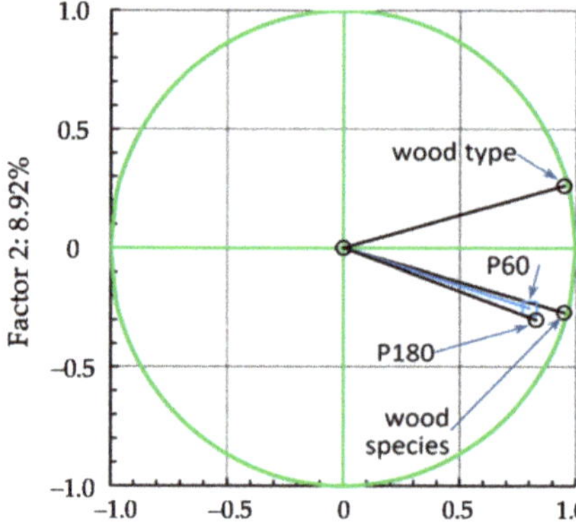

Figure 6. The result of principal component analysis (PCA) for the entire population of dust.

The correlation coefficients between the wood density and the mean arithmetic particle size of dust were calculated. Very high values of these coefficients for the tested wood species were found at the level of 0.9588 for grit size P60 and 0.8794 for grit size P180 based on this analysis. A similar relationship was found for particles <10 μm. The correlation coefficient value was 0.9227 for the P60 grit size and 0.8812 for the P180 grit size. This confirms the highly significant relationship between the grit size of sandpaper and the mean arithmetic particle size and the content of the finest particles in wood dust.

The share of beech wood dust particles with a diameter of ≤80 μm in the range of 89.21–96.29% was described by Očkajová et al. [18]. The finest dust particles that can be found in a such large proportions of small particles are undesirable in working environments. They can penetrate the alveoli and be the source of serious diseases. A similar relationship was obtained when using P180 paper, where the share of this fraction was also significant.

The wood species also has a significant impact on the sanding belt wear. This is due to the specific, mainly microscopic, structure of the wood, as well as different physical and mechanical properties. Beech wood has a uniform structure in spring–summer rings and scattered rings. Oak wood has significant differences in density between spring and summer rings, as well as a relatively high share of extractive substances, even up to 6.1% [36]. Softwood, such as coniferous species, has relatively long fibers, which result in a fraction with larger particles.

4. Conclusions

Based on these studies, we concluded that: The largest mean arithmetic dust particle sizes were obtained for alder, pine, and spruce. The smallest mean arithmetic dust particle sizes were obtained for beech, hornbeam, oak, ash, larch, and walnut. The highest content of the finest particles was found for alder, walnut, oak, beech, and hornbeam. The type of wood (hardwood or softwood) has a significant influence on the mean arithmetic dust size and the content of the dust fraction with the size <10 μm. The particle size analyses of dust from the sanding of different wood species showed that the sanding of walnut, oak, beech, and hornbeam can be a source of a considerable amount of very fine dust particles, which can from a respirable fraction when dispersed in the air.

Author Contributions: Conceptualization, M.S. and T.R.; Data curation, M.P. and K.S.-S.; Formal analysis, M.S.; Funding acquisition, M.S.; Investigation, M.P. and K.S.-S.; Methodology, M.P. and T.R.; Project administration, M.S. and T.R.; Resources, T.R.; Software, K.S.-S.; Supervision, T.R.; Validation, M.S.; Visualization, K.S.-S.; Writing, original draft, T.R.; Writing, review and editing, M.S. All authors have read and agreed to the published version of the manuscript.

Funding: The article processing charge (APC) was financed within the European project POIR.01.02.00 00 00102/17, "The first Polish innovative universal system of furniture fasteners for joining various wood and wood-composite materials in the furniture industry", implemented by Digitouch sp. z o.o. (Suchy Las, Poland). The project is a part of the Polish sectoral programme WoodINN financed by the Polish National Centre for Research and Development (NCRD).

References

1. Douwes, J.; McLean, D.; Slater, T.; Pearce, N. Asthma and Other Respiratory Symptoms in New Zealand Pine Processing Sawmill Workers. *Am. J. Ind. Med.* **2001**, *39*, 608–615. [CrossRef] [PubMed]
2. Schlunssen, V. Asthma and Other Respiratory Diseases among Workers in the Furniture Industry Occupationally Exposed to Wood Dust. *Dan. Med. Bull.* **2001**, *48*, 191.
3. Schlünssen, V.; Sigsgaard, T.; Raulf-Heimsoth, M.; Kespohl, S. Workplace Exposure to Wood Dust and the Prevalence of Wood-Specific Sensitization. *Allergol. Sel.* **2018**, *2*, 101. [CrossRef] [PubMed]
4. Mračková, E.; Krišťák, Ľ.; Kučerka, M.; Gaff, M.; Gajtanska, M. Creation of Wood Dust during Wood Processing: Size Analysis, Dust Separation, and Occupational Health. *BioResources* **2016**, *11*, 209–222. [CrossRef]

5. Asgedom, A.A.; Bråtveit, M.; Moen, B.E. High Prevalence of Respiratory Symptoms among Particleboard Workers in Ethiopia: A Cross-Sectional Study. *Int. J. Environ. Res. Public Health* **2019**, *16*, 2158. [CrossRef]

6. Tureková, I.; Mračková, E.; Marková, I. Determination of Waste Industrial Dust Safety Characteristics. *Int. J. Environ. Res. Public Health* **2019**, *16*, 2103. [CrossRef]

7. Jacobsen, G.; Schaumburg, I.; Sigsgaard, T.; Schlünssen, V. Non-Malignant Respiratory Diseases and Occupational Exposure to Wood Dust. Part II. Dry Wood Industry. *Ann. Agric. Environ. Med.* **2010**, *17*, 29–44.

8. Lorincová, S.; Hitka, M.; Čambál, M.; Szabó, P.; Javorčíková, J. Motivation Factors Influencing Senior Managers in the Forestry and Wood-Processing Sector in Slovakia. *BioResources* **2016**, *11*, 10339–10348. [CrossRef]

9. Igaz, R.; Kminiak, R.; Krišťák, Ľ.; Němec, M.; Gergeľ, T. Methodology of Temperature Monitoring in the Process of CNC Machining of Solid Wood. *Sustainability* **2019**, *11*, 95. [CrossRef]

10. Marková, I.; Mračková, E.; Očkajová, A.; Ladomerský, J. Granulometry of selected wood dust species of dust from orbital sanders. *Wood Res.* **2016**, *61*, 983–992.

11. Marková, I.; Ladomerský, J.; Hroncová, E.; Mračková, E. Thermal Parameters of Beech Wood Dust. *BioResources* **2018**, *13*, 3098–3109. [CrossRef]

12. Mračková, E.; Tureková, I. The Dimensional Characteristics of the Particles of Wood Dust of Selected Deciduous Trees Considering to Explosion. *Key Eng. Mater.* **2016**, *688*, 182–189. [CrossRef]

13. Tureková, I.; Marková, I. Ignition of Deposited Wood Dust Layer by Selected Sources. *Appl. Sci.* **2020**, *10*, 5779. [CrossRef]

14. Branowski, B.; Starczewski, K.; Zabłocki, M.; Sydor, M. Space for Innovation in the Design of Furniture Fasteners. *BioResources* **2020**, *15*, 8472–8495. [CrossRef]

15. Lučić, R.B.; Kos, A.; Antonović, A.; Vujasinović, E.; Šimičić, I. Properties of Chipped Wood Generated during Wood Processing. *Drv. Ind.* **2005**, *1*, 11–19.

16. Dzurenda, L.; Kucerka, M. Change in Structure of Granularity and Aeromechanical Characteristics of Wet Spruce Sawdust in the Process of Drying. *Ann. Wars. Agric. Univ. For. Wood Technol.* **2005**, *56*, 230–236.

17. Dzurenda, L.; Wasielewski, R.; Orłowski, K. Granulometric Analysis of Dry Sawdust from the Sawing Process on the Frame Sawing Machine PRW15M = Granulometrická Analýza Suchej Piliny z Procesu Pílenia Borovicového Dreva Na Rámovej Píle PRW-15M. *Acta Fac. Xylologiae Zvolen* **2006**, *48*, 51–57.

18. Očkajová, A.; Stebila, J.; Rybakowski, M.; Rogozinski, T.; Krišták, L.; Ľuptáková, J. The Granularity of Dust Particles When Sanding Wood and Wood-Based Materials. *Adv. Mater. Res.* **2014**, *1001*, 432–437. [CrossRef]

19. Očkajová, A.; Kučerka, M.; Banski, A.; Rogoziński, T. Factors Affecting the Granularity of Wood Dust Particles. *Chip Chipless Woodwork. Process.* **2016**, *10*, 137–144.

20. Hlásková, L.; Rogoziński, T.; Kopecký, Z. Influence of Feed Speed on the Content of Fine Dust during Cutting of Two-Side-Laminated Particleboards. *Drv. Ind.* **2016**, *67*, 9–15. [CrossRef]

21. Pałubicki, B.; Rogoziński, T. Efficiency of Chips Removal during CNC Machining of Particleboard. *Wood Res.* **2016**, *61*, 811–818.

22. Piernik, M.; Rogoziński, T.; Krauss, A.; Pinkowski, G. The Influence of the Thermal Modification of Pine (*Pinus Sylvestris* L.) Wood on the Creation of Fine Dust Particles in Plane Milling: Fine Dust Creation in the Plane Milling of Thermally Modified Pine Wood. *J. Occup. Health* **2019**, *61*, 481–488. [CrossRef] [PubMed]

23. Dzurenda, L.; Orlowski, K.; Grzeskiewicz, M. Effect of Thermal Modification of Oak Wood on Sawwdust Granularity. *Drv. Ind.* **2010**, *61*, 89–94.

24. Hlaskova, L.; Rogozinski, T.; Dolny, S.; Kopecky, Z.; Jedinak, M. Content of Respirable and Inhalable Fractions in Dust Created While Sawing Beech Wood and Its Modifications. *Drew. Pr. Nauk. Doniesienia Komun.* **2015**, *58*, 135–146. [CrossRef]

25. Korčok, M.; Koleda, P.; Barcík, Š.; Očkajová, A.; Kučerka, M. Effect of Technological and Material Parameters on Final Surface Quality of Machining When Milling Thermally Treated Spruce Wood. *Bioresources* **2019**, *14*, 10004–10013. [CrossRef]

26. Očkajová, A.; Kučerka, M. Granularity of Dust Particles Obtained in the Process of Sanding and Milling of Particleboard. *Woodwork. Tech.* **2011**, *4*, 211–217.

27. Očkajová, A.; Kučerka, M.; Krišťák, L.; Igaz, R. Granulometric Analysis of Sanding Dust from Selected Wood Species. *BioResources* **2018**, *13*, 7481–7495. [CrossRef]

28. Kminiak, R.; Orlowski, K.A.; Dzurenda, L.; Chuchala, D.; Banski, A. Effect of Thermal Treatment of Birch Wood by Saturated Water Vapor on Granulometric Composition of Chips from Sawing and Milling Processes from the Point of View of Its Processing to Composites. *Appl. Sci.* **2020**, *10*, 7545. [CrossRef]

29. Očkajová, A.; Kučerka, M.; Kminiak, R.; Krišťák, Ľ.; Igaz, R.; Réh, R. Occupational Exposure to Dust Produced When Milling Thermally Modified Wood. *Int. J. Environ. Res. Public Health* **2020**, *17*, 1478. [CrossRef]

30. Mazzoli, A.; Favoni, O. Particle Size, Size Distribution and Morphological Evaluation of Airborne Dust Particles of Diverse Woods by Scanning Electron Microscopy and Image Processing Program. *Powder Technol.* **2012**, *225*, 65–71. [CrossRef]

31. Welling, I.; Lehtimäki, M.; Rautio, S.; Lähde, T.; Enbom, S.; Hynynen, P.; Hämeri, K. Wood Dust Particle and Mass Concentrations and Filtration Efficiency in Sanding of Wood Materials. *J. Occup. Environ. Hyg.* **2008**, *6*, 90–98. [CrossRef]

32. Aro, M.D.; Geerts, S.M.; French, S.; Cai, M. Particle Size Analysis of Airborne Wood Dust Produced from Sawing Thermally Modified Wood. *Eur. J. Wood Wood Prod.* **2019**, *77*, 211–218. [CrossRef]

33. Beljo-Lučić, R.; Čavlović, A.O.; Jug, M. Definitions and Relation of Airborne Wood Dust Fractions. In Proceedings of the International Science Conference: Woodworking techniques, Prague, Czech Republic, 7–10 September 2011; pp. 25–32.

34. Rogoziński, T.; Wilkowski, J.; Gorski, J.; Czarniak, P.; Podziewski, P.; Szymanowski, K. Dust Creation in CNC Drilling of Wood Composites. *Biorecources* **2015**, *10*, 3657–3665. [CrossRef]

35. Rogoziński, T.; Wilkowski, J.; Górski, J.; Szymanowski, K.; Podziewski, P.; Czarniak, P. Fine Particles Content in Dust Created in CNC Milling of Selected Wood Composites. *Wood Fiber Sci.* **2017**, *49*, 461–469.

36. Očkajová, A.; Kučerka, M.; Krišťák, L.; Ružiak, I.; Gaff, M. Efficiency of Sanding Belts for Beech and Oak Sanding. *BioResources* **2016**, *11*, 5242–5254. [CrossRef]

Publisher's Note: MDPI stays neutral with regard to jurisdictional claims in published maps and institutional affiliations.

Article

Optimization of Parameters for the Cutting of Wood-Based Materials by a CO_2 Laser

Ivan Kubovský, Ľuboš Krišťák *, Juraj Suja, Milada Gajtanska, Rastislav Igaz, Ivan Ružiak and Roman Réh

Faculty of Wood Sciences and Technology, Technical University in Zvolen, T. G. Masaryka 24, SK-960 01 Zvolen, Slovakia; kubovsky@tuzvo.sk (I.K.); jukas@itenec.sk (J.S.); gajtanska@tuzvo.sk (M.G.); igaz@tuzvo.sk (R.I.); ruziak@tuzvo.sk (I.R.); reh@tuzvo.sk (R.R.)
* Correspondence: kristak@tuzvo.sk

Received: 15 October 2020; Accepted: 12 November 2020; Published: 16 November 2020

Abstract: This article deals with the laser cutting of wood and wood composites. The laser cutting of wood and wood composites is widely accepted and used by the wood industry (due to its many advantages compared to, e.g., saw cutting). The goal of this research was to optimize the cutting parameters of spruce wood (*Pices abies* L.) by a low-power CO_2 laser. The influence of three factors was investigated, namely, the effect of the laser power (100 and 150 W), cutting speed (3, 6, and 9 mm·s^{-1}), and number of annual rings (3–11) on the width of the cutting kerf on the top board, on the width of the cutting kerf on the bottom board, on the ratio of the cutting kerf width on the top and bottom of the board, on the width of the heat-affected area on both sides of the cutting kerf (this applies to the top and bottom of the board), and on the degree of charring. Analysis of variance (ANOVA) and correlation and regression analysis were used for developing a linear regression model without interactions and a quadratic regression model with quadratic interactions. Based on the developed models, the optimization of parameter settings of the investigated process was performed in order to achieve the final kerf quality. The improvement in the quality of the part ranged from 3% to more than 30%. The results were compared with other research dealing with the laser cutting of wood and wood composites.

Keywords: laser cutting; wood; wood composites; cutting parameters

---

1. Introduction

The cutting of materials is one of the most widely used laser machining processes. The principle consists of the movement of a focused laser beam perpendicular to the plane of the machined surface. The absorption of high energy laser radiation causes the material to melt and subsequently evaporate, creating the desired cut. For this purpose, a powerful CO_2 laser is still very frequently used, producing a beam with a wavelength of 10.6 μm [1]. In the past, CO_2 lasers were by far the most widely used devices for laser cutting. In recent years, the laser with the largest market for industry use has been the fiber laser, with a wavelength of 1064 nm. In industrialized countries, such as Japan, material cutting operations performed by beam radiation are very widespread and have grown significantly in recent years [2,3]. Due to their unique efficiency and accuracy, these technologies have become the center of attraction in various fields of application [4]. The cutting process is fast, non-contact, and highly automated and there is only thermal stress in a very limited area, allowing the cutting of a wide range of materials. Additionally, low operating costs are achieved, despite the high initial investment and the requirement for qualified staff [5].

In the case of wood and wood-based products, laser machining is widely accepted by the wood industry [6–8]. The laser processing of wood was one of the earliest applications of laser use in the

1970s [9,10]. The advantage of the laser system is its ability to cut complicated patterns. The application of lasers in the furniture industry for automated cutting processes us also well-known [11,12]. In addition to cutting, CO_2 lasers are also used for the irradiation or engraving of a wood surface [13–17]. The main advantages of the laser cutting of wood in comparison with conventional cutting methods are the highly precise cut, flexibility to start and finish cutting at any point of the board, narrow kerf width (0.1–0.3 mm compared to saw cut of 3–6 mm), and extremely smooth surfaces [18–23]. Other advantages of laser wood cutting are the absence of tool wear, low noise emission and vibration, low sensitivity to very variable processing properties [23–26], reduced amount of sawdust [27–31], and low mechanical stress in the workpiece [32,33].

Factors influencing the process of cutting wood and wood composites by a laser can be divided into three groups [34–36], namely, the properties of the radiation beam, the properties of the laser device and the characteristics of the cutting process, and the properties of the workpiece, specifically, the effects of the beam power, mode, and polarization as aspects of optics; the location of the focal point; the feed speed; the gas-jet assist system and workpiece thickness; the density; and the moisture content [37–44].

In the case of wood composites, Lum et al. [45] investigated the parameters of the process of MDF board cutting with a beam of radiation generated by a CO_2 laser with a power of 520–530 W in pulse and continuous mode. The optimal setting of the parameters for the MDF board cutting was achieved by the experimental change of the cutting speed, the composition of the auxiliary gas and its pressure, and the change of the position of the radiation beam focusing in the material. The results showed that narrow kerf widths are achievable for MDF laser-cut boards, particularly for pulse mode cutting. Striation patterning, although masked by external charring, is evident, but this is of little significance to the overall quality of cut, as evidenced by the low surface roughness values obtained. Burnout was also minimal, even for angular profile cuts of small internal angles. These results were also in agreement with Powell [46]. Eltawahni et al. [47] investigated a means for selecting the process parameters for the laser cutting of MDF based on the design of experiments. They defined a methodology according to which we can partially evaluate the efficiency and quality of radiation beam cutting by the proportion of the cutting joint width on the cut upper workpiece surface to the cutting joint width on the cut workpiece bottom (the so-called ratio). The focal point position and laser power are the principal factors affecting the ratio. In the case of the upper board's surface kerf, the focal point position has the main role, and in the case of the lower kerf, the laser power and cutting speed have the main effect on the lower kerf width. The roughness of the cut section decreases as the focal point position and laser power increase. The roughness increases as the cutting speed and air pressure increase. Smoother cut sections could be processed, but with an increase in the processing operating cost. Barnekov et al. [48] investigated the effect of different focal point positions on the cut surface quality. They found that a smooth surface of the workpiece middle can be achieved with less charring with the focal point at or slightly above the middle. Most of these results are in harmony with previous published research in the case of wood and wood composites [49–51]. The focal point position was also investigated by Barnekov et al. [40]. In their research, they achieved the best quality of particleboard laser cutting when the beam was focused on the surface of the composite material.

In the case of wood, Tayal et al. investigated the significance of the focus position of the CO_2 laser optical system when hardwood cutting [52]. They concluded that the focal point should be at a location Z/2 below the surface to achieve a maximum average laser power density, where Z is the focal length (depth of focus). They verified these results by experimental observations. Nukman et al. [53] used a CO_2 laser to cut a wide range of Malaysian wood and plywood. The processing variables taken into account were the laser power, focal point position, nozzle size, assist gas pressure, types of assist gas, cutting speed, and delay time. The wood properties observed were the thickness, density, and moisture content of wood. The analyses considered the geometric and dimensional accuracy (straight sideline length, diameter of the circle, kerf width, and percent over cut), material removal rate, and severity of burning. A guideline for cutting a wide range of Malaysian wood has been outlined. The influence of the laser power, cutting speed, and shield gas on the cut quality in the case of hard and soft timber was confirmed by Khan et al. [6]. McMillin [54] dealt with the moisture content and its influence on the process of pine wood cutting by a beam of radiation generated by a CO_2 laser in his work. He found that the cutting speed of wetter wood is slower compared to drier wood. The reason for this is that the moisture content in wood increases the thermal conductivity, which results in energy loss in the heating zone. Grad and Mozina [41] showed that the CO_2 laser beam is almost completely absorbed by wood. This result has been confirmed by Hattori [42], who claimed that a CO_2 laser is most suitable for wood processing due to the wavelength used. The authors of the research [55,56] proved that a cut section quality with less roughness results from an increasing beam power.

The optimal conditions for laser cutting were determined by several authors using theoretical models. Zhou and Mahdavian [57] explored the possibilities of cutting wood and particleboards with various levels of laser power and different workpiece cutting speeds. They introduced a theoretical model that estimates the depth of the cut in terms of the material properties and cutting speed. The experimental cutting results were compared with theoretical predictions. Two correction parameters were introduced in the analysis to improve the theoretical model. Numerous other theoretical models for laser machining have been developed by other researchers, among them, Moradi et al. [58], Choudhury et al. [59], Yang et al. [60], Elsheikh et al. [61], Alizadeh and Omrani, and others [62–64]. Various modes of heat transfer, phase changes, workpiece motions, and material properties have been taken into account in these analysis.

There are also several simulations and mathematical and statistical models, in addition to theoretical models. Statistical and experimental analyses of the multiple-pass laser cutting of wet and dry pine wood were presented by Castaneda et al. [65]. The parameters investigated were the laser power, traverse speed, focal plane position, gas pressure, number of passes, direction of cut (normal or parallel to the wood´s tracheids), and moisture content. The experimental results were compared against process responses defining the efficiency (i.e., kerf depth and energy consumption) and quality of the cut section (i.e., kerf width, heat-affected zone, edge surface roughness, and perpendicularity). An energy balance-based simple analytical model was developed and validated with experimental results by Prakash Kumar [66]. The optimal properties of the cutting process can also be determined by suitable simulation methods. Jianying and Yun [67] developed a simulation technique with ANSYS software using the finite element method in order to propose the optimal parameters of the wood beam cutting process. Polak et al. [68] used ANSYS and the numerical finite element method to model the radiation beam cutting and drilling process. Yang et al. [69,70] analyzed the influence of the ablative mechanism of wood processed with a nanosecond laser on the cutting quality and established a prediction model through multiple linear regression equations. Hardalov et al. [71] used the finite element method in their work to quantify the temperature gradient on the surface of ceramic material in the process of irradiation with pulsed and continuous beams using FEMLAB software. In the research by Modest [72], a previously developed three-dimensional conduction model for the scribing of a thick solid was extended to predict the transient temperature distribution inside a finite thickness slab that was irradiated by a moving laser source. The governing equations were solved for both constant and variable thermophysical properties using a finite-difference method on a boundary-fitted coordinate system. An evaluation of the radiation beam cutting quality was also addressed by Rogerro et al. [73], whose method is a combination of neural networks and traditional algorithmic techniques. Bianco et al. [74] used COMSOL Multiphysics 3.2 software to solve a transient two- and three-dimensional temperature field irradiated by a beam of radiation. Babiak et al. [75] dealt with the analysis of temperature profiles created by a moving Gaussian energy beam on a wood surface. Yilbas et al. [76] analyzed both mathematical and numerical solutions of material heating by a beam of radiation.

The aim of this research was to find the optimal combination of parameters of the cutting process of an 8 mm thick spruce board with a moisture content of 12% by a CO_2 laser. The required cut quality was determined by the minimum width of the heat-affected zone, the minimum value of the degree of charring, and the ratio of the cutting joint width on the upper and bottom board´s surface. The parameters were the laser power, cutting speed, and number of annual rings intersecting the area of the cutting kerf.

2. Materials and Methods

The experiments were carried out on Norway spruce (*Picea abies* (L.) H.Karst). Wood was harvested from the Polana region in Slovakia. Cutting kerf was created by the cutting of tangential spruce lumber with dimensions of 8 mm x 100 mm x 1000 mm (tangential x radial x longitudinal), with a relative moisture content $w = 12 \pm 1\%$ and average density $\rho = 428.4 \pm 27.9$ kg·m^{-3}. The moisture content of the samples was determined according to ISO 13061-1 [77]. The wood density was determined according to ISO 13061-2 [78]. The CO$_2$ laser LCS 400-1/W (TST Strojárne Piesok, Piesok, Slovakia) was used for cutting with a wavelength of 10.6 µm operated at a continuous mode output power of 100 and 150 W. Three speeds of cutting were used, with values 3, 6, and 9 mm·s^{-1}. A power of 150 W and speed of cutting of 9 mm·s^{-1} were used as a reference, since this combination is used in practice due to the speed of production. The focal length was 127 mm (5″), beam diameter was 10 mm, and spot diameter was 0.3 mm. The focal point position of the laser beam was set to 1/2 of the sample thickness (measured from the upper surface of the board). The process gas was supplied via a laval contour nozzle with 0.25 MPa pressured air.

All cuts were made parallel to the wood fibers in the tangential direction (Figure 1). The quality of the cut section of the wood was examined using Digital Microscopy (DM).

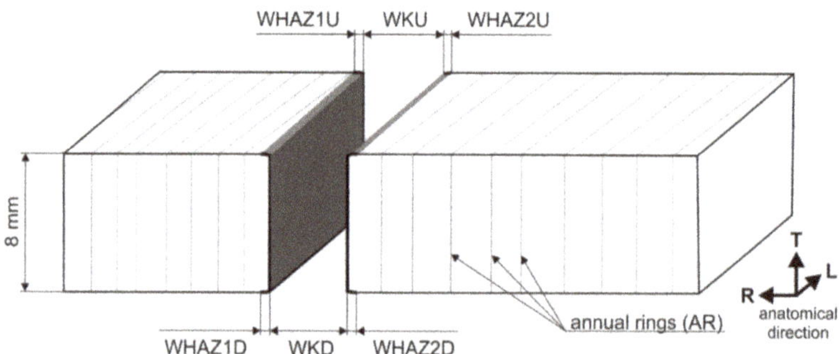

Figure 1. Cutting scheme.

The Response Surface Method (RSM) was used to prepare the experiment. The influence of three factors was investigated, namely, the effect of the laser power P, cutting speed V, and number of annual rings AR (Table 1) on the width of the cutting kerf WKU (width of the cutting kerf on the upper board´s surface) and WKD (cutting kerf width on the board bottom), on their ratio WKR (ratio of the cutting kerf width on the upper and bottom board´s surface), on the width of the heat-affected areas on the spruce board surface WHAZx (width of the heat-affected area on both sides of the cutting kerf is equal, and this applies to the upper board´s surface, as well as to the bottom of the board), and on the degree of charring B.

Table 1. Experimental factors.

Factor	Min. Value	Max. Value	Levels Number	Factor Levels	Unit	Type
AR	3	11	9	3, 4, 5, 6, 7, 8, 9, 10, 11	-	random
P	100	150	2	100, 150	W	controlled
V	3	9	3	3, 6, 9	mm·s^{-1}	controlled

Measuring of the cutting kerf width WKU, WKD, and WHAZx was observed by DM using the K-means clustering segmentation method [79]. The K-means clustering algorithm is an unsupervised algorithm and it clusters or partitions the given data into K-clusters based on the K-centroids. The algorithm is used when we have unlabeled data (without defined categories) and the objective is to minimize the sum of squared distances between all points and the cluster center. In our case, three different color-coded regions were searched in image processing software, namely, the area of the unaffected workpiece, the area of the heat-affected zone, and the area of the cutting kerf. The mean values of the width of the cutting kerf and especially the width of the heat-affected zones were determined from the segmented photos by applying the theorem on the mean value theorem for integrals (Figure 2).

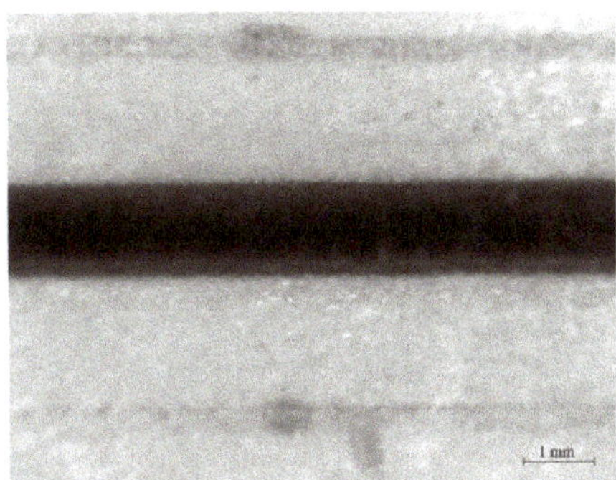
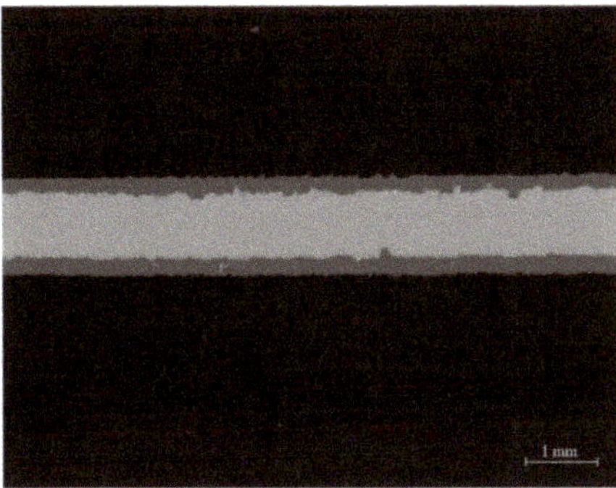

Figure 2. Photo used for measuring the width of the cutting kerf and heat-affected zone (**left**), and photo edited by segmentation (**right**) by Digital Microscopy (DM).

The measuring of the degree of charring consisted of an analysis of cutting kerf in a direction perpendicular to the surface and a subsequent analysis of the morphology (texture) of the examined surface. Surface charred areas are amorphous areas of cutting kerf without a visible texture with a damaged anatomical wood structure due to the thermal degradation of wood. The bimodal histogram was modeled as a mixture of two Gaussian density functions. The use of adaptive particle swarm optimization for the suboptimal estimation of the means and variances of these two Gaussian density functions, and then the computation of the optimal threshold value, was straightforward [80]. The result of this analysis was obtained as the degree of charring in %.

The Shapiro–Wilk test was used to test the normality of measured data, followed by Box-Cox nonlinear normalizing transformation. For greater data clarity of the response, variables were transformed into variables that also passed the normality test. Analysis of variance (ANOVA) and correlation and regression analysis were used for developing a linear regression model without interactions and a quadratic regression model with quadratic interactions. Based on the developed models, the optimization of parameter settings of the investigated process was performed in order to achieve the required response.

3. Results and Discussion

The total number of measurements was 108 in one block, with 864 measured or calculated data. Due to the random factor AR, nine measurements were performed at each monitored cutting speed. The WKR ratio was calculated and the degree of charring was subsequently determined from the measured data. The data were then tested by the Shapiro–Wilk test of good agreement of the measured data, with the normal distribution at a significance level of 0.05. Due to the fact that the measured data did not meet the assumption of a normal distribution, a normal distribution was achieved by a nonlinear Box-Cox transformation (Table 2).

Table 2. Box-Cox transformation.

Reaction	Box-Cox Transformation
WKU*	$1 + (WKU^{-0.569} - 1)/(-0.569 \cdot 0.847226^{-1.569})$
WHAZU*	$1 + (WHAZU^{0.134} - 1)/(0.134 \cdot 0.172185^{-0.866})$
WKD*	$1 + (WKD^{0.206} - 1)/(0.206 \cdot 0.460085^{-0.794})$
WHAZD*	$1 + (WHAZD^{-0.605} - 1)/(-0.605 \cdot 0.169228^{-1.605})$
WKR*	$1 + (WKR^{0.372} - 1)/(0.372 \cdot 0.543049^{-0.628})$
B*	$1 + (B^{2.049} - 1)/(2.049 \cdot 46.4258^{1.049})$

The results of the measured data after the transformation enabled the use of parametric mathematical-statistical methods of experimental evaluation (in the next parts asterisk symbol * is used for transformed values). Descriptive statistics were subsequently produced for all monitored transformed responses (Table 3).

Table 3. Descriptive statistics for transformed values of all parameters.

	WKU*	WHAZU*	WKD*	WHAZD*	WKR*	B*
Average	0.8566	0.1741	0.5524	0.2192	0.6337	0.4688
Standard deviation	0.1761	0.0657	0.2093	0.1471	0.1709	0.0079
Variation coefficient	20.55%	37.75%	37.88%	67.14%	26.97%	16.74%
Displacement coefficient	−0.42	2.344	1.964	4.685	1.452	−1.328
Peak coefficient	−1.256	−0.1885	−0.5222	2.288	−1.079	−0.657
95% confidence interval for the average	<0.8085; 0.9046>	<0.1561; 0.1920>	<0.4953; 0.6095>	<0.1179; 0.2593>	<0.5870; 0.6803>	<0.4474; 0.4902>
95% confidence interval for the standard deviation	<0.148; 0.2174>	<0.0553; 0.0811>	<0.1759; 0.2583>	<0.1237; 0.1816>	<0.1437; 0.2110>	<0.0660; 0.0969>

It follows from the descriptive statistics tables for each monitored transformed response that the cutting kerf width at the board top is larger than the cutting kerf width at the board bottom. The term "upper board´s surface" refers to the side where the primary radiation beam interacts with the material. The values of the variation coefficients are smaller on the upper board´s surface compared to the values of the variation coefficients of the observed responses on the board bottom. This is related to the properties of the wood from the point of view of heat transfer, as well as to the power of the radiation beam required for thermal decomposition of the wood in the cutting process. Larger values of variation coefficients also point to defects in the form of burns on the surface of the board bottom, which are caused by the sudden blowing of burns from the space of the cutting kerf with auxiliary gas (air).

3.1. Analysis of Variance and Regression Analysis

A three-factor experimental model without interactions was used to primarily determine the influence of the investigated factors AR, P, and V for responses WKU*, WHAZU*, WKD*, WHAZD*, WKR*, and B*, with an evaluation of the variability degree of the investigated response (Table 4). Additional information on the influence of interactions between factors was analyzed by a second-degree polynomial model with first-order interactions (Table 5). The coefficients of the factors of the examined responses are marked in gray, which are significant in terms of impact at the significance level of 0.05. The force measures of the models are also marked in gray, which can be considered as sufficiently explanatory of the spruce board cutting process. The coefficients of the regression function were determined by the least squares method. The significance of the investigated factors was evaluated by analysis of variance (F test, p test, R^2, standard deviation, PACH analysis, and percentage error of the model were performed in all cases), graphically by a Pareto diagram, and by a graphical analysis of their influence (Figure 3).

Table 4. Summary of linear models without interactions (e.g., WKU* = 0.9691 + 0.0031·AR + 0.1762·P − 0.0586·V). Background color was used for statistically significant values for 99% confidence interval.

Response	WKU*	WHAZU*	WKD*	WHAZD*	WKR*	B*
Factor	Coeff.	Coeff.	Coeff.	Coeff.	Coeff.	Coeff.
AR	0.0031	0.0021	−0.0112	0.0003	−0.0177	0.00487
P	0.1762	0.0451	0.3501	0.2583	0.2847	−0.05437
V	−0.0586	0.0168	−0.061	−0.0110	−0.0304	−0.00415
Constant	0.9691	0.5613	0.6486	0.5189	0.5699	0.22126
R^2	71.76%	49.14%	84.65%	46.51%	46.74%	2.73%

Table 5. Summary of linear models with interactions (e.g., WKU*= 1.96353 + 0.00695·V^2 + 0.08847·P*V + 0.00032·AR*V − 0.00745·AR*P + 0.00526·AR^2 − 0.25323·V − 0.29207·P − 0.05862·AR). Background color was used for statistically significant values for 99% confidence interval.

Response	WKU*	WHAZU*	WKD*	WHAZD*	WKR*	B*
Factor	Coeff.	Coeff.	Coeff.	Coeff.	Coeff.	Coeff.
AR	−0.05862	0.02398	0.06392	0.01209	0.10382	−0.03579
P	−0.29207	0.19104	0.81739	0.00574	1.17681	−0.51849
V	−0.25323	0.08067	0.03693	−0.05375	0.23656	−0.15057
AR^2	0.00526	−0.00142	0.00034	−0.00349	−0.00263	0.00107
AR*P	−0.00745	0.00257	−0.05651	0.02135	−0.05468	0.01333
AR*V	0.00032	−0.00108	−0.00208	0.00121	−0.00352	0.00172
P*V	0.08847	−0.02749	−0.02015	0.01778	−0.09821	0.06385
V^2	0.00695	−0.00193	−0.00489	0.00099	−0.09964	0.00467
Constant	1.96353	0.23073	−0.13999	−0.78817	−1.01551	1.04444
R^2	85.88%	56.49%	87.93%	50.50%	62.53%	29.40%

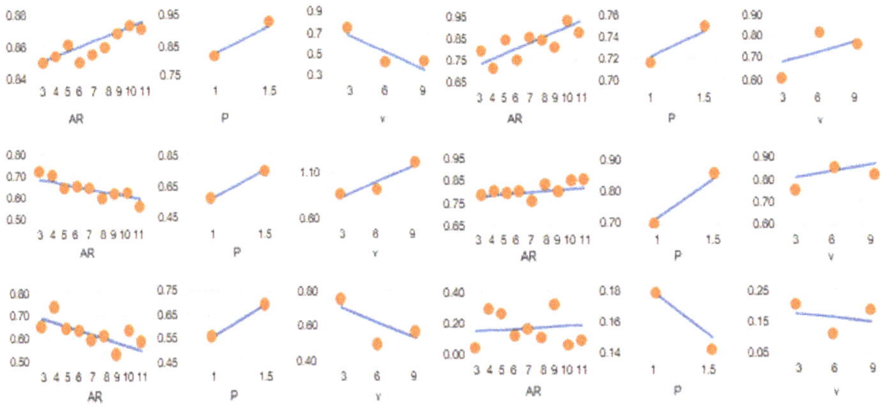

Figure 3. Influence of the main factors: First line (left) WKU* and (right) WHAZU*; second line (left) WKD* and (right) WHAZD*; and last line (left) WKR* and (right) B*.

It follows from the table of results on the analysis of variance for WKU* for the transformed values of the cutting kerf width (Table 4) that the influence of the cutting speed factor V and the laser power P is statistically significant. The influence of the AR factor was not confirmed by the mathematical-statistical analysis, which is related to the wood density. The reason for this may be the fact that the thermal conductivity of wood only plays a major role in heat transfer in the phase of penetration of the energy beam into the material, while on the surface, respectively in the phase of energy contact with the board surface, the surface properties of wood are more pronounced, e.g., color, roughness, reflectivity, and the properties of the laboratory environment. The cutting speed has the greatest influence on the value of the width size of the cutting kerf on the upper board´s surface. However, this effect is disproportionate, and the width of the cutting kerf WKU* decreases with an increasing cutting speed. We can explain this phenomenon by the decreasing amount of energy interacting with the board material per unit time. For this reason, it is necessary to include the interactions between the investigated factors in the model, which correlates with the results of Hernandez [65] in the case of Scots pine, with Ready [81] in the case of Silver fir, and with Liu et al. [82] in case of cherry wood. On the contrary, the width of the cutting kerf increases with an increasing laser power, which was confirmed in the work of Nukman et al. [53] in their research on selected Malaysian wood cutting. If we consider the factor AR, then its effect on WKU*, although negligible, is directly proportional. In the case of the interaction model of factors for WKU*, in addition of the factors P and V, the interactions P*V and V^2 are also statistically significant. This confirmed the assumption of the mutually important influence of V and P on the cutting kerf width, so these factors significantly affect the amount of absorbed and reflected energy in the board structure, as pointed out by Hernandez in his work [65] on pine wood cutting, as well as Barnekov [48]. A statistically significant factor AR in the form of the self-interaction AR^2 already appeared in the interactions model. This is indicated by the fact that the wood density affects the properties of the cutting kerf, which is also stated in the work of Asibu [83].

It follows from the results of the analysis of variance for the transformed values of WHAZU* that the influence of the cutting speed factor V and the laser power factor P is statistically significant. The influence of the factor AR, which is related to the wood density, was not confirmed by the mathematical-statistical analysis. The significant influence of the speed factor V points to the fact that the cutting speed significantly affects the energy distribution in the cutting process by the radiation beam on the workpiece surface at the point of energy interaction with the material. This is consistent with Hernandez's results [65]. This follows from a detailed analysis showing that the cutting speed V and laser power P have the greatest influence on the width value of the heat-affected area on the upper board´s surface in comparison with the other investigated factors. In the case of the cutting speed V, the effect is directly proportional, and the width of the heat-affected area WHAZU* increases with an increasing cutting speed. The result obtained is in contrast to the findings published in the work of Asibu [83], Barcikowsky [7], and Lum [45]. The reason for this is that these authors examined the heat-affected area below the surface of cut wood, while the subject of our research was the size of the heat-affected area on the board surface. This reduces the amount of energy in the process of decomposing the material below its surface by increasing the cutting speed, as well as increasing the amount of energy on the board surface. Therefore, it is clear that it increases the area affected by heat due to energy excess on the board surface, which is consistent with the results of Arai and Hayashi, [84,85]. The width of the heat-affected zone increases as the power of the laser P increases, as Barcikowski [7] also pointed out. If we also take into account the factor AR, then its effect on WHAZU*, although also negligible, is directly proportional. All model quality indicators improved after the inclusion of interactions in the model WHAZU*, but not so significantly that we can consider this model as credible. However, we can use it to explain the influence of the investigated factors and their interactions on the investigated response, in this case, WHAZU*. The mutual interaction between the laser power P and the cutting speed V determines the degree and form of energy distribution, as in the case of the WKU* or WHAZU*; this causes the wood chemical decomposition and thus determines the width of the heat-affected area. The cutting speed also affects the properties at the same time, and in particular, the distribution of the flowing auxiliary gas on the board surface.

It follows from the results of the analysis of variance for the transformed values of the cutting kerf width on the board bottom of the WKD* that the influence of the cutting speed factor V, the laser power P, and the factor AR is statistically significant. These results are in harmony with Ready [79]. The reason for this is again that the wood thermal conductivity plays an important role in heat transfer in the phase of penetration of the energy beam into the material, while on the surface, respectively in the phase of energy contact with the board surface, the surface properties of wood are manifested analogously, as in the case of WKU* and WHAZU*. It is clear that the cutting speed V increase, as well as the density increase, and respectively the number of annual rings AR, results in a reduction in the cutting kerf width on the board bottom. The results correspond to the results of Mahdavian and Zhou [57] or Lum and Black [45,49] in a cutting depth analysis. On the contrary, an increase in the power P of the laser results in an increase in the value of the cutting kerf width WKD* on the board bottom, which is in accordance with the published results, e.g., Eltawahni et al. [47]. It was found, by comparing the effect of the factors for both widths WKU* and WKD*, that the power factor P and the speed factor V act in the same direction on the width values of the WKU* and WKD*. However, the factor AR, respectively the wood density, acts to increase the cutting kerf width on the board surface, in contrast to the effect on the board bottom, where it reduces this width by its influence. The extent value of the variance explanation only improved insignificantly in the case of the interaction model for WKD*, so the model without interactions sufficiently and simply describes the influence of the factors P, V, and AR on WKD*.

The effect of the factor AR could not be confirmed by the results of the variance analysis for WHAZD*; however, a significant effect was confirmed by the cutting speed factors V and laser power P. It is clear that the value of WHAZD* increases with an increasing laser power P and, conversely, the value of the width of the heat-affected zone on the surface of the board bottom decreases with an increasing speed V. It follows from the results of a detailed analysis that a significant improvement in all properties of the model will not be achieved, even when considering the interactions between the investigated factors. Considering this, it is necessary to include other factors in the analysis in the case of continuing research, e.g., the auxiliary gas flow rate.

It follows from the results of the variance analysis for the transformed values of the width ratio of the cutting kerf WKR* that the influence of the cutting speed factor V, the laser power P, and also the factor AR is statistically significant. Analogous to WKD*, the wood thermal conductivity plays an important role in the material, while the wood surface properties are reflected on the surface. A speed increase, as well as density increase, causes the WKR* value to decrease, while a power increase causes the WKR* value to increase. The results are in line with the results obtained for the analysis of WKU* and WKD*, as well as the works dealing with the joint assessment of the cutting kerf width on the upper and bottom board´s surface in the case of other softwoods, e.g., those by Ready [81] and Eltawahni et al. [47]. It improved all the indicators in the model in the case of the model with interactions for WKR* and this was mainly due to the mutual interaction between power P and speed V, while the factor of power P and speed V had the most significant effect on WKR*.

Analysis of variance did not confirm a statistically significant effect of the investigated factors AR, P, and V in the case of B*. Nevertheless, it is clear from a physical point of view that the power factor P, together with the cutting speed V, i.e., feed of the energy source, determines the amount of energy absorbed by the wood, and thus the amount of energy needed to create the burns. Peters and Banas [86] and Barnekov et al. [48] also reached analogous conclusions. This is due to a decrease in the reaction time of the radiation beam interaction on the workpiece material. It evaporates the material at the upper board´s surface just below the surface due to the uneven distribution of energy supplied to the board material. Below the area, where the material evaporates, is the area where there is less energy, and thus the wood is decomposed by burning. The interactions between the laser power P and the cutting speed V are statistically significant, highlighted by the interaction V*V in the case of the interaction model.

3.2. Optimal Parameter Settings of the Investigated Process

It is desirable to minimize the subsequent post processing due to minimizing the cost of the cutting process by the energy beam. It is possible to achieve cost minimization by optimally setting the parameters of the investigated process, in order to achieve the required values of process responses based on a mathematical-statistical model. The goal is to reach a minimum value in the case of WKU*, WHAZU*, WKD*, WHAZD*, and B*, and in the case of WKR*, the goal is to achieve the value 1 and at the same time to find the optimal values of the factors P, V, and AR. There are six polynomial functions based on Table 5, five of which need to be minimized and one of which needs to be normalized, i.e., needs to be equal to one:

$$WKU^* = 1.96353 + 0.00695{\cdot}V^2 + 0.08847{\cdot}P^*V + 0.00032{\cdot}AR^*V - 0.00745{\cdot}AR^*P + 0.00526{\cdot}AR^2$$
$$- 0.25323{\cdot}V - 0.29207{\cdot}P - 0.05862{\cdot}AR = \mathbf{min,}$$

$$WHAZU^* = 0.23073 - 0.00193{\cdot}V^2 - 0.02749{\cdot}P^*V - 0.00108{\cdot}AR^*V + 0.00257{\cdot}AR^*P - 0.00142{\cdot}AR^2$$
$$+ 0.08067{\cdot}V + 0.19104{\cdot}P + 0.02398{\cdot}AR = \mathbf{min,}$$

$$WKD^* = -0.13999 - 0.00489{\cdot}V^2 - 0.02015{\cdot}P^*V - 0.00208{\cdot}AR^*V - 0.05651{\cdot}AR^*P + 0.00034{\cdot}AR^2$$
$$+ 0.03693{\cdot}V + 0.81739{\cdot}P + 0.06392{\cdot}AR = \mathbf{min,}$$

$$WHAZD^* = -0.78817 + 0.00099{\cdot}V^2 + 0.01778{\cdot}P^*V + 0.00121{\cdot}AR^*V + 0.02135{\cdot}AR^*P - 0.00349{\cdot}AR^2$$
$$- 0.05375{\cdot}V + 0.00574{\cdot}P + 0.01209{\cdot}AR = \mathbf{min,}$$

$$B^* = 1.04444 + 0.00467{\cdot}V^2 + 0.06385{\cdot}P^*V + 0.00172{\cdot}AR^*V + 0.01333{\cdot}AR^*P + 0.00107{\cdot}AR^2 - 0.15057{\cdot}V$$
$$- 0.51849{\cdot}P - 0.03579{\cdot}AR = \mathbf{min,}$$

$$WKR^* = -1.01551 - 0.09964{\cdot}V^2 - 0.09821{\cdot}P^*V - 0.00352{\cdot}AR^*V - 0.05468{\cdot}AR^*P - 0.00263{\cdot}AR^2$$
$$+ 0.23656{\cdot}V + 1.17681{\cdot}P + 0.10382{\cdot}AR = \mathbf{1.}$$

The modeling was performed by the method of linear programming, which determined the optimal values of the investigated factors (based on a set of solutions of a system of six linear equations and inequations using the simplex method of solving the general problem of linear programming).

The results of joint optimization of the investigated factor settings reached values of AR = 3, P = 100 W, and v = 9 mm·s^{-1}. The suitability of the joint optimization reached the value of 0.92. We can conclude that we can approach the optimal parameters of the cutting process of a given board by increasing the cutting speed at the optimal performance based on the analysis of the response area (Figure 4 left). It is possible to achieve the same or a similar result by reducing the power of the radiation beam at the optimum cutting speed of a given board (Figure 4 right).

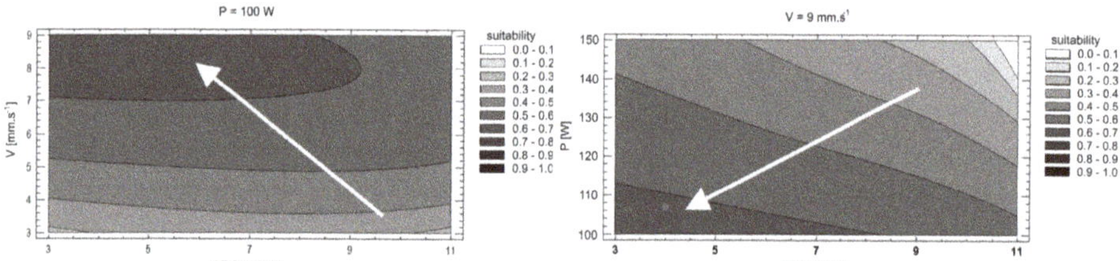

Figure 4. Response areas for joint optimization in the case of fixed power P = 100 W (**left**) and in the case of fixed cutting speed V = 9 mm·s^{-1} (**right**).

The correctness of the optimization was also confirmed by Table 6. From this table, it is clear to see that the chosen parameters of cutting after optimization (AR = 3, P = 100 W, and v = 9 mm·s^{-1}) are improved over the reference parameters of cutting (AR = 3, P = 150 W, and v = 9 mm·s^{-1}) by 3% to 30%.

Table 6. A table presenting the parameters at 150 W power (reference) and 100 W power (optimal), for a cutting speed of 9 mm·s^{-1} and AR = 3.

Parameter	Goal	Cutting Kerf Width (mm) Reference	Cutting Kerf Width (mm) Optimal	Improvement (%)
WKU*	min	0.715	0.627	12.3
WKD*	min	0.591	0.416	29.6
WKR*	1	0.528	0.670	21.1
WHAZU*	min	0.786	0.764	2.7
WHAZD*	min	0.808	0.679	16
B*	min	0.117	0.144	-

In case we need to optimize only one parameter separately, we can achieve even better results, since, in the case of joint optimization, the values are related to each other. The obtained values of separate optimization for all parameters are given in Table 7. In practice, in the case of laser wood cutting, it is not always necessary to achieve the optimized quality of all parameters at the same time. In some cases, with specific requirements for the cutting zone created in laser cutting, optimization of the cutting zone for only one parameter is needed, e.g., the maximum surface flatness is required for subsequent gluing (WKR = 1); in case of increased requirements for the tensile shear strength of the glue joint, minimization of charring is required (B = min); and in the case of subsequent edging and the requirement of one viewing area, minimization of the HAZ is required (WHAZxU alebo WHAZxD = min). In some cases, the perfect surface is counterproductive (in the case of adhesion of the glue, etc.). Alternatively, there is a requirement to combine the optimization of two parameters, e.g., at subsequent edging and the requirement of two viewing surfaces (WHAZxU and WHAZxD = min). These requirements are applied, especially in the case of wood composites (plywood, particleboard, fiberboard, etc.), in the production of furniture elements. In practice, in most cases, the second optimization is more important due to the above reasons.

Table 7. Results of separate optimization of the investigated factor settings.

	Goal	Cutting Kerf Width (mm)	Suitability	AR	P (W)	V (mm·s^{-1})
WKU* (WKU)	min	0.627 (0.565)	0.92	3	100	9
WKD* (WKD)	min	0.327 (0.334)	0.92	11	100	9
WKR* (WKR)	1	0.853 (0.853)	0.92	3	150	3
WHAZU* (WHAZU)	min	0.663 (0.113)	0.92	3	100	3
WHAZD* (WHAZD)	min	0.679 (0.093)	0.88	3	100	9
B* (B)	min	0.117 (33.245)	0.69	3	150	9

The obtained results show a high correlation with experimentally measured results, despite the difficulties largely resulting from the anisotropy of spruce wood, and at the same time, they are comparable with the results of authors investigating cutting by a radiation beam generated by a CO_2 laser for other wood species, as well as wood composites [53,65,81–83].

4. Conclusions

The goal of our research was to find the optimal combination of parameters of the cutting process with a beam of a CO_2 laser energy of a 8 mm thick spruce board with a moisture content of 12% and thus to achieve the required quality of cutting. The targets of past research were mainly equivalent homogeneous wood material. The target of this research was highly anisotropic material. Therefore, parameter AR was considered and discussed, in addition to standard parameters, such as the effect of the cutting speed V and beam power output P, and the influence of these parameters on responses was also analyzed too, including the cutting kerf width on the upper board´s surface WKU, cutting kerf width on the board bottom WKD, the width of the heat-affected area on the upper board's surface WHAZU and the board bottom WHAZD, the ratio of the cutting kerf width on the board bottom to the width on the upper board´s surface WKR, and the degree of char B. Using a mathematical-statistical model, we achieved the optimal parameters (AR = 3, 100 W, and 9 mm·s^{-1}). These parameters of cutting are much better than the non-optimized control parameters (AR = 3, 150 W, and 9 mm·s^{-1}). The improvement in the quality ranged from 3% to more than 30%. The results exhibit a high correlation with experimental measurements and they are also comparable to the results of research on the cutting of more homogeneous wood and wood composite materials cut with a CO_2 laser.

Research has further shown that even inhomogeneous materials such as spruce wood can be cut with a low power laser. However, in order to make this process more predictable, further research should focus on assessing the impact of other factors, such as the spruce moisture content and composition and flow rate of the auxiliary gas, and extending the ranges of factors considered, especially the beam power output and the cutting speed. We can use the method of measuring the cutting kerf width in other areas, where we require a fast and inexpensive method to measure small lengths, which are determined by the resolution of the scanning device. At the same time, this method is suitable for the proposal of electronic equipment that would perform the measurements of small two-dimensional objects. The used method of char measuring can be applied in the mass identification of errors or characteristic properties of the surface of the scanned workpiece by division into characteristic areas.

As mentioned in the introduction, from a comprehensive point of view, the cutting process of wood and wood-based products with a CO_2 laser can be evaluated as a fast, highly-automated, and workpiece-friendly approach that is suitable for cutting homogeneous and inhomogeneous wood and wood composite materials. Other benefits include the highly precise cut, the narrow kerf width, the smooth surface, no tool wear, and the reduced amount of sawdust.

Innovative solutions are essential for sustaining the market position in the competitive business environment. This clearly supports the position of the wood-working industry among the sectors completely fulfilling the requirements of green business products and principles of sustainable development [87–91].

Author Contributions: Conceptualization, I.K. and R.I.; methodology, M.G.; formal analysis, Ľ.K. and J.S.; investigation, J.S. and M.G.; resources, Ľ.K.; data curation, I.R.; writing—original draft preparation, Ľ.K. and R.R.; writing—review and editing, Ľ.K. and R.R. All authors have read and agreed to the published version of the manuscript.

Funding: This research was supported by the Slovak Research and Development Agency under contract no. APVV-18-0378, APVV-19-0269, and VEGA 1/0717/19.

Conflicts of Interest: The authors declare no conflict of interest.

References

1. Masoud, F.; Sapuan, S.; Mohd Ariffin, M.K.A.; Nukman, Y.; Bayraktar, E. Cutting Processes of Natural Fiber-Reinforced Polymer Composites. *Polymers* **2020**, *12*, 1332. [CrossRef] [PubMed]
2. Yoshida, K.; Yahagi, H.; Wada, M.; Kameyama, T.; Kawakami, M.; Furukawa, H.; Adachi, K. Enormously Low Frictional Surface on Tough Hydrogels Simply Created by Laser-Cutting Process. *Technologies* **2018**, *6*, 82. [CrossRef]
3. Yang, L.; Wei, J.; Ma, Z.; Song, P.; Ma, J.; Zhao, Y.; Huang, Z.; Zhang, M.; Yang, F.; Wang, X. The Fabrication of Micro/Nano Structures by Laser Machining. *Nanomaterials* **2019**, *9*, 1789. [CrossRef] [PubMed]
4. Abidou, D.; Yusoff, N.; Nazri, N.; Awang, M.A.O.; Hassan, M.A.; Sarhan, A.A.D. Numerical simulation of metal removal in laser drilling using radial point interpolation method. *Eng. Anal. Bound. Elem.* **2017**, *77*, 89–96. [CrossRef]

5. Mukherjee, K.T.; Grendzwell, P.A.A.; McMillin, C.W. Gas-flow parameters in laser cutting of wood–nozzle design. *For. Prod. J.* **1990**, *40*, 39–42.

6. Khan, P.A.A.; Cherif, M.; Kudapa, S.; Barnekov, V.; Mukherjee, K. *High Speed, High Energy Automated Machining of Hardwoods by Using a Carbon Dioxide Laser: ALPS*; Laser Institute of America: Orlano, FL, USA, 1991; Volume 2, pp. 238–252.

7. Barcikowski, S.; Koch, G.; Odermatt, J. Characterisation and modification of the heat affected zone during laser material processing of wood and wood composites. *Holz Roh Werkst.* **2006**, *64*, 94–103. [CrossRef]

8. Huber, H.E.; McMillin, C.W.; Rasher, A. Economics of cutting wood parts with a laser under optical image analyzer control. *For. Prod. J.* **1982**, *32*, 16–21.

9. Belforte, D.A. Non-metal cutting. *Ind. Laser Rev.* **1998**, *13*, 11–13.

10. Powell, J.; Ellis, G.; Menzies, I.A.; Scheyvaerts, P.F. CO_2 laser cutting of non-metallic materials. In Proceedings of the 4th International Conference Lasers in Manufacturing, Birmingham, UK, 12–14 May 1987.

11. Wieloch, G.; Pohl, P. Use of laser in the furniture industry. In Proceedings of the Laser Technology IV: Research Trends, Instrumentation, and Applications in Metrology and Materials Processing, Szczecin, Poland, 1 March 1995; Volume 2202, pp. 604–607.

12. Pires, M.C. Plywood inlays through CO_2 laser cutting. In Proceedings of the CO_2 Laser and Applications, Los Angeles, CA, USA, 28 July 1989; Volume 1042, pp. 97–102.

13. Kúdela, J.; Kubovský, I.; Andrejko, M. Surface properties of beechwood after CO_2 laser engraving. *Coatings* **2020**, *10*, 77. [CrossRef]

14. Kúdela, J.; Kubovský, I.; Andrejko, M. Impact of different radiation forms on beech wood discolouration. *Wood Res.* **2018**, *63*, 923–934.

15. Kubovský, I.; Kačík, F.; Veľková, V. The effects of CO_2 laser irradiation on color and major chemical component changes in hardwoods. *Bioresources* **2018**, *13*, 2515–2529. [CrossRef]

16. Kubovský, I.; Kačík, F.; Reinprecht, L. The impact of UV radiation on the change of colour and composition of the surface of lime wood treated with a CO_2 laser. *J. Photochem. Photobiol. A Chem.* **2016**, *322*, 60–66. [CrossRef]

17. Kubovský, I.; Kačík, F. Colour and chemical changes of the lime wood surface due to CO_2 laser thermal modification. *Appl. Surf. Sci.* **2014**, *321*, 261–267. [CrossRef]

18. Sinn, G.; Chuchala, D.; Orlowski, K.; Tauble, P. Cutting model parameters from frame sawing of natural and impregnated Scots Pine (*Pinus sylvestris* L.). *Eur. J. Wood Wood Prod.* **2020**, *78*, 777–784. [CrossRef]

19. Očkajová, A.; Kučerka, M.; Kminiak, R.; Krišťák, L'.; Igaz, R.; Réh, R. Occupational Exposure to Dust Produced When Milling Thermally Modified Wood. *Int. J. Environ. Res. Public Health* **2020**, *17*, 1478.

20. Kučerka, M.; Očkajová, A. Thermowood and granularity of abrasive wood dust. *Acta Fac. Xylologiae Zvolen* **2018**, *60*, 43–51.

21. Gaff, M.; Razaei, F.; Sikora, A.; Hýsek, Š.; Sedlecký, M.; Ditommaso, G.; Corleto, R.; Kamboj, G.; Sethy, A.; Vališ, M.; et al. Interactions of monitored factors upon tensile glue shear strength on laser cut wood. *Compos. Struct.* **2020**, *234*, 111679. [CrossRef]

22. Martinez-Conde, A.; Krenke, T.; Frybort, S.; Muller, U. Review: Comparative analysis of CO_2 laser and conventional sawing for cutting of lumber and wood-based materials. *Wood Sci. Technol.* **2017**, *51*, 943–966. [CrossRef]

23. Pinkowski, A.; Krauss, A.; Sydor, M. The effect of spiral grain on energy requirement of plane milling od Scots pine (*Pinus sylvestris* L.) wood. *Bioresources* **2016**, *11*, 1930–2126.

24. Vlckova, M.; Gejdos, M.; Nemec, M. Analysis of vibration in wood chipping process. *Akustika* **2017**, *28*, 106–110.

25. Suchomel, J.; Belanova, K.; Gejdos, M.; Nemec, M.; Danihelova, A.; Maskova, Z. Analysis of Fungi in Wood Chip Storage Piles. *Bioresources* **2014**, *9*, 4410–4420. [CrossRef]

26. Sydor, M.; Rogozinski, T.; Stuper-Szablewska, K.; Starczewski, K. The Accuracy of Holes Drilled in the Side Surface of Plywood. *Bioresources* **2020**, *15*, 117–129.

27. Barcík, Š.; Kvietková, M.; Gašparík, M.; Kminiak, R. Influence of technological parameters on lagging size in cutting process of solid wood by abrasive water jet. *Wood Res.* **2013**, *58*, 627–636.

28. Igaz, R.; Kminiak, R.; Krišťák, L'.; Němec, M.; Gergel', T. Methodology of Temperature Monitoring in the Process of CNC Machining of Solid Wood. *Sustainability* **2018**, *11*, 95. [CrossRef]

29. Eltawahni, H.A.; Rossini, N.S.; Dassisti, M.; Alrashed, K.; Aldaham, T.A.; Benyounis, K.Y.; Olabi, A.G. Evaluation and optimization of laser cutting parameters for plywood materials. *Opt. Lasers Eng.* **2013**, *51*, 1029–1043. [CrossRef]

30. Orlowski, K.; Chuchala, D.; Muzinski, T.; Barariski, J.; Banski, A.; Rogozinski, T. The effect of wood drying method on the granularity of sawdust obtained during the sawing process using the frame sawing machine. *Acta Fac. Xylologiae Zvolen* **2019**, *61*, 83–92.

31. Rogozinski, T.; Wilkowski, J.; Gorski, J.; Szymanowski, K.; Podziewski, P.; Czarniak, P. Technical note: Fine particles content in dust created in CNC milling of selected wood composites. *Wood Fiber Sci.* **2017**, *49*, 461–469.

32. Hlaskova, L.; Orlowski, K.; Kopecky, Z.; Jedinak, M. Sawing Processes as a Way of Determining Fracture Toughness and Shear Yield Stresses of Wood. *Bioresources* **2015**, *10*, 5381–5394. [CrossRef]

33. Marková, I.; Mračková, E.; Očkajová, A.; Ladomerský, J. Granulometry of selected wood dust species of dust from orbital sanders. *Wood Res.* **2016**, *61*, 983–992.

34. Igaz, R.; Gajtanska, M. The influence of water vapour concentration in CO_2 laser active region on output power of emited beam. *Acta Fac. Tech.* **2014**, *19*, 35–40.

35. Gajtanska, M.; Igaz, R.; Krišťák, L'.; Ružiak, I. *Contamination of CO_2 Laser Mixture*; Technical University in Zvolen Publishing: Zvolen, Slovakia, 2014.

36. Gajtanska, M.; Suja, J.; Igaz, R.; Krišťák, L'.; Ružiak, I. *CO_2 Laser Cutting of Spruce Wood*; Technical University in Zvolen Publishing: Zvolen, Slovakia, 2015.

37. Antov, P.; Neykov, N. Costs of occupational accidents in the Bulgarian woodworking and furniture industry. In Proceedings of the 3rd International Scientific Conference Wood Technology & Product Design, Ohrid, Republic of Macedonia, 11–14 September 2017; Volume III, pp. 213–221.

38. Brezin, V.; Antov, P. *Engineering Ecology*, 1st ed.; Publishing House of the UF: Sofia, Bulgaria, 2015; 259p.

39. Li, L.; Mazumder, J. A study of the mechanism of laser cutting of wood. *For. Prod. J.* **1992**, *41*, 53–59.

40. Barnekov, V.G.; Huber, H.A.; McMillin, C.W. Laser machining wood composites. *For. Prod. J.* **1989**, *39*, 76–78.

41. Grad, L.; Mozina, J. Optodynamic studies of Er: YAG laser interaction with wood. *Appl. Surf. Sci.* **1998**, *127*, 973–986. [CrossRef]

42. Hattori, N. Laser processing of wood. *Mokuzai Gakkaishi* **1995**, *41*, 703–709.

43. Riveiro, A.; Quintero, F.; Boutinguiza, M.; del Val, J.; Comesana, R.; Lusquinos, F.; Pou, J. Laser Cutting: A Review on the Influence of Assist Gas. *Materials* **2019**, *12*, 157. [CrossRef]

44. Piili, H.; Hirvikaki, M.; Salminen, A. Repeatability of laser cutting of uncoated and coated boards. In Proceedings of the NOLAMP, Copenhagen, Denmark, 24–26 August 2009.

45. Lum, K.C.P.; Hg, S.L.; Black, I. CO_2 laser cutting of MDF: Determination of process parameter settings. *Opt. Laser Technol.* **2000**, *32*, 67–76. [CrossRef]

46. Powell, J. *CO_2 Laser Cutting*, 2nd ed.; Springer-Verlag: Berlin/Heidelberg, Germany, 1998.

47. Eltawahni, H.A.; Olabi, A.G.; Benyounis, K.Y. Investigating the CO_2 laser cutting parameters of MDF wood composite material. *Opt. Laser Technol.* **2011**, *43*, 648–659. [CrossRef]

48. Barnekov, V.G.; McMillin, C.W.; Huber, H.A. Factors influencing laser cutting of wood. *For. Prod. J.* **1986**, *36*, 55–58.

49. Lum, K.C.P.; Hg, S.L.; Black, I. CO_2 laser cutting of MDF, Estimation of power distribution. *J. Opt. Laser Technol.* **2000**, *32*, 77–87. [CrossRef]

50. Antov, P.; Mantanis, G.I.; Savov, V. Development of Wood Composites from Recycled Fibres Bonded with Magnesium Lignosulfonate. *Forests* **2020**, *11*, 613. [CrossRef]

51. Antov, P.; Savov, V.; Mantanis, G.; Neykov, N. Medium-density fibreboards bonded with phenol-formaldehyde resin and calcium lignosulfonate as an eco-friendly additive. *J. Wood Mater. Sci. Eng.* **2020**. [CrossRef]

52. Tayal, M.; Barnekov, V.; Mukherjee, K. Focal point location in laser machining of thick hard wood. *J. Mater. Sci. Lett.* **1994**, *13*, 644–646. [CrossRef]

53. Nukman, Y.; Saiful, R.I.; Azuddin, M.; Aznijar, A.Y. Selected Malaysian Wood CO_2 Laser Cutting Parameters and Cut Quality. *Am. J. Appl. Sci.* **2008**, *5*, 990–996.

54. McMillin, W.C.; Harry, J.E. Laser Machining of Southern Pine. *For. Prod. J.* **1971**, *21*, 24–37.

55. Quintero, F.; Riveiro, A.; Lusquinos, F.; Comesana, R.; Pou, J. CO_2 laser cutting of phenolic resin boards. *J. Mater. Process. Technol.* **2011**, *211*, 1710–1718. [CrossRef]

56. Quintero, F.; Riveiro, A.; Lusquinos, F.; Comesana, R.; Pou, J. Feasibility study on the laser cutting of phenolic resin boards. *Phys. Procedia* **2011**, *12*, 578–583. [CrossRef]
57. Zhou, B.H.; Mahdavian, S.M. Experimental and theoretical analyses of cutting nonmetallic materials by low power CO_2 laser. *J. Mater. Process. Technol.* **2004**, *146*, 188–192. [CrossRef]
58. Moradi, M.; Mehrabi, O.; Azdast, T.; Benyounis, K.Y. Enhancement of low power CO_2 laser cutting process for injection molded polycarbonate. *Opt. Laser Technol.* **2017**, *96*, 208–218. [CrossRef]
59. Choudhury, I.A.; Shirley, S. Laser cuting of polymeric materials: An experimental investigation. *Opt. Laser Technol.* **2010**, *42*, 503–508. [CrossRef]
60. Yang, C.B.; Deng, C.S.; Chiang, H.L. Combining the Taguchi method with artificial neural network to construct a prediction model of a CO_2 laser cutting experiment. *Int. J. Adv. Manuf. Technol.* **2012**, *59*, 1103–1111. [CrossRef]
61. Elsheikh, A.H.; Shehabeldeen, T.A.; Zhou, J.; Showaib, E.; Elaziz, M.A. Prediction of laser cutting parameters for polymethylmethacrylate sheets using random vector functional link network integrated with equilibrium optimizer. *J. Intell. Manuf.* **2020**. [CrossRef]
62. Alizadeh, A.; Omrani, H. An integrated multi response Taguchi- neural network- robust data envelopment analysis model for CO_2 laser cutting. *Measurement* **2019**, *131*, 69–78. [CrossRef]
63. Dubey, A.K.; Yadava, V. Laser beam machining: A review. *Int. J. Mach. Tools Manuf.* **2008**, *48*, 609–628. [CrossRef]
64. Radovanovic, M.; Madic, M. Experimental investigations of CO_2 laser cut quality: A review. *Nonconv. Technol. Rev.* **2011**, *4*, 35–42.
65. Hernandez-Castaneda, C.J.; Sezer, K.H.; Li, L. The effect of moisture content in fibre laser cutting of pine wood. *Opt. Lasers Eng.* **2011**, *49*, 1139–1152. [CrossRef]
66. Prakash, S.; Kumar, S. Experimental investigation and analytical modeling of multi-pass CO_2 laser processing on PMMA. *Precis. Eng.* **2017**, *49*, 220–234. [CrossRef]
67. Jianying, S.; Yun, Z. FEM Simulation Technologies of Laser Cutting Wood Board Based on ANSYS. *Adv. Mater. Res.* **2010**, *113*, 1629–1631.
68. Polak, M.; Chmelickova, H.; Vasicek, L. Numerical modeling of laser treatment of metal and nonmetal materials. In Proceedings of the 13th Polish-Czech-Slovak Conference on Wave and quantum Aspects of Contemporary Optics, Krzyzowa, Poland, 21 November 2003; Volume 5259, pp. 303–307. [CrossRef]
69. Yang, C.; Jiang, T.; Yu, Y.; Dun, G.; Ma, Y.; Liu, J. Study of surface quality of wood processed by water-jet assisted nanosecond laser. *Bioresources* **2018**, *13*, 3125–3134. [CrossRef]
70. Yang, C.; Jiang, T.; Yu, Y.; Bai, Y.; Song, M.; Miao, Q.; Ma, Y.; Liu, J. Water-jet assisted nanosecond laser microcutting of norheast China ash wood: Experimental study. *Bioresources* **2019**, *14*, 128–138.
71. Hardalov, C.M.; Christov, C.G.; Mihalev, M.S. Numerical modeling of laser machining on the ceramic surface. *AU J. Technol.* **2006**, *9*, 163–171.
72. Modest, F.M. *Laser Machining of Ablating/Decomposing Materials–Through Cutting and Drilling Models*; ICALEO: Quebec, Canada, 1996.
73. Roggero, G.; Scotti, F.; Piuri, V. Quality analysis measurement for laser cutting. In Proceedings of the 2001 International Workshop on Virtual and Intelligent Measurement Systems, Budapest, Hungary, 20–20 May 2001.
74. Bianco, N.; Manca, O.; Nardini, S.; Tamburino, S. Transient Heat Conduction in Solids Irradiated by a Moving Heat Source. *Defect Diffus. Forum* **2009**, *283*, 358–363. [CrossRef]
75. Babiak, M.; Orech, J.; Kleskeňová, M. Temperature distribution in wood heated by scanning gaussian laser beam. *Wood Res.* **1986**, *108*, 1–15.
76. Yilbas, B. Laser Cutting Qualitz Assesment and Thermal Efficiency Analysis. *J. Mater. Process. Technol.* **2004**, *155*, 2106–2115. [CrossRef]
77. ISO 13061-1:2014. *Physical and Mechanical Properties of Wood. Test Methods for Small Clear Wood Specimens. Part 1: Determinantion of Moisture Content for Physical and Mechanical Tests*; International Organization for Standardization: Geneva, Switzerland, 2014.
78. ISO 13061-2:2014. *Physical and Mechanical Properties of Wood. Test Methods for Small Clear Wood Specimens. Part 2: Determination of Density for Physical and Mechanical Tests*; International Organization for Standardization: Geneva, Switzerland, 2014.

79. Dhanachandra, N.; Manglem, K.; Chanu, Y.J. Image Segmentation Using K–means Clustering Algorithm and Subtractive Clustering Algorithm. *Procedia Comput. Sci.* **2015**, *54*, 764–771. [CrossRef]

80. Moallem, P.; Razmjooy, N. Adaptive Particle Swarm Optimization. *J. Appl. Res. Technol.* **2012**, *5*, 703–712.

81. Ready, J.F.; Farson, D.F.; Feeley, T. *LIA Handbook of Laser Materials Processing*; Laser Institute of America, Magnolia Publishing: Orlando, FL, USA; Springer Nature: Berlin/Heidelberg, Germany, 2001.

82. Liu, Q.; Yang, C.; Xue, B.; Miao, Q.; Liu, J. Processing Technology and Experimental Analysis of Gas-assisted Laser Cut Micro Thin Wood. *Bioresources* **2020**, *15*, 5366–5378.

83. Asibu, E.K. *Principles of Laser Materials Processing*; John Wiley and Sons: New York, NY, USA, 2009.

84. Arai, T.; Hayashi, D. Factors affecting the Laser processing of wood. 1. Effects of mechanical laser parameters on machinability. *Mokuzai Gakkaishi* **1992**, *38*, 350–356.

85. Arai, T.; Hayashi, D. Factors affecting the Laser processing of wood. 2. Effects of material parameters on machinability. *Mokuzai Gakkaishi* **1994**, *40*, 497–503.

86. Peters, C.C.; Banas, C.M. Cutting wood and wood-base products with a multikilowatt CO_2 laser. *For. Prod. J.* **1977**, *27*, 41–45.

87. Pędzik, M.; Bednarz, J.; Kwidziński, Z.; Rogoziński, T.; Smardzewski, J. The Idea of Mass Customization in the Door Industry Using the Example of the Company Porta KMI Poland. *Sustainability* **2020**, *12*, 3788. [CrossRef]

88. Potkany, M.; Gejdoš, M.; Debnár, M. Sustainable Innovation Approach for Wood Quality Evaluation in Green Business. *Sustainability* **2018**, *10*, 2984. [CrossRef]

89. Němec, F.; Lorincová, S.; Hitka, M.; Turinská, L. The Storage Area Market in the Particular Territory. *Nase More* **2015**, *62*, 131–138. [CrossRef]

90. Lorincová, S.; Schmidtová, J.; Balážová, Z. Perception of the corporate culture by managers and blue collar workers in Slovak wood-processing businesses. *Acta Fac. Xylologiae Zvolen* **2016**, *58*, 149–163.

91. Potkany, M.; Hitka, M. Utilization of contribution margin in the costing system in the production of components for wood working machines. *Drv. Ind.* **2009**, *60*, 101–110.

Publisher's Note: MDPI stays neutral with regard to jurisdictional claims in published maps and institutional affiliations.

Article

Tropical Wood Dusts—Granulometry, Morfology and Ignition Temperature

**Miroslava Vandličková, Iveta Marková *, Linda Makovická Osvaldová,
Stanislava Gašpercová, Jozef Svetlík and Jozef Vraniak**

Department of Fire Engineering, Faculty of Security Engineering, University of Žilina, Univerzitná 8215/1,
010 26 Žilina, Slovakia; miroslava.vandlickova@fbi.uniza.sk (M.V.); linda.makovicka@fbi.uniza.sk (L.M.O.);
stanislava.gaspercova@fbi.uniza.sk (S.G.); jozef.svetlik@fbi.uniza.sk (J.S.); vraniak.jojo@gmail.com (J.V.)
* Correspondence: iveta.markova@fbi.uniza.sk; Tel.: +421-41-513-6799

Received: 26 September 2020; Accepted: 26 October 2020; Published: 28 October 2020

Abstract: The article considers the granulometric analysis of selected samples of tropical wood
dust from cumaru (*Dipteryx odorata*), padauk (*Pterocarpus soyauxii*), ebony (*Diospyros crassiflora*),
and marblewood (*Marmaroxylon racemosum*) using a Makita 9556CR 1400 W grinder and K36
sandpaper, for the purpose of selecting the percentages of the various fractions (<63; 63; 71; 200;
315; 500 µm) of wood dust samples. Tropical wood dust samples were made using a hand orbital
sander Makita 9556CR 1400 W, and sized using the automatic mesh vibratory sieve machine Retsch
AS 200 control. Most dust particles (between 50–79%) from all wood samples were under 100 µm in
size. This higher percentage is associated with the risk of inhaling the dust, causing damage to the
respiratory system, and the risk of a dust-air explosive mixture. Results of granulometric fractions
contribution of tropical woods sanding dust were similar. Ignition temperature was changed by
particle sizes, and decreased with a decrease in particle sizes. We found that marblewood has the
highest minimum ignition temperature (400–420 °C), and padauk has the lowest (370–390 °C).

Keywords: tropical wood dust; granulometric sieve analysis; morphology shape of particles;
temperature of ignition

---

1. Introduction

During wood processing, dust is created as a by-product [1–7], and plays a negative role in
assessing the risk of fire [8] or explosion [9–14]. Wood fust also poses a significant risk to the health of
the human body [15,16].

The damaging effect of wood dust is determined according to the particle size. Larger size
fractions tend to settle [17,18], whereas if the particle is smaller (such as below 100 µm), the dust
becomes airborne. In the production process, dust is formed that contain particles of various sizes [5].
A granulometric analysis determines the degree of crushing of the base material—which is one of the
characteristic abilities of form airborne dust mixture [8].

Reinprecht et al. [19] classify tropical woods as wood that demonstrates significant resistance
to biological agents and machine wear and tear, together with solid dimensional stability and pretty
aesthetics. These types of wood are commonly used for exterior constructions, tiles, garden furniture,
or special plywood [20–22]. The expansion of their processing brings the creation of their wood dust.

Reinprecht et al. [23] made a detail analysis of seven types of tropical woods: Kusia (*Nauclea
diderichii* Merill), bangkirai (*Shorea obtusa* Wall; Sh. spp.), massaranduba (*Manilkara bidentata* A. Chev.;
M. spp.), jatobá (*Hymenaea courbaril* L.), ipé (*Tebebuia serratifolia* Nichols.; T. spp.), cumaru (*Dipteryx
odorata* (Aubl.) Wild.)—996 kg·m^{-3}, and cumaru rosa (*Dipteryx magnifica* (Ducke))—1014 kg·m^{-3}.

The samples were studies during different weather conditions, using a 36-mount in the exterior, and results showed the lowest lightening cumaru samples than others.

Tropical woods have natural durability [24]. There are resistant to decay fungi [19], insects, and dimensional changes. This is due to extractants that have a biocidal effect, such as coumarins, flavonoids, and tannins, and a hydrophobic effect, such as fats, oils, and waxes [25,26]. Giraldo et al. [27] declare that morphology, together with differences in the inorganic constituents, significantly affects the combustion process of wood.

The major parameter to assess the risk of airborne dust ignition is ignition temperature. Ignition temperature is closely surveyed using standardized equipment [28], where the airborne tropical wood dust is loaded heat. The ignition temperature (SIT) is the lowest temperature at which, under the defined test conditions, ignition occurs by heating, without the presence of any additional flame source [29].

This study aims to examine and seek comparative similarities between the granulometric structure of wood sanding dust from cumaru (*Dipteryx odorata*), padouk (*Pterocarpus soyauxii*), ebony (*Diospyros crassiflora*), and marblewood (*Marmaroxylon racemosum*). The prepared tropical dust was analyzed for the purpose of identifying its morphological structure, and determining the given physical properties (average dust moisture and bulk density). We have focused on microfractions of tropical wood dust (size of particles ≤100 μm), and focused the minimum particle size (<100 μm) required to cause ignition within airborne tropical dusts.

2. Sample Materials and Methods

2.1. Samples—Tropical Woods

Four samples of wood dust from foreign wood species were used for this study. Samples of tropical wood were selected while considering their use in the production of floor coverings, furniture, and interior decorative items (Table 1).

Table 1. Samples used in the experiment.

Common Name [1]	Scientific Name	Density (kg·m^{-3})
Cumaru	*Dipteryx odorata*	1075.69 ± 10.04
Padauk	*Pterocarpus soyauxii*	720.40 ± 9.52
Ebony	*Diospyros crassiflora*	960.00 ± 7.46
Marblewood	*Marmaroxylon racemosum*	1000.08 ± 9.98

[1] Association Technique Internationale des Bois Tropicaux (ATIBT) in France.

Cumaru (*Dipteryx odorata*), together with abiurana (*Pouteria guianensis*), garapeira (*Apuleia molaris*), jequitiba (*Cariniana* sp.), Cedro (*Cedrela odorata*), angelim (*Parkia pendula*), angelim pedra (*Hymenolobium excelsum*), and cerejeira (*Amburana acreana*) belongs to the group of Amazonian woody plants [30,31]. The cumaru tree is very dense (950–1000 kg·m^{-3}), tough, highly durable, and resistant to cracking when exposed to sunlight. Therefore, it is suitable for solid flooring, stair treads, furniture, and pool decks [32]. Moreover, it is frequently found in the states of Acre, Amapá, Amazonas, Pará, Rondônia, and Mato Grosso, as well as in neighboring countries like Guyana, Venezuela, Colombia, Bolivia, Peru, and Suriname [32].

Padauk (*Pterocarpus soyauxii*) is moderately heavy, strong, and stiff, with exceptional stability. It is a popular hardwood among hobbyist woodworkers because of its unique color and low cost. It has a unique reddish-orange coloration, and the wood is sometimes referred to by the name 'Vermillion'. It is commonly used in flooring, musical instruments/objects, tool handles, furniture, other small particular wood objects, and as veneer [33].

Ebony woods have many common names, such as Gabon Ebony, African Ebony, Nigerian Ebony, and Cameroon Ebony, and originate from the equator part of Western Africa. Heartwood (955 kg·m^{-3}) is usually jet-black, with little to no variation or visible grain. Occasionally, dark brown or grayish-brown streaks may be present. Ebony's common uses small items, such as piano keys, musical instrument parts, pool cues, carvings, and other small specialty items [34]. Ebony is highly valued in the Hindu religion as a building material [35], base material [36,37], and for use in other items [38].

Marblewood, also known as Angelim Rajado, is distributed from Northeastern South America. This heartwood (1005 kg·m^{-3}) is yellow to golden brown, with irregular brown, purple, or black streaks [39]. It is commonly used for tuned instruments/objects, flooring, carpentry, sliced veneer, and delicate furniture. Vivek et al. [40] introduced the possibility of using marblewood dust as a partial replacement for cement and sand in concrete.

The basic samples (Figure 1a) of cumaru (152 × 38 × 38 mm), Padouk (131 × 50 × 20 mm), ebony (142 × 36 × 30 mm), and marblewood (120 × 35 × 35 mm) were made by a private wood company in Žilina (Slovakia) by a wood cutting saw (CNC Panel Saw Machine, Shandong, China) (Table 1). The moisture content of the basic samples was approximately 8–10%.

Figure 1. Experimental samples (**a**) in the form of plates; (**b**) in the form of prepared wood dust. The legend: (1) Cumaru (*Dipteryx odorata*); (2) African Parakeet (*Pterocarpus soyauxii*); (3) African Ebony (*Diospyros crassiflora*), and (4) Marblewood (*Marmaroxylon racemosum*).

2.2. Preparation of Tropical Wood Dust Samples

Technical equipment was used to prepare the homogeneous dust particles dust samples (Figure 1b) was Makita 9556CR 1400 W disc sander (Makita Numazu Corp., Branesti Ilfov, Romania) and K36 sandpaper (Topex, Kinekus, Žilina, Slovakia). The grinding was carried out by the Experimental Laboratory at the Faculty of Security Engineering, University of Žilina (Slovakia). Tropical wood dust samples were prepared by a specialist in grinding. We aimed to ensure the grinding process was as close to reality as it is possible (in terms of pressure of the grinding surface of the component, grinding speed, and grinding direction (cross)). The prepared dust particles were amassed in the hopper with a hermetically sealed glass container, because it needed to stop the dust from absorbing any moisture. After the samples were ground three times, the hopper always was cleaned (in between samples). Overall, 300 g of dust was amassed from each board (three boards collection), and served for the granulometric examination. Detailed information about the preparation of dust samples can be found in Reference [29].

2.3. Experimental Methods Utilised for Characteristics of Tropical Wood Dusts and Sieve Analysis

The moisture of the tropical wood samples determined according to Reference [41], and the bulk density determined according to Reference [42], were selected characteristics of samples before sieve analysis (Table 2). The Retsch AS 200 sieve shaker vibration machine (Retsch AS 200 control, Retsch GmbH, Haan, Germany), using the seven fraction sizes (500, 315, 200, 100, 71, 63, and <63 μm), was used for sieve analysis [43]. The sieves, in addition to the tropical wood dust, weighed 30 g (using laboratory scales with precision readings of 0.001 g). The measurement procedures were conducted five times lasting 10 min. The wood dust moisture testing was carried out according to Reference [44]—in a heated oven, at a temperature of 103 ± 2 °C for 24 h (Table 2).

Table 2. Basic physical parameters of samples used in the experiment.

Tropical Dust Samples	Cumaru	Padauk	Ebony	Marblewood
Average bulk density (kg·m−3)	190.3 ± 2.466	167.3 ± 4.355	206.5 ± 2.562	187.9 ± 8.535
Dust moisture (%)	5.93 ± 0.05	5.34 ± 0.08	4.83 ± 0.07	7.34 ± 0.1

2.4. Experimental Methods for Shape of Wood Dust Particles

The size, shape, and form of wood dust particles were studied by microscopic analysis, through a wide-field microscopy system (Nikon Eclipse Ni (Nikon Corp., Tokyo, Japan)) with a Nikon DS-Fi2 camera (Nikon Instruments Inc., Melville, NY, USA) [45]. This microscope possesses two c-mount camera ports, and an electric XY stage. This is the normal configuration for the bright field observation of this microscope.

A Nikon DS-Fi2 full-HD color camera (Nikon Instruments Inc., Melville, NY, USA), that uses one port, 2560 × 1920 pixels, was also used. This camera was used to establish the image by connecting it to the computer using a USB cable. Microscopic analyses of tropical wood dusts were performed using 100 μm fraction precision.

The Institute of Research in Banská Bystrica, Slovakia, has a Nikon SMZ 1270 stereomicroscope (Nikon Corp., Beijing, China), which was used to analyze the shape of the dust particles. Equipment for the research of dust particles shape has a range with a 12.7:1 zoom head, and a magnification range of 0.63× to 8×. The samples of the 500 μm, 315 μm, 200 μm, and 100 μm fractions were measured in this stereomicroscope.

2.5. Experimentation So as to Determine the Ignition Temperature Measurement of Airborne Dusts

Specific test equipment (the detail of which can be found in Reference [29]) (VVUÚ, a.s., Ostrava, Czech Republic), with automatic weighing machines (Steinberg Systems, Łódź, Poland), an air compressor HL 100 ZU EINHELL (Einhell Corp., Landau an der Isar, Germany), and ALMEMO equipment (Ahlborn, Berlin, Germany) was used to the measurement of minimum ignition temperatures of airborne dusts (Figure 1a) according to the standard instructions [28]. The details of this are described in Vandličková et al. [29].

3. Results and Discussion

The average bulk density (kg·m^{-3}) and dust moisture (%) were physical parameters connected with the preparation of dust samples (Table 2). Poorter et al. [46] consider of the density of tropical woods and the influence of climatic conditions on the growth and quality of wood mass. Tropical tree cumaro is considered a hardwood [47–50]. Soriano et al. [51] determined the density of cumoro in the range of 1060–1070 kg·m^{-3}. Marblewood is also ranked among the hardwoods. Everything about working with marblewood revolves around its incredible hardness and density [52].

Ebony is different—it has a lower density (960 kg·m^{-3}), and the highest value of average bulk density (206.5 kg·m^{-3}).

King et al. [53] performed research on the growth of tropical trees in the Amazon and investigated their physical properties. The research samples included padauk trees, namely, *Dipterocarpus cornutus Dyer*. With a density of (680 ± 0.013) kg·m^{-3} and *Dipterocarpus globosus Vesque* with a density of (690 ± 0.017) kg·m^{-3}.

3.1. Results of Particle Sizes (Fractions) of Dust Samples

Sanding dust particulates are minute and of complex composition, and are more hazardous to humans than the cutting dust particulates [54]. Wood dust particles (Figure 2) are normally generated in different sizes [12,55]. The percentages of particle sizes start from the size of 500 µm (Table 3). Larger fractions occurred only at a minimal rate (at 1%). Očkajová and Marková [56]; Očkajová et al. [5,57] presented the same result, which was presented for the chosen domestic tree samples.

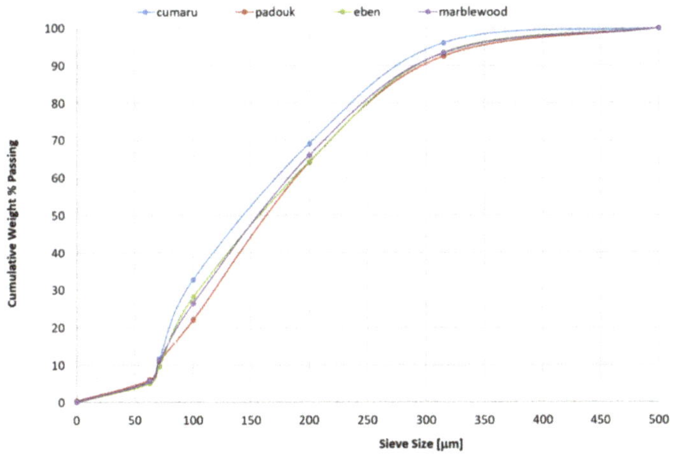

Figure 2. Continuous cumulative curve of cumaru, padauk, ebony, and marblewood dust.

Table 3. Determination of % particle number fractions of the tropical wood dusts.

Fraction	% Particle Number			
Size	Cumaru	Padauk	Ebony	Marblewood
500	1.15 ± 0.076	1.55 ± 0.115	1.36 ± 0.275	1.26 ± 0.244
315	2.78 ± 0.745	6.29 ± 0.621	5.15 ± 0.537	5.33 ± 0.591
200	26.92 ± 0.779	28.37 ± 0.547	29.12 ± 0.464	27.31 ± 0.462
100	36.37 ± 1.2018	42.14 ± 1.049	36.21 ± 1.126	39.61 ± 0.367
71	21.14 ± 0.875	10.97 ± 1.453	18.56 ± 0.180	15.22 ± 0.111
63	6.07 ± 0.591	4.92 ± 0.312	4.44 ± 0.858	5.5 ± 0.383
<63	5.18 ± 0.495	5.81 ± 0.399	5.04 ± 0.420	5.64 ± 0.612

The dust particles prepared from African Padauk had similar composition % particle number, as well as spruce, oak, and beechwood [29].

Očkajová et al. [5] studied the size of fraction particles and share of wood dust particles (beech, oak, and spruce). Their solution formulated that the percentage share of dust particles is also very different, depending on the kind of this tree species. Our results showed similar size of particles (Table 3). Očkajová et al. [6] found a connection between the shares and densities of the wood dust. Their analyses regarding the wood dust in the sanding process showed that more dust is produced as the density of wood increases [54].

The results of dust sieving are presented by cumulative curves (Figure 2). The results are presented in passed weight percent of the individual fractions collected on a sieve with the appropriate mesh size. The results of the sieve analysis should be presented according to the appropriate standards [58,59]. Prepared continuous cumulative curves are presented in Figure 2 with complete mesh size.

3.2. Shape Analysis of Dust Particles

The characterization of wood dust samples from a morphological (Figures 3–6), and a dimensional point of view, yields information that can help epidemiologists and toxicologists to understand the causes of respiratory illnesses [60].

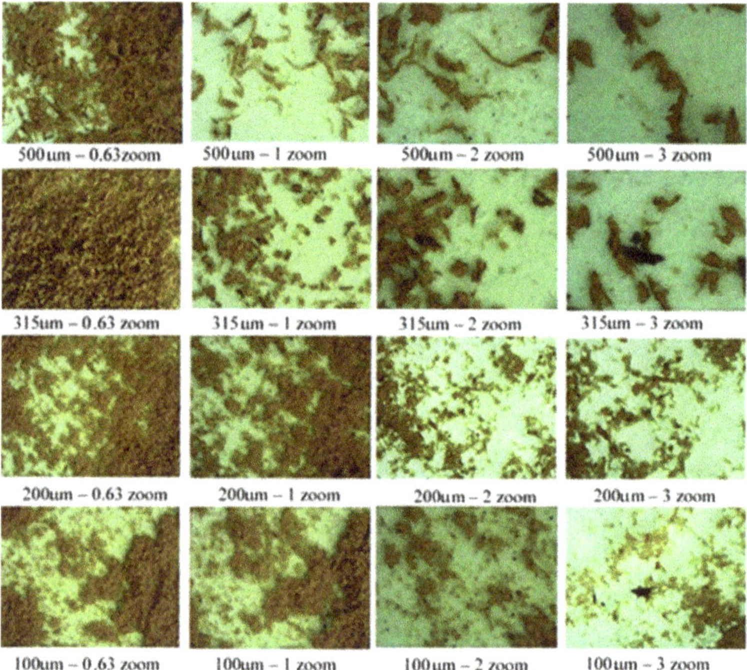

Figure 3. Cumaru dust particles and their shape.

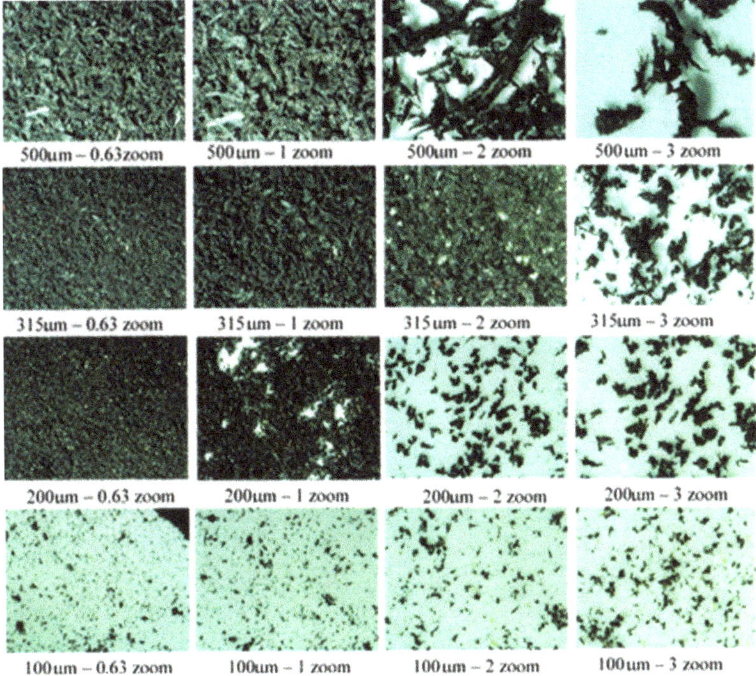

Figure 4. Ebony dust particles and their shape.

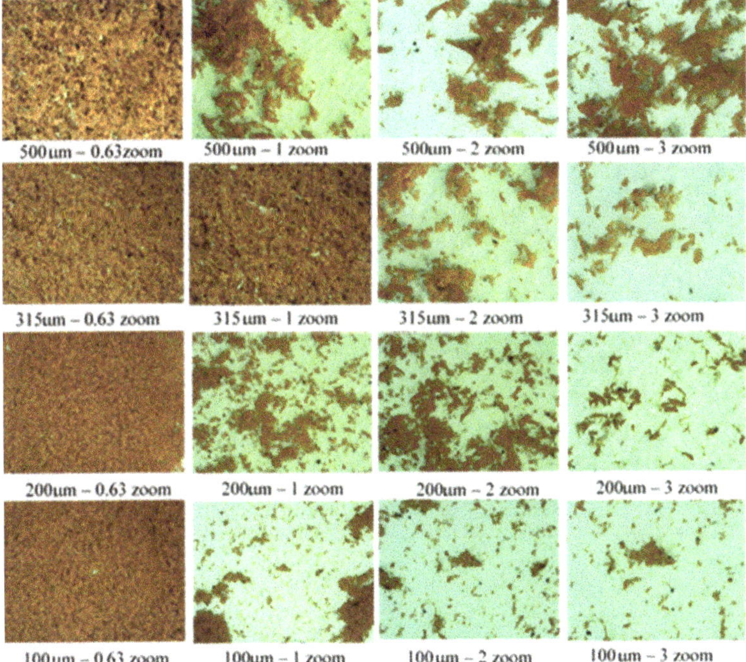

Figure 5. Padauk dust particles and their shape.

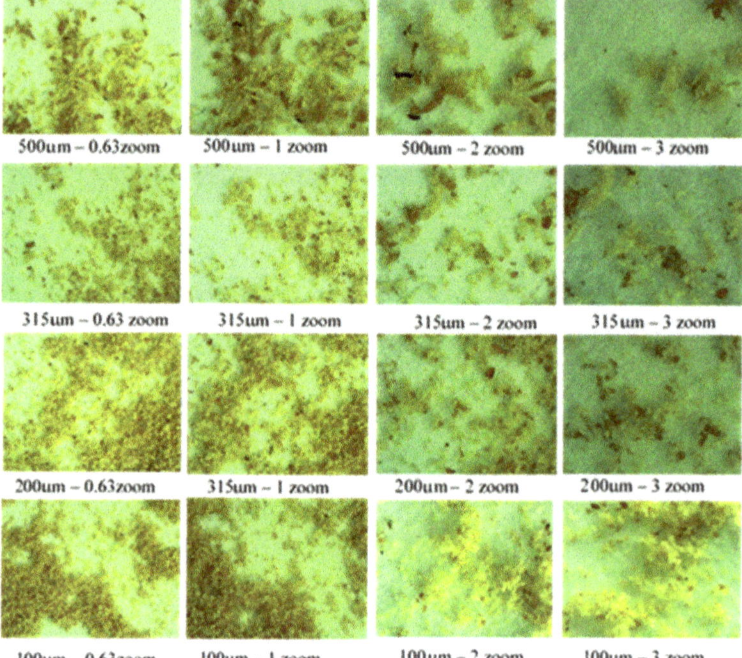

Figure 6. Marblewood dust particles and their shape.

The particles in our sample show a range of shapes and sizes of particles [61]. The size 100 μm is the boundary value of a particle size, where dust is expected to becomes airborne and is potentially explosive [62,63]. Within the basic range of size particles, dust can be classified as coarse (particle diameter of >100 μm) or fine (particle diameter of <100 μm) [64].

The anatomical structure is preserved in analyzed particles [65]. When magnified, the fractions of 500, 315, 200, and 100 μm (Figures 3–6) appeared differently—they had their own specific shapes. Figure 3 shows the fractions of cumaru, and Figure 6 fractions of marblewood.

Selected tropical woods produce dust particles within the whole spectrum. The differences in morphology are shown in Figures 3–6. Cumaru and marblewood are hardwoods, with a density of 1000 gm-m-3, and it is possible to state the similarity of the formed particles. The relationship between the particle shape and the initiation temperature can be shown in marblewood dust, which has the highest ignition temperature. Ebony offers a limited or bounded particle shape, and its ignition temperature is lower.

3.3. Microscopic Dust Analysis

The particles of tropical dust samples (Figure 7) have the size of <100 μm. From the safety and occupational hygiene perspective, particles below 100 μm are the most dangerous in the working environment [2]. The scans clearly show that the fibers of the dust particles maintain their anatomical structure with the fibrous character of particles [66]. The two-dimensional pictures of particles can be used for determining the smallest particle sizes.

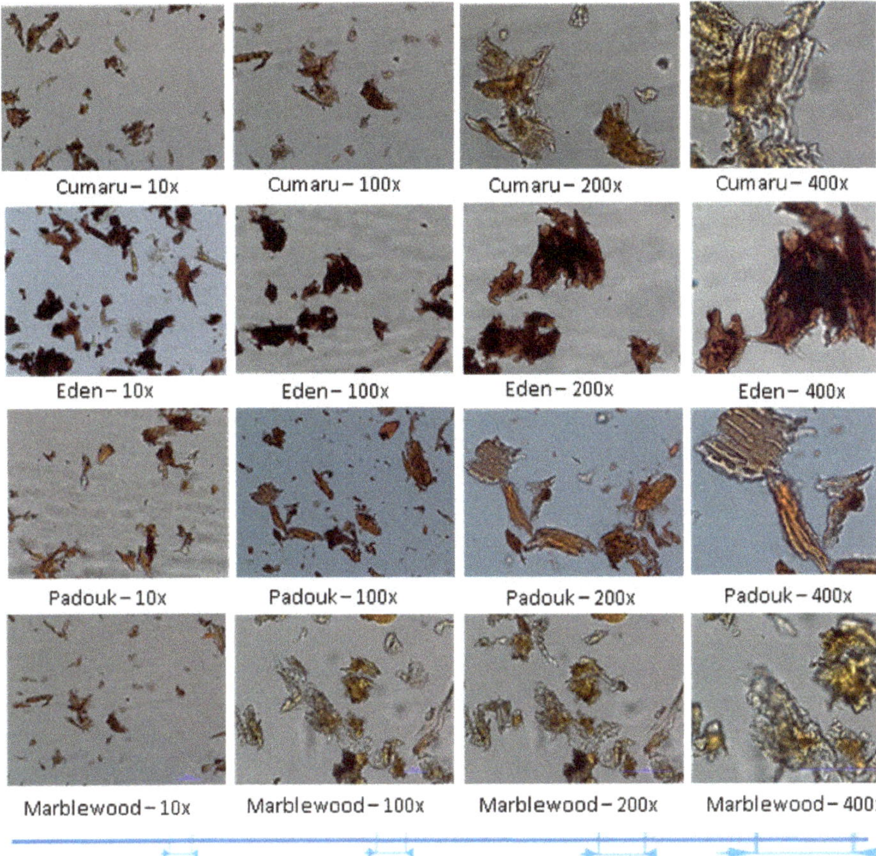

Figure 7. Light microscopic images of tropical dust fibers with 100×, 200×, and 400× zoom. Legend: Blue line presents a size of 100 microns (μm) in a 2D layout.

Picture of wood samples, after sanding, taken by an electron microscope, show different and complex shapes of particles of wood dust (Figure 7). As described in the Particle Atlas [67], the diverse geometric expression could be observed, such as cylinders, cones, rectangular prisms, and spheres [60].

Mazzoli et al. [60] provided microscope analyses of dusts from two hardwoods (sessile oak, oak-tree), two tropical hardwoods (padouk, iroko), and three softwoods (pine, spruce, and larch). These were obtained using a grinding machine with 360-grit sanding paper.

Gómez Yepes and Cremades [61] analyzed the particle characteristics in Quindio (size distributions, aerodynamic equivalent diameter (*Da*), elemental composition, and shape factors), and particles were then characterized via scanning electron microscopy (*SEM*) in conjunction with energy dispersive *X-ray* analysis. Results from their analysis of particulate matter showed that the cone-shaped particle ranged from 2.09 to 48.79 μm *Da*; the rectangular prism-shaped particle from 2.47 to 72.9 μm *Da*; the cylindrically-shaped particle from 2.5 to 48.79 μm *Da*; and the spherically-shaped particle from 2.61 to 51.93 μm *Da*.

Oak dust provides a similar comparison with marblewood dust, as oak is the hardest Slovak wood. In all investigated oak fractions, the isometric shape of particles with sharp edges and rounded corners, which is more typical for fraction <100 μm [3].

3.4. Ignition Temperature of Airborne Dust

Vandličkova et al. [29] studied the processes of ignition within airborne wood dusts. They [29] studied the specificities of the dust ignition process in different stages. The first stage is the beginning of an explosion: The dust was sprinkled into a ceramic tube furnace, with dust carbon residue. In the second stage, the unburned wood dust was ignited with the most incredibly intense flame. Subsequently, in the third state, flame slowly diminished with side effects.

The ebony dust of 500 μm, 315 μm, and 71 μm fractions were obtained the maximum flame (Figure 8). It was noted that when using larger particles, the flame had a lower intensity.

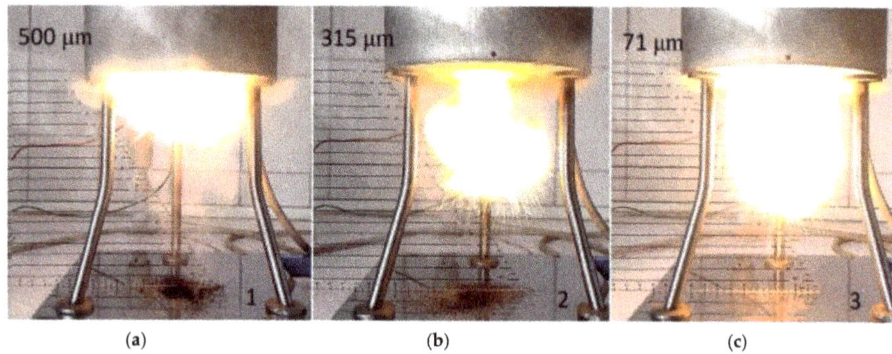

(a)   (b)   (c)

Figure 8. The flame on ignition (conditions: 0.2 g of dust, air pressure of 30 kPa, the temperature of the ceramic tube furnace 500 °C). Images were prepared using a Basler a602fc-2 high-speed camera (Basler AG, Ahrensburg, Germany). Legend: (**a**) Point of ignition for fraction 500 μm of ebony dust; (**b**) point of ignition for fraction 315 μm of ebony dust; (**c**) point of ignition for fraction 71 μm of ebony dust.

These results determined that the ignition temperature of airborne dust (Figure 9) showed a reduction in values relating to the smaller size fraction [68].

The minimum ignition temperature of cumaru for the 500 μm fraction was 410 °C, and this temperature decreased with changing particle size.

Particles of 100 μm or less had the most significant effect on the change in the minimum temperature. The minimum ignition temperature of the ebony dust particles 500 μm, and 200 μm was identical (400 °C). The 100 μm particle fraction had a minimum ignition temperature of 380 °C. The last three size fractions had minimum ignition temperatures, due to strong dispersion in the heating furnace. An assumption was made applied to a larger dispersion of particles in a space for larger particles (of 500 μm), compared to 100 μm with the same weight of the batch.

Marblewood had the highest tested minimum ignition temperatures of all samples. The minimum ignition temperatures started from 400 °C at a particle size of 500 μm to 100 μm. Subsequently, the minimum ignition temperatures decreased with the particle size change to the level of 400 °C, at a particle size of 63 μm.

Flame propagation behaviors and temperature characteristics of four types of biomass (poplar, pine, peanut, and corn sawdust particles) with two different particle sizes (50–70 μm and 100–200 μm) distributions were studied experimentally by Jiang et al. [69]. The average flame propagation velocity and the amplitude of the velocity fluctuation are functions of the mass density of the biomass particles and depend on the particle size distributions [69].

Tropical woods came to a global market for their use in various products. Information about tropical wood fire parameters is poor. Fire parameters are not described in Safety Data Sheets [70,71]. Carrasco et al. [72] studied the heat transfer in Brazilian woods, using a sample that was thermally loaded to examine the potential for fire. Their results showed that there are curves of temperature on the time experiments that create the thickness *f* chair layer corresponding with Carrasco's numerical models.

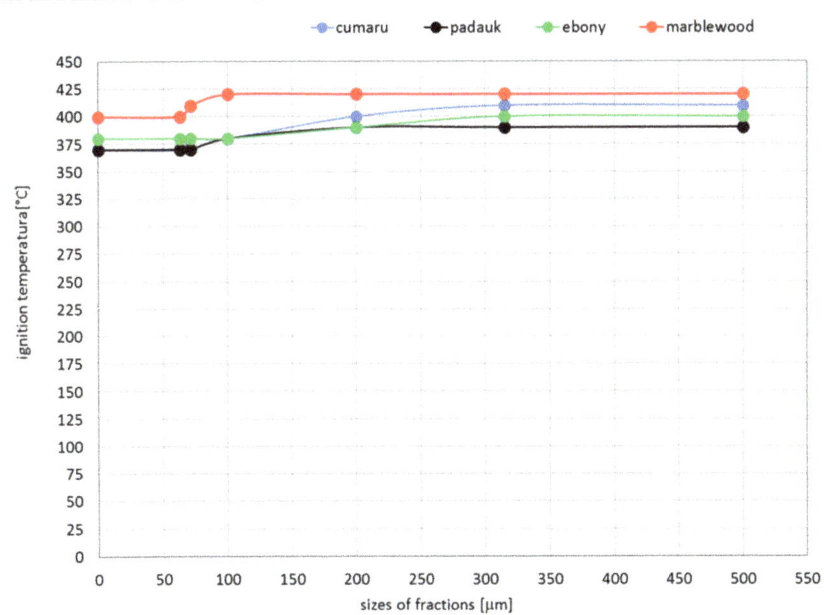

Figure 9. Ascertaining of the ignition temperature fractions of the tropical wood dusts.

4. Conclusions

From the perspective of this study, we can claim that we have obtained original results within the field of morphology of dust particles regarding the wood sanding process of tropical four species, as well as results concerning the determination of minimal ignition temperature airborne dust.

Our results show that fraction sizes of tropical wood dust are an important factor for fire ignition. However, the shape and morphology of tropical wood dust particles may also exert influence ignition process. All these aspects deserve to be studied further.

Marblewood has the highest minimum ignition temperature. Thermal stability can be found in its hardness and density.

Cumaru has specific behavior. It is a hardwood with a density comparable to marblewood, but the minimum initiation temperature decreases significantly with decreasing particle size (fraction <63 µm has a temperature of 370 °C).

The monitored dust particles retain their anatomical structure; the shapes of individual samples are different. These results offer different particle shapes such as: Rectangular prism-shaped particle, cylindrically-shaped particle and spherically-shaped particle.

Author Contributions: Conceptualization, M.V. and I.M.; methodology, investigation and resources S.G., J.S. and J.V.; writing—original draft preparation, I.M.; writing—review and editing V.M., L.M.O. and I.M. All authors have read and agreed to the published version of the manuscript.

Funding: This article was supported by the Cultural and Educational Grant Agency of the Ministry of Education, Science, Research and sport of the Slovak Republic on the basis of the project KEGA 0014UKF-4/2020 Innovative learning e-modules for safety in dual education.

Acknowledgments: This article was supported by the Project KEGA 0014UKF-4/2020 Innovative learning e-modules for safety in dual education.

Conflicts of Interest: The founding sponsors had no role in the design of the study, in the collection, analyses, or interpretation of data; in the writing of the manuscript and in the decision to publish the results.

References

1. Top, Y. Relationship between Employees' Perception of Airborne Wood Dust and Ventilation Applications in Micro-Scale Enterprises Producing Furniture. *BioRes* **2020**, *15*, 1252–1264.
2. Očkajová, A.; Beljo Lučić, R.; Čavlović, A.; Tereňová, J. Reduction of dustiness in sawing wood by universal circular saw. *Drv. Ind.* **2006**, *57*, 119–126.
3. Ockajova, A.; Beljakova, A.; Luptakova, J. Selected properties of spruce dust generated from sanding operations. *Drv. Ind.* **2008**, *59*, 3–10.
4. Piernik, M.; Rogozinski, T.; Krauss, A.; Pinkowski, G. The influence of the thermal modification of pine (Pinus sylvestris L.) wood on the creation of fine dust particles in plane milling Fine dust creation in the plane milling of thermally modified pine wood. *J. Occup. Health* **2019**, *61*, 481–488. [CrossRef] [PubMed]
5. Očkajová, A.; Kučerka, M.; Kminiak, R.; Krišťák, L.; Igaz, R.; Réh, R. Occupational Exposure to Dust Produced When Milling Thermally Modified Wood. *Int. J. Environ. Res. Public Health* **2020**, *17*, 1478. [CrossRef] [PubMed]
6. Očkajová, A.; Barcík, Š.; Kučerka, M.; Koleda, P.; Korčok, M.; Vyhnáliková, Z. Wood dust granular analysis in the sanding process of thermally modified wood versus its density. *BioRes* **2019**, *14*, 8559–8572.
7. Igaz, R.; Kminiak, R.; Kristak, L.; Nemec, M.; Gergel, T. Methodology of Temperature Monitoring in the Process of CNC Machining of Solid Wood. *Sustainability* **2019**, *11*, 95. [CrossRef]
8. Li, H.T.; Chen, X.K.; Deng, J.; Shu, C.M.; Kuo, C.H.; Yu, Y.C.; Hu, X.Y. CFD analysis and experimental study on the effect of oxygen level, particle size, and dust concentration on the flame evolution characteristics and explosion severity of cornstarch dust cloud deflagration in a spherical chamber. *Powder Technol.* **2020**, *372*, 585–599. [CrossRef]
9. Huang, C.; Chen, X.; Yuan, B.; Zhang, H.; Shang, S.; Zhao, Q.; Dai, H.; He, S.; Zhang, Y.; Niu, Y. Insight into suppression performance and mechanisms of ultrafine powders on wood dust deflagration under equivalent concentration. *J. Hazard. Mater.* **2020**, *394*, 122584. [CrossRef]
10. Demers, P.A.; Weinrich, A.J. Wood Dusts. In *Encyclopedia of Toxicology*, 3th ed.; Elsevier: New York, NY, USA, 2014; pp. 981–983.
11. Guo, L.; Xiao, Q.P.; Zhu, N.F.; Wang, Y.; Chen, X.L.; Xu, C.Y. Comparative Studies on the Explosion Severity of Different Wood Dusts from Fiberboard Production. *BioRes* **2019**, *14*, 3182–3199.
12. Rogoziński, T.; Wilkowski, J.; Górski, J.; Czarniak, P.; Podziewski, P.; Szymanowski, K. Dust creation in CNC drilling of wood composites. *BioRes* **2015**, *10*, 3657–3665. [CrossRef]
13. Liu, A.H.; Chen, J.Y.; Huang, X.F.; Lin, J.J.; Zhang, X.C.; Xu, W.B. Explosion parameters and combustion kinetics of biomass dust. *Bioresour. Technol.* **2019**, *294*, 122168. [CrossRef] [PubMed]
14. Gaspercova, S.; Osvaldova Makovicka, L. Fire protection in various types of wooden structures. *Civ. Environ. Eng.* **2015**, *11*, 51–57. [CrossRef]
15. Kamal, A.; Malik, N.R.; Martellini, T.; Cincinelli, A. Source, profile, and carcinogenic risk assessment for cohorts occupationally exposed to dust-bound PAHs in Lahore and Rawalpindi cities (Punjab province, Pakistan). *Environ. Sci. Pollut. Res.* **2015**, *22*, 10580–10591. [CrossRef]
16. Ojima, J. Generation rate and particle size distribution of wood dust by handheld sanding operation. *J. Occup. Health* **2016**, *58*, 640–643. [CrossRef]
17. Mračková, E.; Tureková, I. The dimensional characteristics of the particles of wood dust of selected deciduous trees considering to explosion. *Key Eng. Mater.* **2016**, *688*, 182–189. [CrossRef]
18. Nasir, V.; Cool, J. A review on wood machining: Characterization, optimization, and monitoring of the sawing process. *Wood Mater. Sci. Eng.* **2020**, *15*, 1–16. [CrossRef]
19. Reinprecht, L.; Mamonova, M.; Panek, M.; Kacik, F. The impact of natural and artificial weathering on the visual, colour and structural changes of seven tropical woods. *Eur. J. Wood Wood Prod.* **2018**, *76*, 175–190. [CrossRef]
20. Panek, M.; Reinprecht, L. Effect of vegetable oils on the colour stability of four tropical woods during natural and artificial weathering. *J. Wood Sci.* **2016**, *62*, 74–84. [CrossRef]
21. Reinprecht, L.; Vidholdova, Z. Rot Resistance of Tropical Wood Species Affected by Water Leaching. *BioRes* **2019**, *14*, 8664–8677.
22. Vidholdova, Z.; Reinprecht, L. The Colour of Tropical Woods Influenced by Brown Rot. *Forests* **2019**, *10*, 322. [CrossRef]

23. Reinprecht, L.; Vidholdova, Z.; Izdinsky, J. Bacterial and Mold Resistance of Selected Tropical Wood Species. *Bioresources* **2020**, *15*, 5198–5209.

24. Corassa, J.; Tiesen, C.; Dall'Oglio, O.; Melo, R. Durabilidade natural de dez madeiras amazônicas sob condicoes de campo. *Nativa* **2019**, *7*, 758. [CrossRef]

25. Imai, T.; Inoue, S.; Ohdaira, N.; Matsushita, Y.; Suzuki, R.; Sakurai, M.; Henriques De Jesus, J.; Ozaki, S.; Finger, Z.; Fukushima, K. Heartwood extractives from the Amazonian trees Dipteryx odorata, Hymenaea courbaril, and Astronium lecointei and their antioxidant activities. *J. Wood Sci.* **2008**, *54*, 470–475. [CrossRef]

26. Liu, Y.; Shao, L.; Gao, J.; Guo, H.; Chen, Y.; Cheng, Q.; Via, B.K. Surface photo-discoloration and degradation of dyed wood veneer exposed to different wavelengths of artificial light. *Appl. Surf. Sci.* **2015**, *331*, 353–361. [CrossRef]

27. Lacasta, A.M.; Haurie, L.; Monton, J.; Navarro Ezquerra, A.; Giraldo, P.; Sotomayor, J.; Palumbo, M. Characterization of the Fire Bbehavior of Tropical Wood Species for Use in the Construction Industry. In *A: World Conference on Timber Engineering. "WCTE 2016: World Conference on Timber Engineering, Vienna, Austria, 22–25 August 2016 Vienna, Austria: e-book"*; Vienna Technischen Universität: Graz, Austria, 2016.

28. *EN 50281-2-1: 2002. Electrical Apparatus for Use in the Presence of Combustible Dust. Part 2-1: Test Methods. Methods for Determining the Minimum Ignition Temperatures of Dust*; European Committee for Standardization: Brussels, Belgium, 2002.

29. Vandličková, M.; Marková, I.; Osvaldová, L.M.; Gašpercová, S.; Svetlík, J. Evaluation of African padauk (Pterocarpus soyauxii) explosion dust. *BioRes* **2020**, *15*, 401–414.

30. Pereira, K.M.; Garcia, R.A.; do Nascimento, A.M. Surface roughness of Amazonian woods. *Sci. For.* **2018**, *46*, 347–356.

31. Stangerlin, D.M.; Cavalcante, C.F.P.; da Costa, C.A.; Pariz, E.; de Melo, R.R.; Dall'oglio, O.T. Mechanical properties of Amazonian woods estimated by ultrasound waves propagation methods. *Nativa* **2017**, *5*, 628–633.

32. De Oliveira Araújo, S.; Rocha Vital, B.; Oliveira, B.; Oliveira Carneiro, A.; Lourenço, A.; Pereira, H. Physical and mechanical properties of heat treated wood from Aspidosperma populifolium, dipteryx odorata and mimosa scabrella. *Maderas. Cienc. Tecnol.* **2016**, *18*. [CrossRef]

33. African Padauk. In the Wood Database. Available online: https://www.wood-database.com/african-padauk/ (accessed on 20 March 2020).

34. Ebony. In the Wood Database. Available online: https://www.wood-database.com/gaboon-ebony/ (accessed on 22 March 2020).

35. Book Group Author(s). *IOP Conference Series: Earth and Environmental Science, Volume 126, Friendly City 4 'From Research to Implementation For Better Sustainability' 11–12 October 2017, Medan, Indonesia*; IOP Publishing Ltd.: Bristol, UK, 2018; Volume 126.

36. Iringova, A.; Idunk, R. Assessment and usability of historic trusses in terms of fire protection—A case study. *Int. Wood Prod. J.* **2017**, *8*, 80–87. [CrossRef]

37. Mari, M.; Filippidis, G. Non-Linear Microscopy: A Well-Established Technique for Biological Applications towards Serving as a Diagnostic Tool for in situ Cultural Heritage Studies. *Sustainability* **2020**, *12*, 1409. [CrossRef]

38. Kirksey, S.; Helmreich, S. The emergence of multispecies ethnography. *Cult. Anthropol.* **2010**, *25*, 545–576. [CrossRef]

39. Marblewood. The Wood Database. Available online: https://www.wood-database.com/marblewood/ (accessed on 21 March 2020).

40. Vivek, V.; Vinod, K.; Sonthwal, K. Effect of Marble Dust Powder & Wood Sawdust Ash on UCS and CBR Values of Soil. *Int. J. Innov. Res. Sci. Eng. Technol.* **2017**, 17442–17446.

41. *STN 49 0103: 1979. Wood. Determination of Moisture Content at Physical and Mechanical Testing*; Slovak Technical Normalisation: Bratislava, Slovakia, 1979. (In Slovak)

42. *ISO 23145-1:2007. Determination of Bulk Density of Ceramic Powders—Part 1: Tap Density*; International Organization for Standardization: Geneva, Switzerland, 2007.

43. *ISO 3310-1:201. Test Sieves—Technical Requirements and Testing—Part 1: Test Sieves of Metal Wire Cloth*; International Organization for Standardization: Geneva, Switzerland, 2016.

44. *CEN Standard EN 13183-1: 2002. Workplace Atmospheres. Size Fraction Definitions for Measurement of Airborne Particles. Moisture Content of a Piece of Timber-Part 1: Determination by Oven Dry Method*; European Committe for Standartion: Brussels, Belgium, 2002.

45. Goldsberry, A.; Hanke, C.W.; Countryman, N.B. A comparison of super wide field microscopy systems in Mohs surgery. *J. Drugs Dermatol.* **2014**, *13*, 1463–1465.

46. Poorter, L.; Rozendaal, D.; Bongers, F.; de Almeida-Cortez, J.S.; Zambrano, A.M.A.; Alvarez, F.S.; Andrade, J.L.; Villa, L.F.A.; Balvanera, P.; Becknell, J.M.; et al. Wet and dry tropical forests show opposite successional pathways in wood density but converge over time. *Nat. Ecol. Evol.* **2019**, *3*, 928–934. [CrossRef]

47. Stangerlin, D.M.; da Costa, A.F.; Goncalez, J.C.; Pastore, T.C.M.; Garlet, A. Monitoring of biodeterioration of three Amazonian wood species by the colorimetry technique. *Acta Amaz.* **2013**, *43*, 429–438. [CrossRef]

48. De Melo, B.A.; Molina-Rugama, A.J.; Leite, D.T.; de Godoy, M.S.; de Araujo, E.L. Bioactivity of powders from plant species on reproduction of Callosobruchus maculatus (FEBR. 1775) (*Coleoptera: Bruchidae*). *Biosci. J.* **2014**, *30*, 346–353.

49. Chipaia, F.D.; Reis, A.R.S.; Reis, L.P.; de Carvalho, J.C.; da Silva, E.F.R. Description anatomical macroscopic wood forest species of eigth market in the municipality of Altamira-PA, Brazil. *J. Bioenergy Food Sci.* **2015**, *2*, 18–24.

50. Eleoterio, J.R.; da Silva, C.M.K. Comparision of dry kiln schedules for Cumaru (Dipteryx odorata), Jatoba (Hymenaea spp) and Muiracatiara (Astronium lecointei) obtained by different methods. *Sci. For.* **2012**, *40*, 537–545.

51. Soriano, J.; da Veiga, N.S.; Martins, I.Z. Wood density estimation using the sclerometric method. *Eur. J. Wood Wood Prod.* **2015**, *73*, 753–758. [CrossRef]

52. Marblewood, Wood Turning Pens. Available online: https://www.woodturningpens.com/marblewood/ (accessed on 21 April 2020).

53. King, D.A.; Davies, S.J.; Tan, S.; Nur Supardi, M.D.; Noor, R. The role of wood density and stem support costs in the growth and mortality of tropical trees. *J. Ecol.* **2006**, *94*, 670–680. [CrossRef]

54. Yuan, N.; Zhang, J.; Lu, J.; Liu, H.; Sun, P. Analysis of inhalable dust produced in manufacturing of wooden furniture. *BioRes* **2014**, *9*, 7257–7266. [CrossRef]

55. Ratnasingam, J.; Ramasamy, G.; Ioras, F.; Thanesegaran, G.; Mutthiah, N. Assessment of dust emission and working conditions in the bamboo and wooden furniture industries in Malaysia. *BioRes* **2016**, *11*, 1189–1201. [CrossRef]

56. Očkajová, A.; Marková, I. Particular size analysis of selected wood dust species particles generated in the wood working environment. *Acta Univ. Matthiae Belii Ser. Environ. Manag.* **2016**, *18*, 24–31. (In Slovak)

57. Očkajová, A.; Kučerka, M.; Krišťák, L.; Igaz, R. Granulometric analysis of sanding dust from selected wood species. *Bioresources* **2018**, *13*, 7481–7495. [CrossRef]

58. ISO 9276-1:1998. *Representation of Results of Particle Size Analysis—Part 1: Graphical Representation*; International Organization for Standardization: Geneva, Switzerland, 1988.

59. Wolfrom, R.L. The Language of Particle size. As published in GXP. *Spring* **2011**, *15*, 2.

60. Mazzoli, A.; Favoni, O. Particle size, size distribution and morphological evaluation of airborne dust particles of diverse woods by Scanning Electron Microscopy and image processing program. *Powder Technol.* **2012**, *225*, 65–71. [CrossRef]

61. Gómez Yepes, M.E.; Cremades, L.V. Characterization of Wood Dust from Furniture by Scanning Electron Microscopy and Energy-dispersive X-ray Analysis. *Ind. Health* **2011**, *49*, 492–500. [CrossRef]

62. Dado, M.; Lamperova, A.; Kotek, L.; Hnilica, R. An evaluation of on-tool system for sanding dust collection: Pilot study. *Manag. Syst. Prod. Eng.* **2020**, *28*, 184–188. [CrossRef]

63. Sisler, J.D.; Mandler, W.K.; Shaffer, J.; Lee, T.; McKinney, W.G.; Battelli, L.A.; Orandle, M.S.; Thomas, T.A.; Castranova, V.C.; Qi, C.L. Toxicological assessment of dust from sanding micronized copper-treated lumber in vivo. *J. Hazard. Mater.* **2019**, *373*, 630–639. [CrossRef] [PubMed]

64. Dado, M.; Mikusova, L.; Schwarz, M.; Hnilica, R. Effect of selected factors on mass concentration of airborne dust during wood sanding. *MM Sci. J.* **2019**, *2019*, 3679–3682. [CrossRef]

65. Mamonova, M.; Reinprecht, L. The impact of natural and artificial weathering on the anatomy of selected tropical hardwoods. *IAWA J.* **2020**, *41*, 333–355. [CrossRef]

Appl. Sci. **2020**, *10*, 7608

66. Alfonso, V.A.; Carlquist, S.; Chimelo, J.P.; Rauber Coradin, V.T.; Détienne, P.; Grosser, D.; Ilic, J.; Wheeler, E.A.; Baas, P.; Gasson, P.E.; et al. IAWA list of microscopic features for hardwood identification with an appendix on non-anatomical information. *IAWA Bull.* **1989**, *10*, 219–232.

67. Stewart, I. The Particle Atlas Electronic Edition. *Microsc. Today* **1993**, *1*, 3. [CrossRef]

68. Calle, S.; Klaba, L.; Thomas, D.; Perrin, L.; Dufaud, O. Influence of the size distribution and concentration on wood dust explosion: Experiments and reaction modelling. *Powder Technol.* **2005**, *157*, 144–148. [CrossRef]

69. Jiang, H.P.; Bi, M.S.; Li, B.; Gan, B.; Gao, W. Combustion behaviors and temperature characteristics in pulverized biomass dust explosions. *Renew. Energy* **2018**, *122*, 45–54. [CrossRef]

70. Safety Data Sheet Cumaru. Available online: https://www.porta.com.au/wp-content/uploads/2019/03/Porta-SDS-Cumaru-Hardwood-5303-90SDS.pdf (accessed on 20 April 2020).

71. Safety Data Sheet Cumaru. Available online: https://tropix.cirad.fr/FichiersComplementaires/-EN/Africa/PADOUK.pdf (accessed on 30 April 2020).

72. Carrasco, E.V.M.; Caldas, R.B.; Oliveira, A.L.C.; Fakury, R.H. Numerical analysis of heat transfer in Brazilians wood species exposed to fire. *Cerne. Lavras* **2010**, *16*, 58–65.

Publisher's Note: MDPI stays neutral with regard to jurisdictional claims in published maps and institutional affiliations.

applied sciences

Article

Effect of Thermal Treatment of Birch Wood by Saturated Water Vapor on Granulometric Composition of Chips from Sawing and Milling Processes from the Point of View of Its Processing to Composites

Richard Kminiak [1],*, Kazimierz A. Orlowski [2], Ladislav Dzurenda [1], Daniel Chuchala [2] and Adrián Banski [1]

1   Faculty of Wood Sciences and Technology, Technical University in Zvolen, 960 53 Zvolen, Slovakia; dzurenda@tuzvo.sk (L.D.); banski@tuzvo.sk (A.B.)
2   Department of Manufacturing and Production Engineering, Faculty of Mechanical Engineering, Gdansk University of Technology, Narutowicza 11/12, 80-233 Gdansk, Poland; korlowsk@pg.edu.pl (K.A.O.); danchuch@pg.edu.pl (D.C.)
*   Correspondence: richard.kminiak@tuzvo.sk; Tel.: +421-0904-827-285

Received: 1 October 2020; Accepted: 23 October 2020; Published: 27 October 2020

Abstract: The goal of this work is to investigate the impact of thermal modification of birch wood with saturated steam on the particle size distribution of the sawing and milling process. Birch wood (Betula pendula Roth) is an excellent source to produce plywood boards. Wastes from mechanical processing of birch wood are suitable to produce composite materials. Granulometric analyses of chips from sawing processes on the PRW 15M frame saw, as well as on the 5-axis CNC machining centre SCM TECH Z5 and the 5-axis CNC machining centre AX320 Pinnacle, proved that more than 95% of chips are chips of coarse and medium coarse chip fractions with dimensions above 0.125 mm. Depending on the shape, coarse and medium-thick chips belong to the group of fiber chips, the length of which is several times greater than the width and thickness. Fine fractions with dimensions smaller than 125 μm are isometric chips that are approximately the same size in all three dimensions. Thoracic dust fractions below 30 μm were not measured. The performed analyses showed that the heat treatment of birch wood with saturated steam did not affect the grain size of chips formed in sawing and milling processes on CNC machining centre and can be used as a raw material for the production of composite materials. Fabric filters are suitable for separating chips extracted from frame saws, PRW-15M or machining centre. Environmental criteria for the separation of chips from transport air in textile filters are met by filters with a fabric classified in class G4.

Keywords: birch wood; chips; wood composites; granulometric composition of sawdust and chips; air handling; ecological filtration

---

1. Introduction

Thermal treatment of wood with saturated water vapor, in addition to the targeted physico-mechanical changes of wood, is often used in the production of veneer and plywood, bent furniture, or pressed wood. The mentioned treatment is also frequently accompanied by chemical reactions causes the colour changing of wood [1–7]. While in the past, the colour changes of the darkening of thermally treated wood were used to eliminate undesirable colour differences between light beige and dark kernel, or to remove unwanted colour spots caused by evaporation, browning or molding, in recent times the research focuses on targeted changes in wood colour to more or less pronounced colour shades, namely wood imitations of domestic trees as exotic trees [8–12]. The influence of the hydro-thermal modification process on wood machinability was also investigated by Sandak et al. [13] with four minor species such as black poplar (*Populus nigra* L.), deodar cedar (*Cedrus deodara* Roxb.), black pine (*Pinus nigra* Arnold.) and alder (*Alnus cordata* Loisel) the sharpness of the tool has a lower importance for the final surface smoothness.

Chips from sawing and milling processes were characterized in the literature as a polydisperse bulk material consisting of coarse and medium coarse fractions with a grain size of 0.5–3.5 mm, while the proportion of fine (dusty) fractions with smaller particle sizes below 300 μm is not excluded [14–30]. Dzurenda and Orlowski [31] examined sawdust of thermally modified ash wood obtained during sawing on a sash gang saw PRW-15M, and they revealed increasing of chip homogeneity in the range of granularity $a = 250$ μm–2.4 mm. A combination of wood chips and pine sawdust (*Pinus Sylvestris* L.) with a percentage of sawdust of up to 50% positively affects the board quality by making its structure more homogeneous [32].

The geometry of wood elements used in composites production is one of the important factors affecting board properties. Therefore, the research of birch wood after its thermal treatment on granulometric composition of wood particles from sawing and milling processes is substantial when we consider how large volumes of birch wood are processed in the sawmill industry and in the formatting of plywood boards around the world (Youngquist [32], Maloney [33], and Istek et al. [34]). According to these studies, bending and modulus of elasticity as well as dimensional stability of composites improve with appropriate elements length used for composite production.

The aim of this work was to determine the effect of thermal treatment-modification of the colour of birch wood (*Betula pendula* Roth) by saturated water vapor upon grain size from cutting processes of thermally modified birch wood on different machine tools as follows: the sash gang saw PRW-15M, the 5-axis CNC machining centre SCM TECH Z5 and AX320 Pinnacle. Furthermore, specification of separation technique requirements for the mentioned processes have to be defined.

2. Materials and Methods

2.1. Material

Birch wood (*Betula pendula* Roth) in the form of tangential boards with dimensions: 40 (radial direction) × 80 (tangential direction) × 600 mm (longitudinal direction), in total 180 samples, were divided into 3 groups consisting of 60 samples each. The initial moisture content MC of wet birch was in the range MC = 54.7–58.2%. The samples in the Group 1 were not heat treated. The boards of the second group were heat treated with the MODE I and the blanks of third group were heat treated with the MODE II. Thermal treatment of birch wood with saturated water vapor was carried out in a pressure autoclave APDZ 240 (Himmasch AD, Haskovo, Bulgaria) installed at Sundermann Ltd. in Banska Stiavnica (Slovakia).

2.2. Thermal Treatment

The process of thermal treatment of birch timber with saturated water vapor (steam) is shown in Figure 1, and the technical parameters of each mode are given in Table 1.

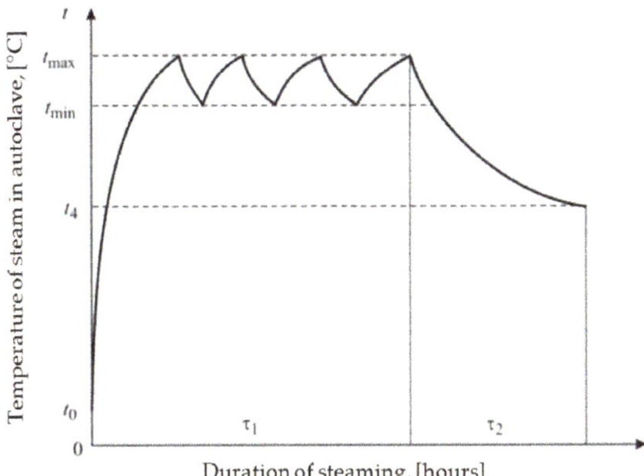

Figure 1. Mode of colour modification of birch wood with saturated water steam.

Table 1. Regimes for thermal treatment of the birch wood using saturated water vapor.

Thermal Treatment Regimes	Temperature of Steam in Autoclave, [°C]			Duration of Steaming τ, [h]		
	t_{min}	t_{max}	t_4	τ_1- stage I.	τ_2-stage II.	Total $\tau_1 + \tau_2$
MODE I	122.5	127.5	100	6.0	1.5	5.5
MODE II	132.5	137.5	100	6.0	1.5	7.5

The temperatures t_{max} and t_{min} in Figure 1 are the intervals between which saturated water vapor is fed into the autoclave to carry out the technological process. Temperature t_4 is a parameter of the saturated water vapor pressure in the autoclave to which the autoclave vapor pressure must be reduced before the pressure equipment is safely opened.

Subsequently, thermally treated and not-thermally treated blanks were dried by a low-temperature regime without changing the colour of the wood to a moisture content MC = 12 ± 0.5% in a conventional hot-air dryer: KC 1/50 (SUSAR s.r.o). After the blanks were dried, a portion of the blanks from each group was used to produce ST = 5 mm thick slats on a PRW-15M, a portion of the blanks was machined on the 5 axis machining centre AX320 Pinnacle and a part of the blanks was milled on a machining centre SCM Tech Z5.

2.3. Characteristics of Machine Tools and Milling Cutters

2.3.1. Narrow-Kerf Frame Sawing Machine (Sash Gang Saw) PRW–15M

The lamellae were made from birch thermally untreated and treated wood on the frame sawing machine PRW–15M with the hybrid dynamically balanced drive system and elliptical tooth trajectory movement [35] at the Department of Manufacturing and Production Engineering (Gdańsk University of Technology, PL).

The sawing process for both type of materials was carried out with average cutting speed v_c = 3.69 m·s^{-1} and feed per tooth f_z = 0.14 mm. In the case of the frame sawing process, can be assumed that the value of feed per tooth is equal to the value of average uncut chip thickness, $f_z = h_{av}$ = 0.14 mm (Figure 2). The saw blades were sharp, with Stellite tipped teeth. The other basic parameters of frame saw and saw blades are shown in Table 2.

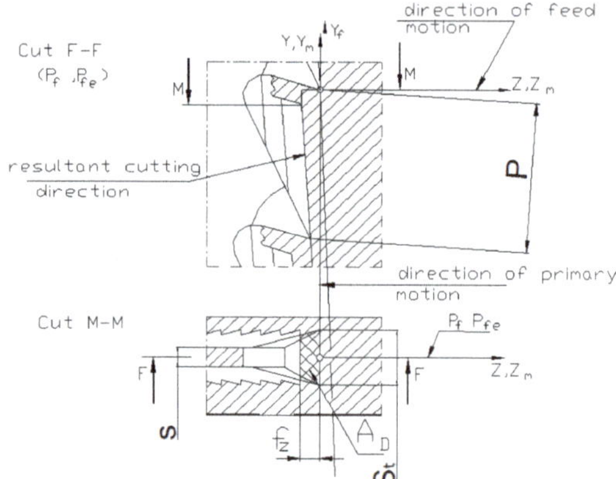

Figure 2. Mode sawing kinematics on the sash gang saw: fz–feed per tooth, s–saw blade thickness, AD–area of the cut, P–pitch, Y, Z and YM, ZM–machine coordinate and setting axes, Yf–f–set coordinate axis, Pfe–working plane.

Table 2. Machine tool and tool settings for frame sawing process.

Machine Tool Settings			
Name of Parameter	**Symbol**	**Value**	**Unit**
Cutting speed	v_c	3.69	m·s^{-1}
Number of strokes of saw frame per min	n_F	685	spm
Saw frame stroke	H_F	162	mm
Number of saws in the gang	n	5	pcs
Feed speed	v_f	1183	mm·min^{-1}
Feed per tooth	f_z	0.14	mm
Uncut chip thickness	h_{av}	0.14	mm
Tool Settings			
Name of Parameter	**Symbol**	**Value**	**Unit**
Overall set (kerf width)	S_t	2	mm
Saw blade thickness	s	0.9	mm
Free length of the saw blade	L_0	318	mm
Blade width	b	30	mm
Tooth pitch	P	13	mm
Tool side rake angle	γ_f	9	°
Tool side clearance angle	α_f	14	°
Tension stresses of saws in the gang	σ_N	300	MPa

2.3.2. Milling Centre AX320 Pinnacle

After the sawing process on the frame machine, the bigger parts of sawed samples were subjected to a milling process on 5 axis milling centre AX320 Pinnacle also located in GUT laboratory. The main parameters of milling process such as cutting speed v_c = 3.69 m·s^{-1} (rotational speed n = 4405 min^{-1}) and average uncut chip thickness h_{avg} = 0.14 mm (feed speed v_f = 2848 mm·min^{-1}) were used the same values as for frame sawing process. The end milling cutter with blades from cemented carbide was used during experimental cutting tests. This cutter was manufactured by ASPI company, Suwalki, Poland and the main geometry dimensions of milling cutter (Figure 3) and main parameters of the milling process are shown in Table 3. The work movements of the tool were performed by the machine tool in accordance with the CNC program on the Heidenhain TNC 640 control system. The standard vice with jaws length 100 mm was used to fixture of samples (Figure 4). The height of the milling samples H = 25 mm was cut depth a_p also (Figure 4).

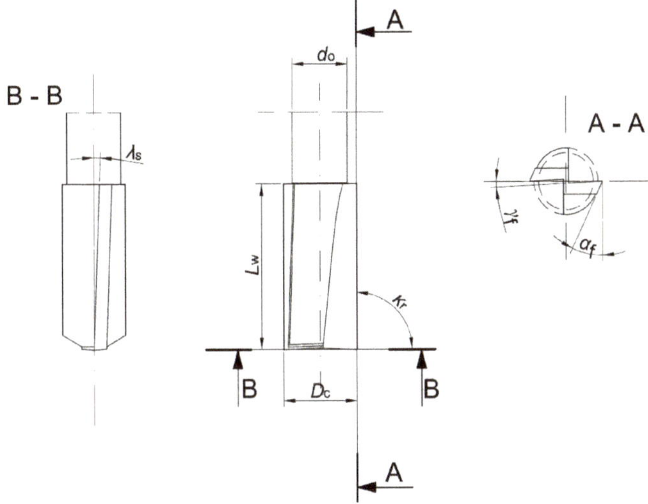

Figure 3. Cutting edge geometry of mill cutter used for wood milling process at the AX320 milling centre.

Table 3. Main parameters of milling cutters and milling processes.

Milling Cutters Settings				
Name of Parameter	**Symbol**	**Unit**	**Value**	
			AX 320	**SCM Tech Z5**
Tool diameter [mm],	D_c	mm	16	14
Shank diameter [mm],	d_o	mm	12	14
Number of teeth [-], z	z	-	2	3
Tool length [mm],	L_t	mm	70	110
Working length of the tool	L_c	mm	35	58
Side rake angle	γ_f	°	3	35
Side clearance angle	α_f	°	25	15
Tool cutting edge angle	κ_r	°	90	90
Cutting edge inclination angle	λ_s	°	2	30
Material of edges			HM	VHM

Milling Processes Settings				
Name of Parameter	**Symbol**	**Unit**	**Value**	
			AX 320	**SCM Tech Z5**
Cutting speed	v_c	m·s^{-1}	3.69	13.19
Spindle speed	n	min^{-1}	4405	18000
Feed speed	v_f	mm·min^{-1}	2848	6000
Feed per tooth	f_z	mm	0.323	0.111
Average uncut chip thickness	h_{av}	mm	0.14	0.066
Cut depth	a_p	mm	25	22
Cut width	a_e	mm	3	5

Figure 4. Machine tool AX320 Pinnacle with equipment for machining tests.

2.3.3. Milling Centre SCM Tech Z5

The birch blanks as well as after the heat treatment by the individual modes were milled on a 5 axis CNC machining centre SCM Tech Z5 manufactured by SCM-group, Rimini, Italy. A positive spiral milling cutter manufactured by IGM under the designation IGM 193 was used in the experiment (Figure 5). The working part of spiral cutter IGM 193 was manufactured with high-quality carbide VHM (Integral HM). The base technical data of this cutter are shown in Table 3.

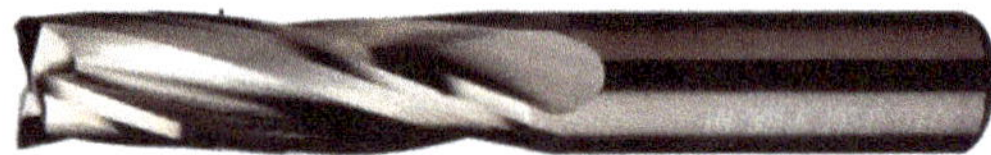

Figure 5. Mill cutter used for wood milling process at the SCM Tech Z5 milling centre.

2.4. Granulometric Analysis of Chips

For granulometric analyses of sawdust and shavings from the sawing and milling process of thermally untreated and treated birch wood, sawdust and chips samples were taken by isokinetic procedure from the exhaust pipe of the individual machine tools: PRW-15M frame saw, AX 320 Pinnacle and SCM Tech Z5 CNC machining centres, according to ISO 9096 [36].

A granulometric composition of the chips was evaluated by sifting. For this purpose, it was used a special set of sieves arranged one above the other (mesh size: 2, 1, 0.5, 0.25, 0.125, 0.063, 0.032 mm, and the bottom), the sieves are placed on a vibration stand of the sifting machine Retsch AS 200c (Retsh GmbH, Haan, Germany). The parameters of sifting were as follows: frequency of sifting interruption of 20 s, amplitude of sieves deflection: 2 mm·g^{-1}, sifting time: $\tau = 15$ min, weighed sample: 50 g. The granulometric composition was obtained by weighing of the portions remaining on the sieves after sifting on an electronic laboratory scale Radwag 510/C/2 (Radwag Balances and Scales, Radom, Poland), weighing to an accuracy of 0.001 g. The sifting was realized with 3 samples for each combination of parameters.

With the purpose of specifying information about the size of the smallest particles of fine fraction of dry chips a microscopic analysis of granules of fraction of dry chips with the size lower than 500 μm was realized. The proposed analysis of chips was carried out by an optical method–analysis of the picture obtained from the microscope Nikon Optiphot–2 with the objective Nikon 4×. Granules of chips were scanned by three low-cost television CCD cameras HITACHI HV-C20

(RGB 752 × 582 pixel), with horizontal resolution 700 TV lines and evaluated by a software LUCIA-G 4.0 (Laboratory Universal Computer Image Analysis), installed on a PC with the processor Pentium 90 (RAM 32 MB) with the graphic card VGA Matrox Magic under the operation system Windows NT 4.0 Workstation. The program of analysis of picture LUCIA-G enables to identify the individual particles of disintegrated wood material, quantitative determination of individual particles situated in the analyzed picture, and basic information such as width and length of particles, and circularity expressing the measure of deviation of projection of a given chip shape from the projection of the shape of a circle ψ according to equation:

$$\psi = \frac{4\pi \cdot S}{P_p^{\,2}} \tag{1}$$

where: S—surface of particle [m^2], P_p—perimeter of particle [m].

3. Results and Discussion

The results of the sieve analysis of sawdust sucked off from the frame saw PRW-15M and the chips sucked off from the machining centre SCM Tech Z5, and AX320 Pinnacle from birch shavings with and without heat treatment, are shown in Tables 4–6.

Table 4. Sawdust from the saw frame PRW-15M, average values with standard deviations.

Measure of Sieve Mesh, [mm]	Mark of Fraction	Fractions Representation of Birch Wood [%]		
		Untreated	Modified by Mode I	Modified by Mode II
2.000	coarse	3.77 ± 0.19	3.11 ± 0.16	3.98 ± 0.20
1.000		2.82 ± 0.14	4.21 ± 0.21	1.80 ± 0.09
0.500	medium coarse	16.20 ± 0.81	14.84 ± 0.74	9.18 ± 0.46
0.250		52.30 ± 2.62	46.27 ± 2.31	47.13 ± 2.36
0.125		22.49 ± 1.12	28.30 ± 1.42	33.13 ± 1.66
0.063	fine	1.82 ± 0.09	3.20 ± 0.16	4.54 ± 0.23
0.032		0.61 ± 0.03	0.09 ± 0.00	0.25 ± 0.01
<0.032		0.00 ± 0.00	0.00 ± 0.00	0.00 ± 0.00

Table 5. Chips from AX320 Pinnacle, average values with standard deviations.

Measure of Sieve Mesh, [mm]	Mark of Fraction	Fractions Representation of Birch Wood [%]		
		Untreated	Modified by Mode I	Modified by Mode II
2.000	coarse	2.86 ± 0.14	2.52 ± 0.13	3.66 ± 0.18
1.000		4.74 ± 0.24	11.72 ± 0.59	4.44 ± 0.22
0.500	medium coarse	30.95 ± 1.50	35.02 ± 1.75	30.15 ± 1.51
0.250		40.92 ± 2.05	28.25 ± 1.41	33.59 ± 1.68
0.125		18.38 ± 0.92	18.14 ± 0.91	22.60 ± 1.13
0.063	fine	1.89 ± 0.09	4.22 ± 0.21	5.23 ± 0.26
0.032		0.27 ± 0.01	0.13 ± 0.01	0.33 ± 0.02
<0.032		0.00 ± 0.00	0.00 ± 0.00	0.00 ± 0.00

Table 6. Chips from the CNC machining centre SCM Tech Z5, average values with standard deviations.

Measure of Sieve Mesh, [mm]	Mark of Fraction	Fractions Representation of Birch Wood [%]		
		Untreated	Modified by Mode I	Modified by Mode II
2.000	coarse	69.43 ± 3.47	66.25 ± 3.31	63.72 ± 3.19
1.000		6.28 ± 0.31	9.12 ± 0.46	8.67 ± 0.33
0.500	medium coarse	10.14 ± 0.51	11.47 ± 0.57	10.23 ± 0.51
0.250		9.13 ± 0.46	9.50 ± 0.48	8.71 ± 0.44
0.125		3.16 ± 0.16	2.02 ± 0.29	5.75 ± 0.29
0.063	fine	1.67 ± 0.08	1.52 ± 0.08	2.63 ± 0.13
0.032		0.19 ± 0.01	0.12 ± 0.01	0.29 ± 0.01
<0.032		0.00 ± 0.00	0.00 ± 0.00	0.00 ± 0.00

By comparing the proportions of the individual sawdust fractions of the PRW-15M frame saw (Table 4), it can be stated that there are no differences in the proportions of the sawdust fractions of natural birch wood and thermally treated birch wood. The distribution of chip fractions according to the distribution is symmetric, with the largest fraction in the grain size range of 0.250–0.500 μm.

Analyses of the shavings of individual sawdust fractions indicate that coarse and medium coarse fractions over 0.5 mm in size from the sawing process of both untreated and treated birch wood on the PRW-15M frame saw belong to the category of polydisperse fibrous materials, rod-shaped with significant elongation in one dimension. The agreement between the granulometric composition of the steam-saturated wood vapor at a temperature of $t = 125$–135 °C and the granulometric composition of the native birch wood chips predicts that the thermal treatment of the birch wood does not affect the chip formation process. This finding is similar to the results of temperature analysis in the process of drying beech timber by low and high temperature regimes on sawdust grain size presented by Orlowski et al. [37].

Microscopic analysis of the chip shape of the fractions below 500 μm indicates that the chip size of these fractions by their shape belongs to the group of isometric chips. Chips having approximately the same dimension in all three directions. Circularity is in the range $\Psi = 0.7$–1.0.

The proportion of fractions of chip extracted from the machining centre AX320 Pinnacle (Table 5), is similar to sawdust from the PRW-15M frame saw. Chips from machining centre AX320 Pinnacle did not show any significant effect of birch wood thermal treatment on the particle size except for the increased proportion of fine fraction of thermally treated birch wood, which increased from 2.1% to 4.4% for thermally treated birch wood mode I, and to 5.5% for the thermally treated wood mode II.

From the chip shape aspect, only chips of a coarse fraction above 1 mm are fibrous chips. Medium and fine fraction chips are isometric chips with a circularity value in the interval: $\Psi = 0.7$–1.0.

The chips formed at the SCM Tech Z5 machining centre (Table 6) differ according to the proportion of the individual fractions compared to the sawdust formed by sawing the birch wood on the PRW 15M sash gang saw. The coarse fraction chips above 2 mm make up to 2/3 of the total chips produced. Similar representation of the fractions of beech, maple, and oak wood chips at the SCM Tech Z5 machining centre at a removal of $a_e = 3$ mm and a feed speed $v_f = 3$ m·min^{-1}, and at feed speed $v_f = 5$ m·min^{-1} reported by: Kminiak and Banski [38]. With the feed speed decreasing at this machining centre when milling wood as the authors stated [39], the share of the coarse fraction is increasing at the expense of the medium coarse fraction.

The chip shape of the coarse fraction corresponds to the shape and size of the layer to be cut. Medium and fine fractions of birch wood chips with or without thermal treatment are isometric chips.

A comparison of the dust fraction of the extracted sawdust from the PRW-15M frame sawing machine and the chips extracted from the CNC machining centres suggests that the smallest chips in the extracted bulk wood mass are chips of size $a = 32$ μm (Figure 6). Fractions below 30 μm (thoracic dust or respirable dust) were not detected by measurements.

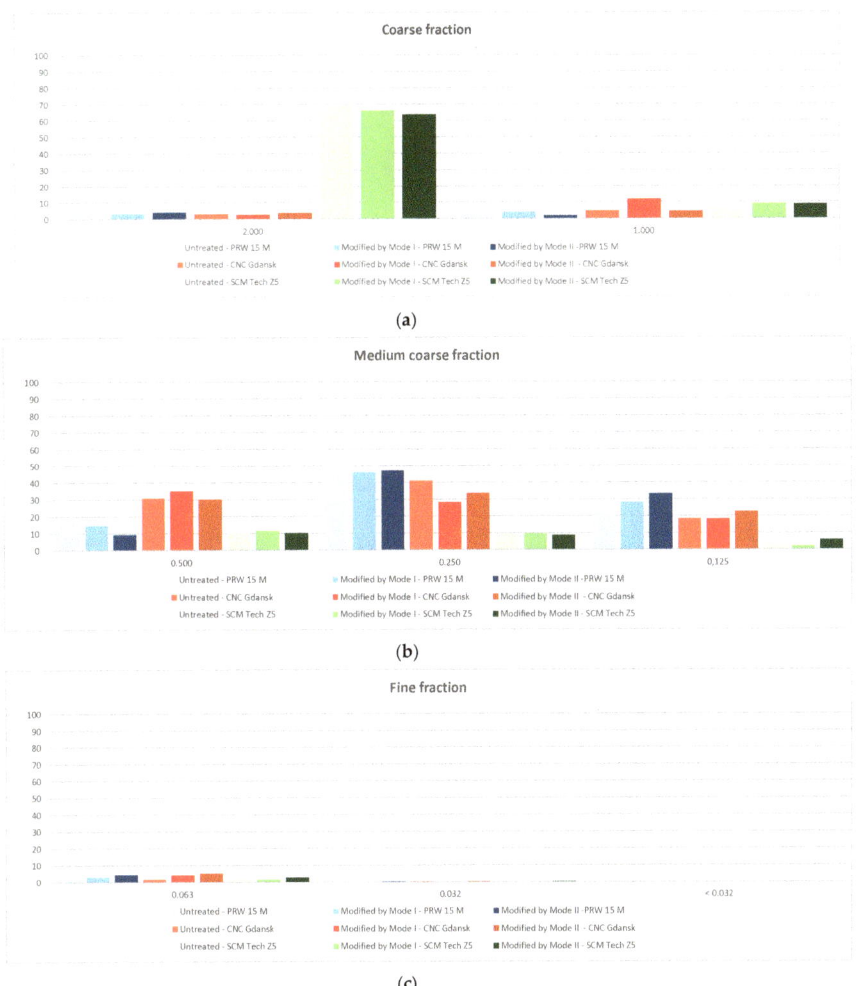

Figure 6. Comparison of sawdust and chips fractions from CNC machining centres (**a**) coarse fraction (**b**) medium coarse fraction (**c**) fine fraction.

Criterion for establishing a separation technique requirement for the separation of extracted bulk solids, including dust chips from conveyed air in exhaust systems is the limit of separation of separation technology. By comparing the smallest chips in the conveyed bulk material from individual machines $a = 32$ µm with the separability limit of the separators in Figure 7, the possibility of meeting environmental criteria is demonstrated. Separation limits of selected filters and separators are shown in Table 7.

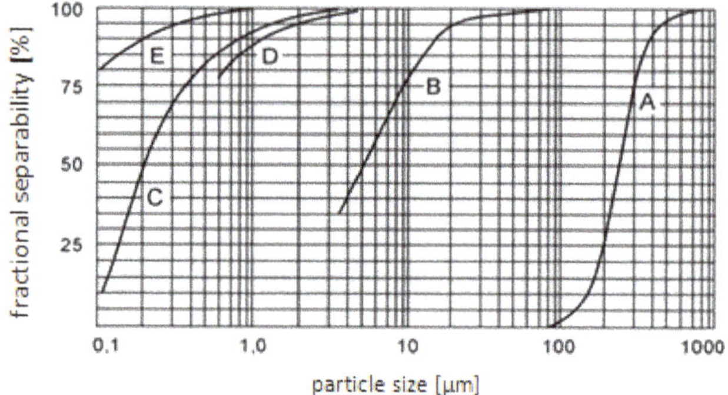

Figure 7. Fractional separation diagram of disintegrated wood mass of individual types of separation technique: A—settling chambers, B—dry mechanical separators, C—fabric filters, D—wet separators, E—electrofiltrators.

Table 7. Separation limits for separation technique.

Type of Separation Device	Separation Limits
dry mechanical separators	$a = 80$ [μm]
fabric filters with fabric class G4	$a = 10$ [μm]
electro separators	$a = 1$ [μm]

Based on data in Table 4 and Figure 6, it follows that for separating chips with dimensions above and ≥ 32 μm, suitable separation techniques are: wet scrubbers, fabric filters, and electro-scrubbers.

For the purpose of separating loose wood from transport air in wood processing plants, the best available technology are filters with G4 fabric having a separation limit value of $a_{SL} = 10$ μm, thus meeting the requirements of EN 779/2012 [40].

Fabric filters are fully sufficient to catch the chips extracted from the sawing and milling process native birch wood, and thermally modified wood with saturated water vapor [38–41].

Analyses of the grain size of chips formed in the processes of sawing dried birch wood on PRW-15M frame saws and milling on CNC machining centers showed that heat treatment of birch wood with saturated steam did not affect the grain size of the chips. Based on this fact, it can be concluded that, just as chips from raw birch wood are used for the production of composite materials, chips from heat treatment birch wood are also a suitable raw material for the production of composite materials. The use of a given chip to produce agglomerated materials prolongs the life cycle of wood and thus contributes to increasing the degree of sustainability.

4. Conclusions

The following conclusions are drawn from the performed granulometric analyses of the extracted sawdust from the frame saw PRW 15M and the chips of the extracted CNC machining centre in order to evaluate the inlet of the thermal modification of the birch wood by the saturated water steam to modify the wood colour:

- The influence of thermal treatment of birch wood did not affect the grain size of sawdust from the PRW 15 M frame saw or the grain chips from CNC centre.
- In the process of sawing birch wood on the frame saw PRW-15M and milling on CNC machining centres, chips are formed with a dominant representation of medium and coarse fractions larger than 125 μm.
- The dust fraction in the grain size range of 125–32 μm does not exceed 5%.

- Fractions of thoracic or respirable dust with particle sizes below 30 μm have not been detected.
- These values are fully in line with the particle sizes used for the production of wood composites and it is not possible to consider reducing the quality of the input raw material for the production of wood composites.
- Environmental criteria for the separation of chips from transport air are met by fabric filters with a fabric classified in G4 filtration class.

Author Contributions: Conceptualization, L.D., K.A.O., D.C. and R.K.; methodology, L.D. and D.C.; software, D.C. and R.K.; validation, L.D. and K.A.O.; formal analysis, L.D. and R.K.; investigation, L.D., K.A.O., D.C. and R.K.; resources, L.D., K.A.O., D.C., R.K. and A.B.; data curation, L.D. and K.A.O.; writing—original draft preparation, L.D., K.A.O., D.C., R.K. and A.B.; writing—review and editing, L.D., K.A.O., D.C. and R.K.; visualization, L.D., K.A.O., D.C., R.K. and A.B.; supervision, L.D. and K.A.O.; project administration, L.D., K.A.O. and D.C.; funding acquisition, R.K. and D.C. Please turn to the CRediT taxonomy for the term explanation. Authorship must be limited to those who have contributed substantially to the work reported. All authors have read and agreed to the published version of the manuscript.

Funding: This research was funded by Slovak Research and Development Agency, grant number APVV-17-0456 and by Polish Ministry of Science and Higher Education, decision number 21/E-359/SPUB/SP/2019.

Acknowledgments: This experimental research was prepared within the grant project: APVV-17-0456 "Termická modifikácia dreva sýtou vodnou parou za účelom cielenej a stabilnej zmeny farby drevnej hmoty" as the result of work of authors and the considerable assistance of the APVV Agency. The authors gratefully acknowledge the Polish Ministry of Science and Higher Education for funding the maintenance of scientific and research equipment–PRW-15M frame saw (decision no. 21/E-359/SPUB/SP/2019).

Conflicts of Interest: The authors declare no conflict of interest.

References

1. Fengel, D.; Wegener, G. *Wood: Chemistry, Ultrastructure, Reactions*; Walter de Gruyter: Berlin, Germany, 1989; p. 613.
2. Hon, N.S.D.; Shiraishi, N. *Wood and Cellulosic Chemistry*; CRC Press: New York, NY, USA, 2001; p. 928.
3. Kačíková, D.; Kačík, F.; Čabalová, I.; Ďurkovič, J. Effects of thermal treatment on chemical, mechanical and colour traits in Norway spruce wood. *Bioresour. Technol.* **2013**, *144*, 669–674. [CrossRef] [PubMed]
4. Timar, M.C.; Varodi, A.M.; Hacibektasoglu, M.; Campean, M. Color and FT-IR analysis of chemical changes in beech wood (*Fagus sylvatica* L.) after light steaming and heat treatment in two different environments. *BioResources* **2016**, *11*, 8325–8343. [CrossRef]
5. Sikora, A.; Kačík, F.; Gaff, M.; Vondrová, V.; Bubeníková, T.; Kubovský, I. Impact of thermal modification on color and chemical changes of spruce and oak wood. *J. Wood Sci.* **2018**, *64*, 406–416. [CrossRef]
6. Geffert, A.; Výbohová, E.; Geffertová, J. Characterization of the changes of colour and some wood components on the surface of steamed beech wood. *Acta Fac. Xylologiae Zvolen* **2017**, *59*, 49–57. [CrossRef]
7. Dzurenda, L.; Geffert, A.; Geffertová, J.; Dudiak, M. Evaluation of the process thermal treatment of maple wood saturated water steam in terms of change of pH and color of wood. *BioResources* **2020**, *15*, 2550–2559. [CrossRef]
8. Tolvaj, L.; Nemeth, R.; Varga, D.; Molnar, S. Colour homogenisation of beech wood by steam treatment. *Drewno* **2009**, *52*, 5–17.
9. Deniz, A.; Gokhan, G.; Seray, O. The influence of thermal treatment on color response of wood materials. *Color Res. Appl.* **2012**, *37*, 148–153. [CrossRef]
10. Hadjiski, M.; Deliiski, N. Advanced control of the wood thermal treatment processing. Cybernetics and information technologies. *Bulg. Acad. Sci.* **2016**, *16*, 176–197.
11. Dzurenda, L. The shades of color of quercus robur L. Wood obtained through the processes of thermal treatment with saturated water vapor. *BioResources* **2018**, *13*, 1525–1533. [CrossRef]
12. Banski, A.; Dudiak, M. Dependence of color on the time and temperature of saturated water steam in the process of thermal modification of beech wood. *AIP Conf. Proc.* **2019**, *2118*, 030003. [CrossRef]
13. Sandak, J.; Goli, G.; Cetera, P.; Sandak, A.; Cavalli, A.; Todaro, L. Machinability of minor wooden species before and after modification with thermo-vacuum technology. *Materials* **2017**, *10*, 121. [CrossRef] [PubMed]

14. Kos, A.; Beljo-Lučić, R.; Šega, K.; Rapp, A.O. Influence of woodworking machine cutting parameters on the surrounding air dustiness. *Holz Als Roh Und Werkst.* **2004**, *62*, 169–176. [CrossRef]

15. Očkajová, A.; Beljo Lučić, R.; Čavlović, A.; Teraňová, J. Reduction of dustiness in sawing wood by universal circular saw. *Drv. Ind.* **2006**, *57*, 119–126.

16. Kopecký, Z.; Rousek, M. Dustiness in high-speed milling. *Wood Res.* **2007**, *52*, 65–76.

17. Orlowski, K.A.; Sandak, J.; Negri, M.; Dzurenda, L. Sawing frozen wood with narrow kerf saws: Energy and quality effects. *Forest Prod. J.* **2009**, *59*, 79–83.

18. Dzurenda, L.; Orlowski, K.A.; Grzeskiewicz, M. Effect of thermal modification of oak wood on sawdust granularity. *Drv. Ind.* **2010**, *61*, 89–94.

19. Hlásková, L.; Rogozinski, T.; Dolny, S.; Kopecký, Z.; Jedinák, M. Content of respirable and inhalable fractions in dust created while sawing beech wood and its modifications. *Drewno* **2015**, *58*, 135–146.

20. Mračková, E.; Krišťák, Ľ.; Kučerka, M.; Gaff, M.; Gajtanska, M. Creation of wood dust during wood processing: Size analysis, dust separation, and occupational health. *BioResources* **2015**, *11*, 209–222. [CrossRef]

21. Hlásková, L.; Rogozinski, T.; Kopecký, Z. Influence of feed speed on the content of fine dust during cutting of two-side-laminated particleboards. *Drv. Ind.* **2016**, *67*, 9–15. [CrossRef]

22. Pałubicki, B.; Rogoziński, T. Efficiency of chips removal during CNC machining of particleboard. *Wood Res.* **2016**, *61*, 811–818.

23. Marková, I.; Hroncová, E.; Tomaškin, J.; Tureková, I. Thermal analysis of granulometry selected wood dust particles. *Bioresources* **2018**, *13*, 8041–8060. [CrossRef]

24. Piernik, M.; Rogozinski, T.; Krauss, A.; Pinkowski, G. The influence of the thermal modification of pine (*Pinus sylvestris* L.) wood on the creation of fine dust particles in plane milling. *J. Occup. Health* **2019**, *61*, 481–488. [CrossRef] [PubMed]

25. Tureková, I.; Marková, I. Ignition of deposited wood dust layer by selected sources. *Appl. Sci.* **2020**, *10*, 5779. [CrossRef]

26. Mirski, R.; Derkowski, A.; Dziurka, D.; Wieruszewski, M.; Dukarska, D. Effects of chip type on the properties of chip–sawdust boards glued with polymeric diphenyl methane diisocyanate. *Materials* **2020**, *13*, 1329. [CrossRef]

27. Očkajová, A.; Kučerka, M.; Kminiak, R.; Krišťák, Ľ.; Igaz, R.; Réh, R. Occupational exposure to dust produced when milling thermally modified wood. *Int. J. Environ. Res. Public Health* **2020**, *17*, 1478. [CrossRef]

28. Pałubicki, B.; Hlásková, L.; Rogoziński, T. Influence of exhaust system setup on working zone pollution by dust during sawing of particleboards. *Int. J. Environ. Res. Public Health* **2020**, *17*, 3626. [CrossRef]

29. Rogozinski, T. Pilot-scale study on the influence of wood dust type on pressure drop during filtration a pulse-jet baghouse. *Process Saf. Environ. Prot.* **2018**, *119*, 58–64. [CrossRef]

30. Očkajová, A.; Kučerka, M.; Kminiak, R.; Rogoziński, T. Granulometric composition of chips and dust produced from the process of working thermally modified wood. *Acta Fac. Xylologiae Zvolen* **2020**, *62*, 103–111. [CrossRef]

31. Dzurenda, L.; Orlowski, K.A. The effect of thermal modification of ash wood on granularity and homogeneity of sawdust in the sawing process on a sash gang saw PRW 15-M in view of its technological usefulness. *Drewno* **2011**, *54*, 27–37.

32. Youngquist, J.A. Wood-based composites and panel products. In *Wood Handbook—Wood as an Engineering Material*; United States Department of Agriculture, Ed.; USDA: Madison, WI, USA, 1999; pp. 10–31.

33. Maloney, T.M. *Modern Particleboard & Dry-Process Fiberboard Manufacturing*, 2nd ed.; Miller Freemann Inc.: San Francisco, CA, USA, 1993; 689p.

34. Istek, A.; Aydin, U.; Özlüsoylu, I. The effect of chip size on the particleboard properties. In Proceedings of the International Congress on Engineering and Life Science (ICELIS), Kastamouno, Turkey, 26–29 April 2018; pp. 439–444.

35. Wasielewski, R.; Orłowski, K.A. Hybrid dynamically balanced saw frame drive. *Holz als Roh-Und Werkst.* **2002**, *60*, 202–206. [CrossRef]

36. ISO 9096. *Stationary Source Emissions—Manual Determination of Mass Concentration of Particulate Matter*; The International Organization for Standardization: Geneva, Switzerland, 2017.

37. Orłowski, K.A.; Chuchała, D.; Muziński, T.; Barański, J.; Banski, A.; Rogoziński, T. The effect of wood drying method upon the granularity of sawdust obtained during the sawing process on the frame sawing machine. *Acta Fac. Xylologiae Zvolen* **2019**, *61*, 83–92. [CrossRef]

38. Kminiak, R.; Banski, A. Separation of exhausted chips from a CNC machining center in filter FR-SP 50/4 with finet PES 4 fabric. *AIP Conf. Proc.* **2018**, *2000*, 30011. [CrossRef]
39. Kminiak, R.; Dzurenda, L. Impact of sycamore maple thermal treatment on a granulometric composition of chips obtained due to processing on a CNC machining centre. *Sustainability* **2019**, *11*, 718. [CrossRef]
40. EN 779. *Particulate Air Filters for General Ventilation—Determination of the Filtration Performance*; The International Organization for Standardization: Geneva, Switzerland, 2012.
41. Dolny, S.; Rogozinski, T. Air flow resistance nonwoven filter fabric covered with Microfiber layer used in wood dust separation. *Drewno* **2014**, *57*, 125–134.

applied sciences

MDPI

Article

TOF-SIMS Molecular Imaging and Properties of pMDI-Bonded Particleboards Made from Cup-Plant and Wood

Petr Klímek [1,2], Rupert Wimmer [2,3,*] and Peter Meinlschmidt [4]

[1] Tescan Orsay Holding a.s., 62300 Brno, Czech Republic; Petr.Klimek@tescan.com
[2] Department of Wood Science and Technology, Mendel University in Brno, 61300 Brno, Czech Republic
[3] Institute of Wood Technology and Renewable Materials, University of Natural Resources and Life Sciences, 3430 Vienna, Austria
[4] Fraunhofer-Institut für Holzforschung—Wilhelm-Klauditz-Institut, 38108 Braunschweig, Germany; peter.meinlschmidt@wki.fraunhofer.de
* Correspondence: Rupert.Wimmer@boku.ac.at

Abstract: Cup-plant (*Silphium perfoliatum* L.) stalks were investigated as a potential wood-replacement in particleboards (PBs). Two types of PBs were produced—(1) single-layer and (2) three-layer boards. In the three-layer cup-plant PB, the core layer was made from cup-plant, while the surface layer consisted of spruce particles. The cup-plant as well as spruce control panels were produced with polymeric methylene diphenyl diisocyanate (pMDI) as the adhesive, with the physical and mechanical properties measured to meet class P1 of the European EN 312 standard. For the intrinsic morphology of the particleboards, scanning electron microscopy was applied. Wood-based and cup-plant-based particleboards indicated significant differences in morphology that affect the resulting properties of particleboards. Furthermore, an innovative approach was used in the determination of the pMDI bondline morphology. With a compact Time-of-Flight Secondary Ion Mass analyser, integrated in a multifunctional focused-ion beam scanning-electron-microscope, it was possible to show that the Ga$^+$ ion source could be detect and visualize in 3D ion molecular clusters specific to pMDI adhesive and wood. Mechanical performance data showed that cup-plant particleboards performed well, even though their properties were below the spruce-made controls. Especially the modulus of rupture (MOR) of the cup-plant PB was lowered by 40%, as compared to the spruce-made control board. Likewise, thickness swelling of cup-plant made boards was higher than the control. Results were linked to the specific porous structure of the cup-plant material. In contrast, it was shown that three-layer cup-plant PB had a higher MOR and also a higher modulus of elasticity, along with lower thickness swelling, compared to its single-layer cup-plant counterpart. The industry relevant finding was that the three-layer PB made from cup-plant stalks fulfilled the EN 312 standard, class P1 (usage in dry conditions). It was shown that raw material mixtures could be useful to improve the mechanical panel performance, also with an altered vertical density profile.

Keywords: particleboard; three-layer particleboard; cup plant; TOF-SIMS; biomass; bioresources

Citation: Klímek, P.; Wimmer, R.; Meinlschmidt, P. TOF-SIMS Molecular Imaging and Properties of pMDI-Bonded Particleboards Made from Cup-Plant and Wood. *Appl. Sci.* **2021**, *11*, 1604. https://doi.org/10.3390/app11041604

Academic Editor: Ľuboš Krišťák
Received: 6 January 2021
Accepted: 3 February 2021
Published: 10 February 2021

Publisher's Note: MDPI stays neutral with regard to jurisdictional claims in published maps and institutional affiliations.

1. Introduction

Wood is the traditional and prime raw material in particleboards production since 1887, and annual production volumes in Europa exceed 30 million m^3 [1]. Considering the high production volumes, declining stocks of natural resources [2], i.e., possible future wood shortage situations, could play important roles. In addition, as the use of potentially contaminated waste wood in particleboards reaches 90% in some European countries, it could create environmental concerns, with a higher request for alternative non-contaminated materials.

Non-wood materials could also be utilized in particleboard (PB) production, which have the advantages of achieving higher resource-effectiveness, at ecologically and economically viable conditions. While agricultural crops are primarily cultivated for food, for various chemical products, or for biogas production [3], unutilized plant parts could be potentially processed to PBs. Using agricultural residues for PBs might have economic benefits for manufacturers, as the expenses for residues and wastes might be below market prices for wood [4]. Likewise, the utilization of waste materials in industrial production is also reducing environmental burdens, as residues such as stalks, husks, or straw are often left on the fields, or even burned.

Cup-plants seem to be a reasonable candidate for replacing wood in PBs. Dry mass yield between 11 t/ha and 20 t/ha per harvest is high and might compete with the yield achieved in forests, which can be ~16 t/ha per harvest [5]. Cup-plant (*Silphium perfoliatum* L.) originates in Eastern North America [6], but is now well-established across Central Europe. Although it was grown in gardens as an ornamental plant during the 18th century, today it is widely cultivated for energy production [7]. Cup-plant characteristics, including aspects of cultivation and utilization, including particleboard manufacturing have been demonstrated [8].

Particleboards from rice straw or rice husks [9–12], wheat straw [13], sunflower stalks [14–17], from vine prunings [18], cotton stalks [19], apple and plum orchard prunings [20], or even teal oil camellia [21], were already shown, i.e., produced. Balducci et al. [22] and Dix et al. [23] introduced residues of several Central European agricultural plants as a raw material for low density PBs, and Selinger and Wimmer [24] introduced light-weight sandwich PBs from hemp shives and fibers. It is obvious that agricultural resources could provide materials to replace wood in PBs, even if their property profiles are generally below conventional PBs. As the mechanical properties of PBs made with alternative materials are lowered, the anatomical structure and morphology of the utilized particles are of ultimate importance due to the close connections to the relevant properties [25].

Various microscopic techniques were applied to describe the anatomical and structural composition of wood-based composites. As an example, scanning electron microscopy was used for describing anatomical structures in PBs [26]. However, due to the lignocellulosic nature of the samples, there is insufficient compositional contrast to distinguish clearly between wood fractions, and adhesive bondlines [27]. To this end, elemental mapping by means of electron dispersive X-rays might be a feasible method. However, Electron Dispersive X-rays (EDX) techniques are also limited by their spatial resolution, sensitivity, or the ability to detect and map molecules that are indicative of wood-adhesive bondlines, particularly when wood and the used adhesive are both represented by the same chemical elements, albeit different molecular clusters. Here, the alternative method Time-of-Flight Secondary Ion Mass Spectrometry (TOF-SIMS) is suggested. Lately, TOF-SIMS was integrated into focused-ion-beam scanning-electron-microscope (FIB-SEM) TESCAN instruments [28], which brings additional advantages such as high spatial resolution (>5 nm) to elemental mapping, as well as simultaneous ion and SEM imaging. It was shown that TOF-SIMS (single instrument) can describe the distribution of molecules in biological systems [29] and also in wood tissues [30–33], but the use of TOF-SIMS for mapping the molecular distribution in adhesive-wood phases was not demonstrated so far. Consequently, the following hypotheses are stated:

Hypothesis 1 (H1). *Compact time-of-flight ion mass analyser (C-TOF) is capable of detecting elements and molecules specific to pMDI adhesion status.*

Hypothesis 2 (H2). *The penetration of the adhesive can be visualized in 3D using TOF-SIMS data.*

Hypothesis 3 (H3). *The mechanical properties of PBs made from the cup-plant fulfil industrial standards (EN 312, class 1).*

Hypothesis 4 (H4). *A three-layer PB made from cup-plant particles in the core layer, and spruce particles in the surface layer, could show improved properties over a single-layer produced panel.*

2. Materials and Methods

Single-layer as well as three-layer particleboards (PBs) were produced using cup-plant stalk particles (*Silphium perfoliatum* L.). Stalks were 1.8 m long and square-sized with 25 × 25 mm^2. As a control, single-layer and three-layers PBs were also produced using spruce particles (*Picea abies [L.] Karst*). The material was first chipped in a Klöckner chipper 120 × 400 H2W.T (Klöckner Maschinenfabrik, Lauenburg, Germany), at a cutting speed of 725 rpm, and a feeding speed of 1 m/s. The obtained chips approximately sized 20 × 10 × 5 mm^3 were then milled in a Condux-Werk HS 350 (Condux Maschinenbau GmbH & Co. KG, Hanau—Wolfgang, Germany) hammer mill. Afterwards, the particles were screened in the cascade vertical drum screener Allgaier D7336 (Allgaier-Werke GmbH, Uhingen, Germany). The sieve screens had mesh size openings of 5.0 mm, 3.15 mm, 1.24 mm, and 0.60 mm, respectively. Particles used to manufacture PBs were taken from the sieves with openings between >3.15 mm and <5 mm, which were then manually mixed at a weight ratio of 50:50. Afterwards, the particles were oven-dried at 74 °C for 4 days, reaching moisture contents between 5% and 7%. For the three-layer PBs, spruce particles at dimensions <1.24 mm were used for the surface layers, while cup-plant particles formed the core layer (3LCP). Same procedure was done for the three-layer PB, with both the core layer and the surface layers made from spruce (3LSP). For both three-layer PB types, the shelling ratio, which is the ratio of the surface layer thickness to the total thickness of the panel, was set at 0.3.

All PBs were produced with the target density of 600 kg/m^3, and a panel thickness of 12 mm, and bonded with polymeric methylene diphenyl diisocyanate (pMDI) resin (Huntsman I-BOND® PM4390, Huntsman GmbH, Hamburg, Germany). Two resin dosages were applied—pMDI was applied in amounts of 4% (MDI4), and 6% (MDI6), respectively. Particles were resinated in a drum blender for 5 min, using a pneumatic spraying nozzle. Consequently, the resonated particles were manually distributed in a wooden forming box (550 × 550 mm^2), and pre-pressed. The pre-pressed mat was then hot-pressed at 200 °C, at 3.2 MPa for 100 s, using a hydraulic Siempelkamp press (Siempelkamp Ma-schinen und Anlagenbau GmbH, Krefeld, Germany). The target thickness of the panels was checked at random positions. In total, one PB per type was manufactured (Figure 1).

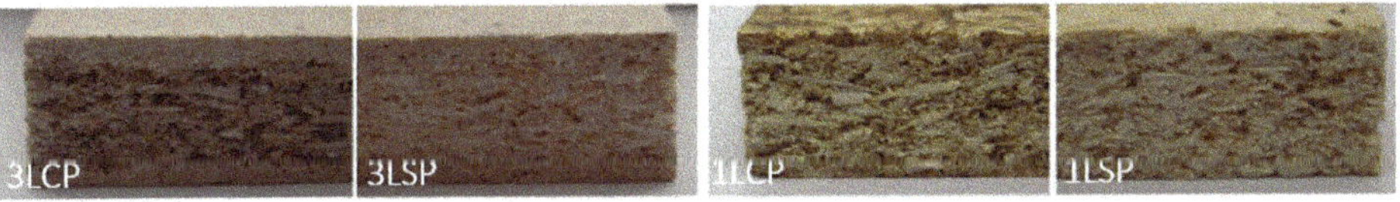

Figure 1. Cross-sectional views of the produced particleboards. 3LCP—three-layer particleboard (PB) with a cup-plant core layer, and spruce surface layers; 3LSP—three-layer spruce PB, 1LCP—single layer cup-plant PB, and 1LSP—single layer spruce PB.

For scanning electron microscopy (SEM), a Tescan S8000 (Tescan Brno, s.r.o., Brno, Czech Republic) was used to study the surface morphology of the various PB types. Likewise, morphology and interactions between cup-plant particles and wood particles were observed as well. The ultra-high-resolution mode was used, by means of an Everhart Thornley secondary electron detector. Low accelerating voltages (between 500 V—1 kV) were used to avoid surface charging. Sample surface was cut with a sliding microtome [34].

The multifunctional focused-ion beam SEM, TESCAN LYRA3 (Tescan Brno, s.r.o., Brno, Czech Republic), was used, which also had an integrated compact time of flight secondary ion mass analyser (C-TOF-SIMS) (TofWerk AG, Thun, Switzerland). An area of interest (AOI) of 50 × 50 μm^2 was scanned with a focused Ga$^+$ ion beam (4092 pA, 30,000 V), while time-of-flight of secondary ions, and their clusters (molecules) were continuously analysed by C-TOF-SIMS. C-TOF was operated at 10 μs dwell time, which provided the mass range of 0–170 m/Q. In parallel with the AOI scanning using focused ion beam

(FIB), C-TOF was used to record spectra at negative ion polarity and to capture elemental distribution maps at resolution of 1024 × 1024 pixels, by always binning 4 × 4 pixels. Data were derived from 100 scanned frames, which resulted in a crater with 1.7 μm depth in the sample (Figure 2, particularly B and D). With TOF-SIMS, the distribution of typical elements in wood, along with molecules indicating pMDI adhesives, were possible to visualize. Here, the distribution of carbon (C), oxygen (O), hydrogen (H), and hydroxyl (OH) groups were displayed to indicate wood, while CNO, CNH, and CN ion clusters were taken to display the pMDI adhesive distributions. Sample surfaces were coated with platinum in a sputter coater, prior to measurements, which avoided charging of the sample when exposed to the primary ion beam.

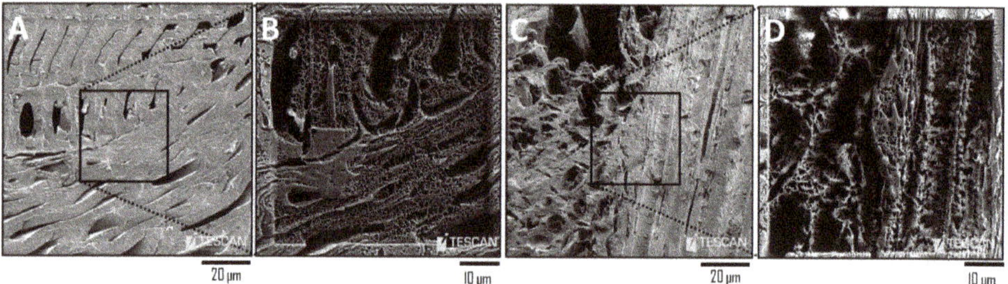

Figure 2. Selected area of interest (AOI) SEM of a spruce particle interphase prior (**A**), and crater-formations due to the TOF-SIMS analysis (**B**). AOI of a cup-plant particle interphase prior (**C**) and crater formations due to the TOF-SIMS analysis (**D**).

Mechanical testing was carried out on a Zwick® 1474 universal testing machine using the testXpert II software (Zwick GmbH & Co. kg, Ulm, Germany). Three-point bending tests according to EN 310 [35] were employed for the bending properties. Samples sized $12 \times 50 \times 290$ mm^3 were subjected to a loading rate of 7 mm·min^{-1}, until failure. Internal bonding (IB) strength following EN 319 [36] was measured with squared samples (50×50 mm^2). Prior to testing the samples were sanded and then glued to the stainless-steel blocks. The blocks were then positioned in gimbal-mounted holders and pre-loaded with 5 N in tension. Subsequently, a loading rate of 1 mm/min was applied until failure was reached.

Thickness swelling was determined according to EN 317, with conditioned samples sized $12 \times 50 \times 50$ mm^2 fully immersed in 20 °C distilled water. Thickness swelling was determined at two-time intervals, i.e., after 2 and 24 h. After the immersion time had elapsed, the test samples were removed from the water and excess water was removed with a paper cloth. Then, the thickness swelling was measured manually, using a thickness gauge, at the center of the samples. Vertical density profiles (VDP) were measured with the x-ray density analyzer GreCon RG44 (GreCon, Germany). Five samples were measured from each type, with samples sized $12 \times 50 \times 50$ mm^2. The obtained data were processed with Statistica v.12 (StatSoft, inc., Tulsa, OK, USA) software. Normality of the data were checked by the Shapiro-Wilk test. The Statistical significance was set at $p < 0.05$ for the analysis of variance (ANOVA), with Scheffé post-hoc tests.

3. Results and Discussion

3.1. Scanning Electron Microscopy

SEM images indicate that spruce fines located in the surface layer of a particleboard (PB) have a better adherence with each other than the core-layer particles (Figure 3A). Further, an apparent porosity was seen at the transition of the spruce surface layer migrating into the cup-plant core layer (Figure 3D), a fact that potentially affected the mechanical properties. The anatomical structure of the core layer vs. surface layer was clearly different.

While the wood showed a rather regular and compact cellular structure dominated by the tracheids (Figure 3B,C), the cup-plant structure was more diverse, showing annular thickenings (Figure 3E) and pitted perforations (Figure 3F), constituting a wider range of pore sizes, all potentially influencing resulting properties.

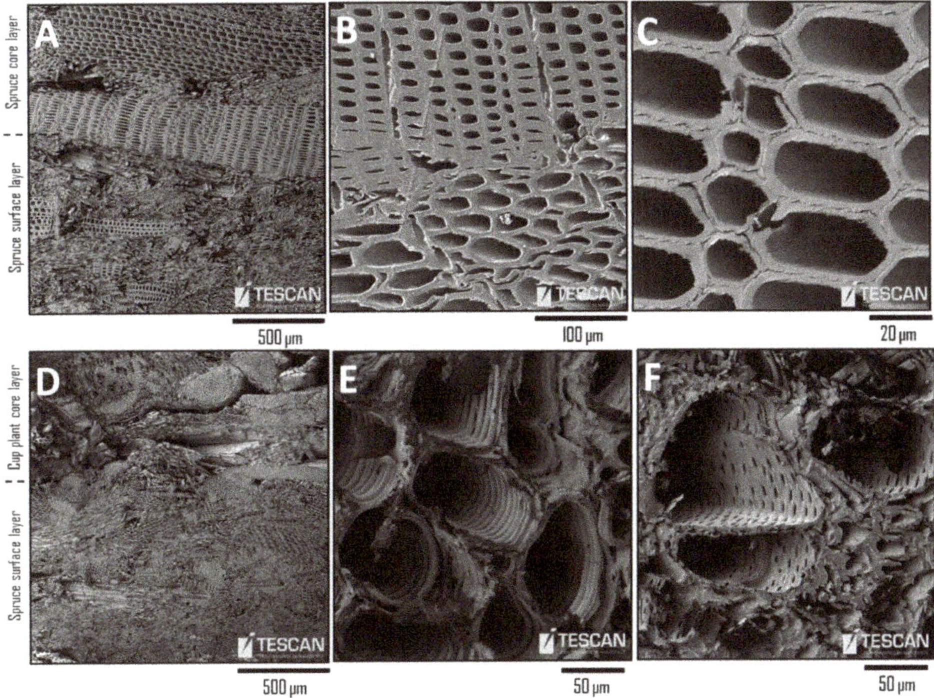

Figure 3. Cross-sectional SEM images of the spruce (**A–C**) and cup-plant core layer—spruce surface layer (**D–F**) PBs.

3.2. TOF-SIMS Analysis

For the spruce wood (Figure 4) and the cup-plant PB (Figure 5), cross sections were prepared for the detection of carbon at mass-to-charge m/Q of 12, hydrogen at a m/Q of 1, oxygen at a m/Q of 16, and hydroxyl ion cluster OH at a m/Q of 17. Secondary ion molecular clusters typical for pMDI adhesives were also detected. The cyanide ion anion (CN) was identified at the m/Q peak of 26, hydrogen isocyanide (CNH) at the m/Q peak 27, while cyanate (CNO) was detected at a m/Q of 42. Due to the detection of molecules associated with pMDI (CN, CNH, and CNO), the resin distribution within the composite could be displayed. Results showed that in the pMDI-bonded PB, the pMDI-wood bondlines were not spot-like, but appeared rather even and non-regular, with a penetration deep into the wood structure. This finding corresponded to data presented by Mahrdt et al. [37]. We are showing that C-TOF attached to FIB-SEM with Ga$^+$ ion source could detect ion molecular clusters specific to pMDI adhesive and wood. Additionally, with the Ga$^+$ ion source, it should be possible to detect G-lignin at peak Q/m 137 [30], however, the applied C-TOF setup delivered only a low secondary ion signal (although visible), which did not allow an elemental mapping. This could be further elaborated in a future study. The relevance of CN, CNH, CNO being related to the pMDI distribution in the wooden structure was also confirmed by the FIB-SEM image (Figures 4 and 5).

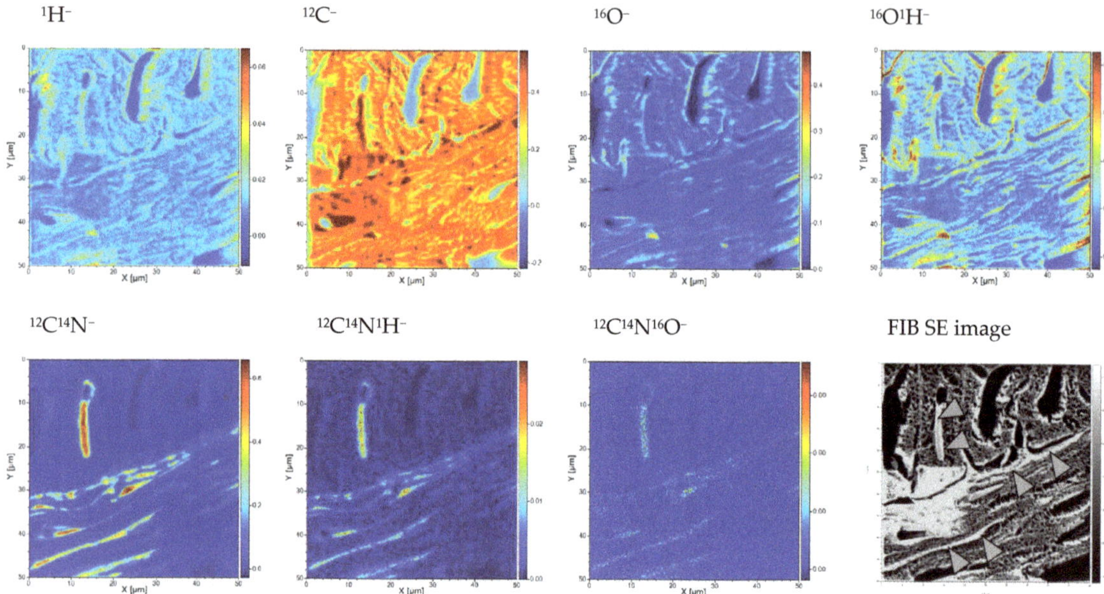

Figure 4. Elemental distribution maps of the selected isotopes and their molecular clusters for a spruce particle. The element exponent indicates a detected mass of ions or their clusters used for displaying the maps. Brighter regions in the FIB SE image refers to pMDI (marked); see Figure 2B for the selected area of interest SEM.

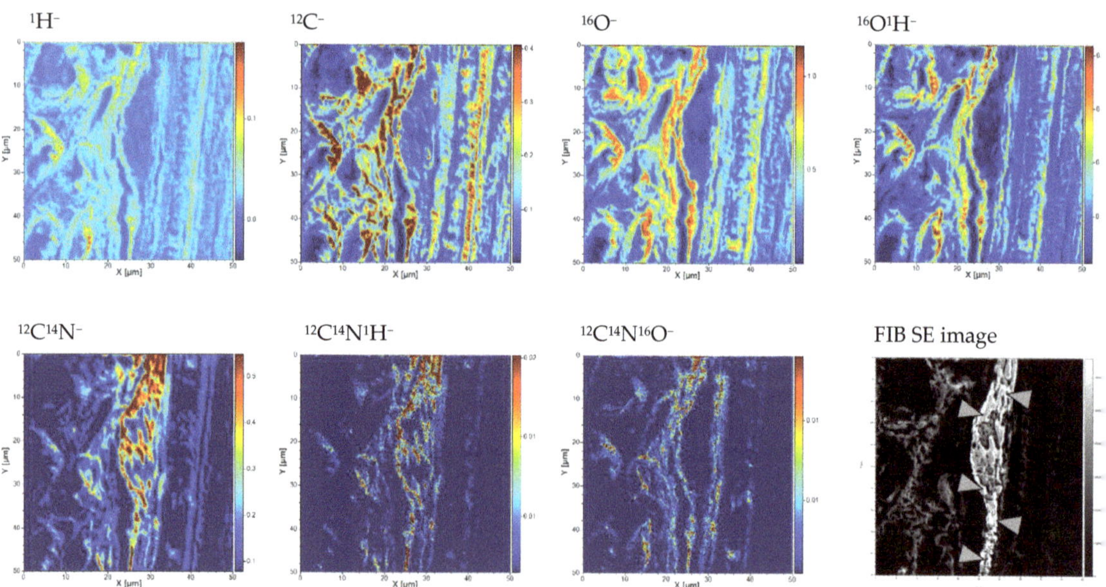

Figure 5. Elemental distribution maps of the selected isotopes and their molecular clusters for a cup-plant particle. The element exponent indicates a detected mass of ions or their clusters used for displaying of maps. Brighter regions in the FIB SE image refers to pMDI (marked); see Figure 2D for the selected area of interest SEM.

Particle–particle bondlines for both PB types were visualized by TOF-SIMS. The Q/m 12 peak (carbon) was adopted to represent the genuine wood structure, while for the particle–particle bondline, the molecular cluster CN (Q/m 26) was taken. The results showed that the pMDI adhesive was more dispersed in the cup-plant PB than in the spruce PB (Figures 6 and 7). While the spruce PB showed very narrow particle–particle bondlines, with penetration in non-compressed cell lumina regions (Figure 6), the adhesive in cup-plant PBs appeared to be more spread-out (Figure 7). This could be related to the greater and more dispersed porosity present in the cup-plant. An essential and novel outcome of this research was also that individual particle–particle bondlines could be visualized through molecular C-TOF SIMS identification. Here, no additional sample preparation such as staining was required, with the samples not getting modified in any way, as it is the case with other bondline identification methods [27]. With C-TOF, it is possible to visualize bondlines in 3D, different to regular electron or light microscopy imaging. The capability of 3D imaging also approved Hypothesis 2. The obtained approach delivered data for in-depth analysis of bondline mechanics and substrate interaction, since the dataset could be transformed into the finite element model, with the stress and strain distributions of the structural components to be further assessed [38].

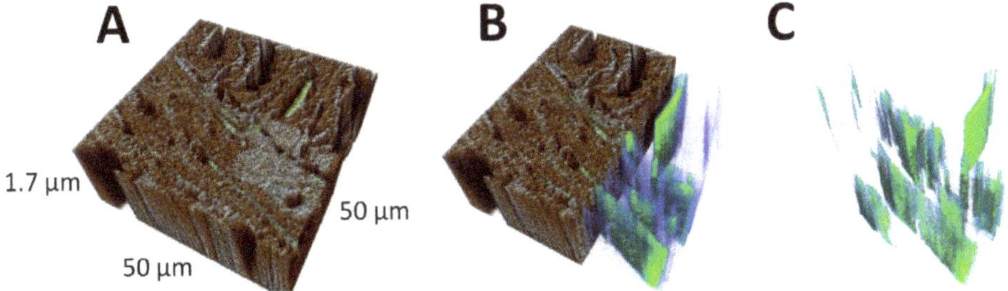

Figure 6. 3D-visualisation of the spruce particle–particle bondlines, using TOF SIMS data, reconstructed by the ORS software (Object Research Systems, Montreal, QC, Canada); (**A**,**B**) spruce PB with visible particle–particle bondline, (**C**) pMDI bondline imaging (Q/m 26).

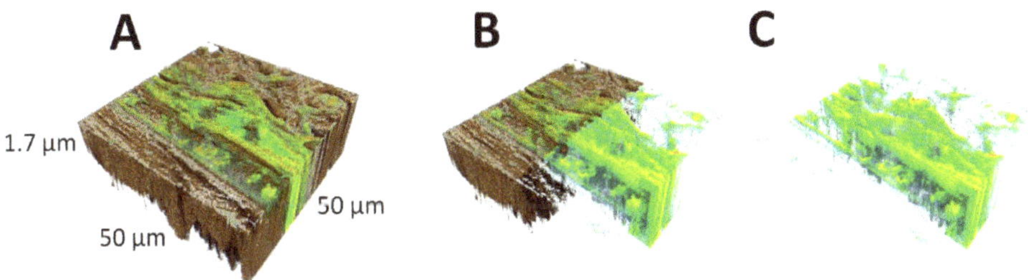

Figure 7. 3D-visualisation of cup-plant particle–particle bondlines, TOF SIMS data reconstructed by the ORS software (Object Research Systems, QC, Canada); (**A**,**B**) cup-plant PB with visible particle–particle bondline, and (**C**) the pMDI bondline imaging (Q/m 26).

3.3. Mechanical Properties

The results showed that the mean MOR values of the cup-plant PBs (3LCP, 1LCP) were below the mean measured for spruce PBs. Further, mean MOR of the three-layer cup-plant PB (3LCP) was above the one-layer cup-plant PB (1LCP), although not statistically significant. Additionally, MOR of three-layer cup-plant PB (3LCP) was not different to the one-layer spruce PB (1LSP, $p > 0.05$; Figure 8). The spruce PB types and the three-layer cup-plant PB did meet the EN312 class P1 standard [39], which is for general use in dry conditions. In contrast, average MOR of the one-layer cup-plant PB did not reach the P1 standard.

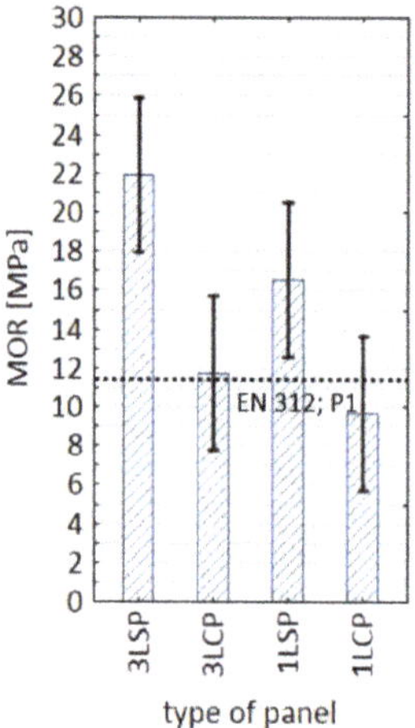

Figure 8. Modulus of rupture (MOR) of measured particleboards (PBs). 3LCP—three-layer cup-plant PB, 3LSP—three-layer spruce PB, 1LCP—single-layer cup-plant PB, and 1LSP—single-layer spruce PB ($n = 10$).

Average MOE measured for both cup-plant PBs types (3LCP, 1LCP) did not differ, ($p > 0.05$), and were also not statistically different from the one-layer spruce PB (1LSP, Figure 9). In addition, MOE of both cup-plant PBs were significantly ($p < 0.05$) lower than the three-layer spruce PB. Even with all seen MOE variability (Figure 9), the cup-plant PBs could be classified as P1 of EN 312, which refers to PBs used in dry conditions for interior fitments, including furniture.

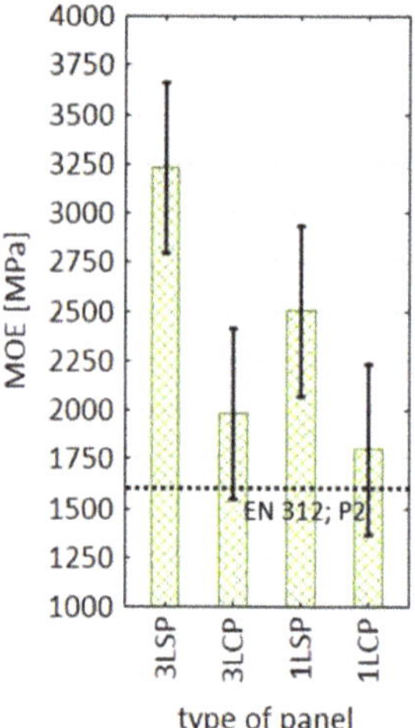

Figure 9. Modulus of elasticity (MOE) of measured particleboards (PBs). 3LCP—three-layer cup-plant PB, 3LSP—three-layer spruce PB, 1LCP—single-layer cup-plant PB, and 1LSP—single-layer spruce PB ($n = 10$).

The measured MOR and MOE of the one-layer spruce PB, and the three-layer spruce PB were consistent with data reported by Rofii et al. [40]. It was also documented that MOE and MOR of three-layer PBs were above the one-layer PBs [41]. MOE and MOR were both affected by particle alignments, surface layer density, as well as by the nature of the used raw material [42]. As cup-plant particles differ in their anatomical structure [8] and sizes when compared to spruce particles, this has evidently an effect on the properties of the produced PBs. Juliana et al. [43] reported a rise in MOE and MOR of three-layer PB made from Kenaf stalk cores, when the surface layers were made of rubberwood. MOR and MOE of the produced one-layer cup-plant PB showed similar values than PBs made from waste tea leaves [44], bleached straw [13], eggplant stalks [45], or rice straw [9].

Internal bonding values (IB) of the cup-plant PBs (3LCP, 1LCP) were also below the spruce PB types (Figure 10). One-layer cup-plant PB showed an average IB of 0.30 MPa, which was not significantly different from the IB of the three-layer cup-plant PB (0.34 MPa). Overall, the PBs made from cup-plant were again found suitable as general usage panels in dry conditions, as defined in EN 312 P1 [39]. In PBs, the core layers strongly determine the internal bonding, which explains why there are no significant differences for IB between single-layer and three-layer PBs. Rofii et al. [40] reported that surface layer characteristics of PBs have no significant effect on the IB. Additionally, Balducci et al. [22] showed that the surface layer of three-layer Miscanthus PBs had no significant influence on the measured IB. A reduced IB was found for rice husks PBs [46], hazelnut husks [47], or waste tea leaves [36]. An IB lower than 0.2 MPa was measured for rice straw [9], and for waste grass clipping PBs [48].

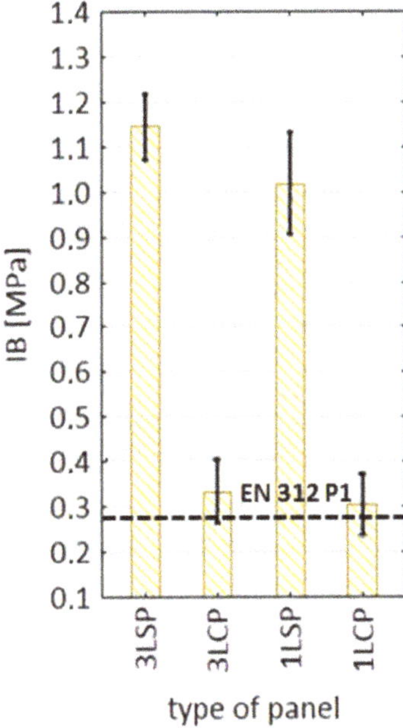

Figure 10. Internal bonding strength (IB) of the measured particleboards (PBs). 3LCP—three-layer cup-plant PBs, 3LSP—three-layer spruce PB, 1LCP—single-layer cup-plant PB, and 1LSP—single-layer spruce PB ($n = 20$).

3.4. Thickness Swelling and Water Uptake

The three-layers PBs made from cup-plant, and the three-layer PB made from spruce wood, both showed similar thickness swelling after 2 h (TS2h, Figure 11). Data also show that the spruce particles in the surface layer did significantly lower TS2h, as it is the case with the cup-plant PBs (3LCP). The TS2h of single-layer cup-plant PB was almost twice the TS2h of the three-layer cup-plant PB. No TS2h difference was found between 3LSP and 3LCP. A lower thickness swelling of three-layer PB than single-layer PB was also measured by [22], where Miscanthus or topinambour stalks were utilized for the core layer in three-layer PB.

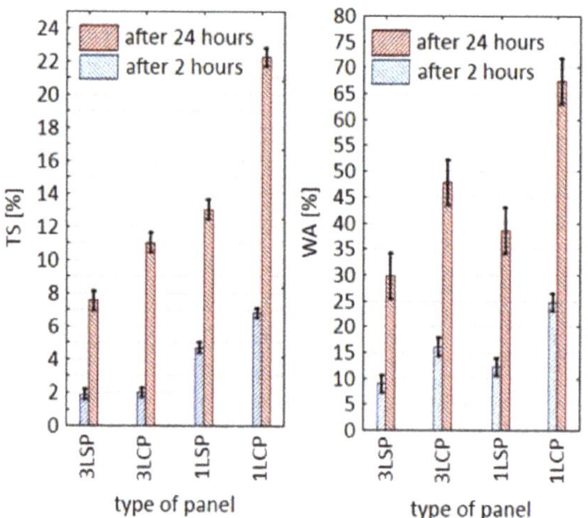

Figure 11. Thickness swelling (TS) and water absorption (WA) of produced particleboards (PBs). 3LCP—three-layer cup-plant PB, 3LSP—three-layer spruce PB, 1LCP—single-layer cup-plant PB, and 1LSP—single-layer spruce PB.

It is further shown that TS24h of both cup-plant PBs were significantly higher than the three-layer spruce comparison. It can be noted that in the 3LCP the spruce surface layer has a positive effect on TS24h. In addition, TS24h of the three-layer cup-plant PB was significantly lower ($p < 0.05$) than the TS24h measured for the single-layer cup-plant PB.

Water uptake results after 24 h were different in a way with the three-layer spruce particleboards absorbing the least water, while the single-layer cup-plant had the highest water uptake. Spruce particles as the surface layer is reducing the water-uptake, meaning that these particles seemingly reduce water access. This could be linked to a less accessible pore structure in spruce, compared to cup-plant particles (see Figure 3).

3.5. Vertical Density Profile

As seen in Figure 12, vertical density profiles of the cup-plant PB and the spruce PB were quite different. Single-layer cup-plant PB showed a flat density profile without distinct surface peaks. With spruce particles present in the surface layers, in an otherwise cup-plant PB, the density profile was also altered. It was found that the core layer density of the three-layer cup-plant PB was higher than that in the surface layers. It was evident that the density profile altered the mechanical performance of the cup-plant PBs. In Wong et al. [49] it was found that the density profile was like the one measured for spruce PBs, which is beneficial to MOE and MOR. Likewise, a flat density profile measured for the single-layer cup-plant PB was commonly connected to reduced bending properties, as shown with own data. The three-layer cup-plant PB had a higher density in the core layer, which at first glance should provide better IB for the three-layer cup-plant PB. Interestingly, this is not shown with our data. It must be noted that commonly not only the core layer, but the transition zone (TZ) between the surface and core layer is more prone to fail during an internal bonding test [50] (Figure 6). As seen in Figure 3, the three-layer cup-plant PB's transition zone had a higher proportion of pores and could be seen as a "weak layer", with a density similar to the core layer of the single-layer cup-plant PB. Thus, internal bonding of the 3LCP did not improve over the other panel types. The three-layer spruce particleboard had the highest surface layer density, which explains the high MOR and MOE values of this PB type.

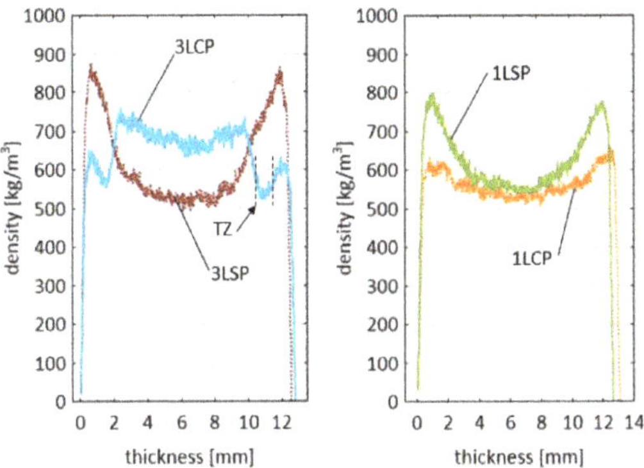

Figure 12. Vertical density profile of the produced particleboards (PBs). 3LCP—three layer-cup-plant PB, 3LSP—three-layer spruce PB, 1LCP—single-layer cup-plant PB, and1LSP—single-layer spruce PB; TZ—transition zone between surface and core layer.

4. Conclusions

In this research, we successfully produced single-layer as well as three-layer cup-plant particleboards (PBs). We approved Hypothesis 1, as such that C-TOF and a combination of SEM-FIB with Ga^+ ion source was able to detect ion molecular clusters specific to pMDI adhesive and wood. The actual setup of the primary beam did not provide enough secondary ion signals to capture G-lignin cluster, although a specific peak at m/Q 137 was visible. This needs to be further elaborated. Mechanical and physical properties of PBs were compared with spruce-made particleboards. Hypothesis 2 was also confirmed, as the bondline was visualized in 3D with datasets acquired by C-TOF. Hypothesis 3 was only partly approved, as the three-layer cup-plant PBs fulfilled the requirement for a general usage panel (EN312, P1). However, MOR of the single-layer cup-plant PB needs to be increased to meet EN 312, P1 requirements. Hypothesis 4 was proven with the restriction that the IB of the cup-plant three-layer PB was not statistically different from the cup-plant single-layer PB. Nevertheless, three-layer cup-plant PBs delivered better MOE and MOR values than the single-layer cup-plant PBs. Raw material mixtures with spruce might be useful to raise MOR and MOE values. The density profile of the three-layer cup-plant PB has been altered in a way the core layer had higher densities than the surface layer.

Author Contributions: Conceptualization, P.K. and R.W.; methodology, P.K. and P.M..; software, P.K.; validation, P.K., R.W. and P.M.; writing-original draft preparation, P.K. and R.W.; writing-review and editing, R.W. and P.M.; visualization, P.K. All authors have read and agreed to the published version of the manuscript.

Funding: This work was supported by Tescan Orsay Holding a.s., 62300 Brno, by providing the relevant analytical resources. Further, the Fraunhofer-Institut für Holzforschung—Wilhelm-Klauditz-Institut provided all material characterisations.

Institutional Review Board Statement: Not applicable.

Informed Consent Statement: Not applicable.

Data Availability Statement: Not applicable.

Conflicts of Interest: The authors declare no conflict of interest.

References

1. European Panel Federation. *European Padixnel Federation—Annual Report 2013/2014*; EPF: Brussels, Belgium, 2014.
2. Giljum, S.; Martin, H.F.B.; Burger, E.; Frühmann, J.; Lutter, S.; Elke, P.; Christine, P.; Hannes, W.; Lisa, K.; Michael, W. Overcosumption? Our Use of the World's Natural Resources. 2009. Available online: https://www.foe.co.uk/sites/default/files/downloads/overconsumption.pdf (accessed on 5 January 2021).
3. Mast, B.; Lemmer, A.; Oechsner, H.; Reinhardt-Hanisch, A.; Claupein, W.; Graeff-Hönninger, S. Methane yield potential of novel perennial biogas crops influenced by harvest date. *Ind. Crops Prod.* **2014**, *58*, 194–203. [CrossRef]
4. Khachatryan, H.; Casavant, K.; Jessup, E. *Waste to Fuels Technology: Evaluating Three Technology Options and the Economics for Converting Biomass to Fuels*; Final Report; Ecology Publication Number 09-07-058; Washington State University: Pullman, WA, USA, 2009; 203p.
5. Pretzsch, H. *Forest Dynamics, Growth and Yield: From Measurement to Model*; Springer: Berlin, Germany, 2009; ISBN 978-3-540-88306-7.
6. Stanford, G. Silphium perfoliatum (cup-plant) as a new forage. In Proceedings of the Twelfth North American Prairie Conference, Cedar Falls, IA, USA, 5–9 August 1990; Volume 1, pp. 33–37.
7. Haag, N.L.; Nägele, H.-J.; Reiss, K.; Biertümpfel, A.; Oechsner, H. Methane formation potential of cup plant (Silphium perfoliatum). *Biomass Bioenergy* **2015**, *75*, 126–133. [CrossRef]
8. Gansberger, M.; Montgomery, L.F.R.; Liebhard, P. Botanical characteristics, crop management and potential of Silphium perfoliatum L. as a renewable resource for biogas production: A review. *Ind. Crops Prod.* **2015**, *63*, 362–372. [CrossRef]
9. Li, X.; Cai, Z.; Winandy, J.E.; Basta, A.H. Selected properties of particleboard panels manufactured from rice straws of different geometries. *Bioresour. Technol.* **2010**, *101*, 4662–4666. [CrossRef]
10. Gerardi, V.; Minelli, F.; Viggiano, D. Steam treated rice industry residues as an alternative feedstock for the wood based particleboard industry in Italy. *Biomass Bioenergy* **1998**, *14*, 295–299. [CrossRef]
11. Yasin, M.; Waheed, A.; Ahmed, A.; Karim, S. Efficient Utilization of Rice-wheat Straw to Produce Value—Added Composite Products. *Int. J. Chem. Enviromental Eng.* **2010**, *1*, 136–143.
12. Kariuki, S.W.; Wachira, J.; Kawira, M.; Murithi, G. Crop residues used as lignocellulose materials for particleboards formulation. *Heliyon* **2020**, *6*, e05025. [CrossRef] [PubMed]
13. Mo, X.; Cheng, E.; Wang, D.; Sun, X.S. Physical properties of medium-density wheat straw particleboard using different adhesives. *Ind. Crops Prod.* **2003**, *18*, 47–53. [CrossRef]
14. Khristova, P.; Yossifov, N.; Gabir, S. Particle Board from Sunflower Stalks: Preliminary Trials. *Bioresour. Technol.* **1996**, *58*, 319–321. [CrossRef]
15. Mati-Baouche, N.; De Baynast, H.; Lebert, A.; Sun, S.; Lopez-Mingo, C.J.S.; Leclaire, P.; Michaud, P. Mechanical, thermal and acoustical characterizations of an insulating bio-based composite made from sunflower stalks particles and chitosan. *Ind. Crops Prod.* **2014**, *58*, 244–250. [CrossRef]
16. Bektas, I. The Manufacture of Particleboards using Sunflower Stalks (helianthus annuus l.) and Poplar Wood (Populus alba L.). *J. Compos. Mater.* **2005**, *39*, 467–473. [CrossRef]
17. Guler, C.; Bektas, I.; Kalaycioglu, H. The experimental particleboard manufacture from sunflower stalks (Helianthus annuus L.) and Calabrian pine (Pinus brutia Ten). *For. Prod. J.* **2006**, *56*, 56–60.
18. Ferrandez-Villena, M.; Ferrandez-Garcia, C.E.; Garcia-Ortuño, T.; Ferrandez-Garcia, A.; Ferrandez-Garcia, M.T. Analysis of the thermal insulation and fire-resistance capacity of particleboards made from vine (Vitis vinifera l.) prunings. *Polymers* **2020**, *12*, 1147. [CrossRef] [PubMed]
19. Guler, C.; Ozen, R. Some properties of particleboards made from cotton stalks (Gossypium hirsitum L.). *Holz als Roh- und Werkst.* **2004**, *62*, 40–43. [CrossRef]
20. Kowaluk, G.; Szymanowski, K.; Kozlowski, P.; Kukula, W.; Sala, C.; Robles, E.; Czarniak, P. Functional assessment of particleboards made of apple and plum orchard pruning. *Waste Biomass Valoriz.* **2020**, *11*, 2877–2886. [CrossRef]
21. Chaydarreh, K.C.; Lin, X.; Guan, L.; Yun, H.; Gu, J.; Hu, C. Utilization of tea oil camellia (Camellia oleifera Abel.) shells as alternative raw materials for manufacturing particleboard. *Ind. Crop Prod.* **2021**, *161*, 113221. [CrossRef]
22. Balducci, F.; Harper, C.; Meinlschmidt, P.; Dix, B.; Sanasi, A. Development of Innovative Particleboard Panels. *Drv. Ind.* **2008**, *59*, 131–136.
23. Dix, B.; Meinlschmidt, P.; Van De Flierdt, A.; Thole, V. Leichte Spanplatten für den Möbelbau aus Rückständen der landwirtschaftlichen Produktion—T.1: Verfügbarkeit der Rohstoffe. *Holztechnologie* **2009**, *50*, 5–10.
24. Selinger, J.; Wimmer, R. A novel low-density sandwich panel made from hemp. In Proceedings of the InWood2015: Innovations in Wood Materials and Processes, Brno, Czech Republic, 19–22 May 2015; Volume 19, pp. 29–31.
25. Klímek, P.; Meinlschmidt, P.; Wimmer, R.; Plinke, B.; Schirp, A. Using sunflower (Helianthus annuus L.), topinambour (Helianthus tuberosus L.) and cup-plant (Silphium perfoliatum L.) stalks as alternative raw materials for particleboards. *Ind. Crops Prod.* **2016**, *92*, 157–164.
26. Klímek, P.; Morávek, T.; Ráhel, J.; Stupavská, M.; Děcký, D.; Král, P.; Kúdela, J.; Wimmer, R. Utilization of air-plasma treated waste polyethylene terephthalate particles as a raw material for particleboard production. *Compos. Part B Eng.* **2016**, *90*, 188–194. [CrossRef]
27. Kamke, F.A.; Lee, J.N. Adhesive penetration in wood: A review. *Wood Fiber Sci.* **2007**, *39*, 205–220.

28. Whitby, J.A.; Östlund, F.; Horvath, P.; Gabureac, M.; Riesterer, J.L.; Utke, I.; Hohl, M.; Sedláček, L.; Jiruše, J.; Friedli, V.; et al. High Spatial Resolution Time-of-Flight Secondary Ion Mass Spectrometry for the Masses: A Novel Orthogonal ToF FIB-SIMS Instrument. *Adv. Mater. Sci. Eng.* **2012**, *2012*, 180437. [CrossRef]

29. Singh, A.V.; Jungnickel, H.; Leibrock, L.; Tentschert, J.; Reichardt, P.; Katz, A.; Luch, A. ToF-SIMS 3D imaging unveils important insights on the cellular microenvironment during biomineralization of gold nanostructures. *Sci. Rep.* **2020**, *10*, 1–11. [CrossRef] [PubMed]

30. Saito, K.; Kato, T.; Takamori, H.; Kishimoto, T.; Yamamoto, A.; Fukushima, K. A new analysis of the depolymerized fragments of lignin polymer in the plant cell walls using ToF-SIMS. *Appl. Surf. Sci.* **2006**, *252*, 6734–6737. [CrossRef]

31. Tokareva, E.N.; Fardim, P.; Pranovich, A.V.; Fagerholm, H.; Daniel, G.; Holmbom, B. Imaging of wood tissue by ToF-SIMS: Critical evaluation and development of sample preparation techniques. *Appl. Surf. Sci.* **2007**, *253*, 7569–7577. [CrossRef]

32. Fardim, P.; Dura, N. Modification of fibre surfaces during pulping and refining as analysed by SEM, XPS and ToF-SIMS. *Colloids Surf. A Physicochem. Eng. Asp.* **2003**, *223*, 263–276. [CrossRef]

33. Kuroda, K.; Imai, T.; Saito, K.; Kato, T.; Fukushima, K. Applied Surface Science Application of ToF-SIMS to the study on heartwood formation in Cryptomeria japonica trees. *Appl. Surf. Sci.* **2008**, *255*, 1143–1147. [CrossRef]

34. Gärtner, H.; Nievergelt, D. The core-microtome: A new tool for surface preparation on cores and time series analysis of varying cell parameters. *Dendrochronologia* **2010**, *28*, 85–92. [CrossRef]

35. European Committee for Standardization. *EN 310, Wood-Based Panels—Determination of Modulus of Elasticity in Bending and of Bending Strength*; European Committee for Standardization: London, UK, 1993.

36. *European Committee for Standardization, EN 319, Particleboards and Fibreboards—Determination of Tensile Strength Perpendicular to the Plane of the Board*; European Committee for Standardization: London, UK, 1993.

37. Mahrdt, E.; Stöckel, F.; Van Herwijnen, H.W.G.; Müller, U.; Kantner, W.; Moser, J.; Gindl-Altmutter, W. Light microscopic detection of UF adhesive in industrial particle board. *Wood Sci. Technol.* **2015**, *49*, 517–526. [CrossRef]

38. Klímek, P. Focused ion beam tomography as a tool for bio-inspired structural design. In Proceedings of the XVI International Conference on Electron Microscopy, Jachranka, Poland, 10–13 September 2017; pp. 37–39.

39. European Committee for Standardization. *EN 312, Particleboards. Specifications—Requirements for Flat-Pressed or Calendar-Pressed Unfaced Particleboards*; European Committee for Standardization: London, UK, 2010.

40. Rofii, M.N.; Yumigeta, S.; Suzuki, S.; Prayitno, T.A. Mechanical properties of three-layered particleboards made from different wood species. In Proceedings of the 3rd International Symposium of Indonesian Wood Research Society (IWoRS), Yogykarta, Indonesia, 3–4 November 2011; pp. 152–161.

41. Geimer, R.L.; Lehmann, W.F. Effects of Layer Characteristics on the Properties Of Three-Layer Particleboards. *For. Prod. J.* **1975**, *25*, 19–29.

42. Nasser, R.A. Physical and Mechanical Properties of Three-Layer Particleboard Manufactured from the Tree Pruning of Seven Wood Species. *World Appl. Sci. J.* **2012**, *19*, 741–753. [CrossRef]

43. Juliana, A.H.; Paridah, M.T.; Rahim, S.; Nor Azowa, I.; Anwar, U.M.K. Properties of particleboard made from kenaf (Hibiscus cannabinus L.) as function of particle geometry. *Mater. Des.* **2012**, *34*, 406–411. [CrossRef]

44. Yalinkilic, K.M.; Imamuraa, Y.; Takahashi, M.; Kalaycioglub, H.; Nemlib, G.; Demircib, Z.; Ozdemirb, T. Physical and mechanical properties of particleboard manufactured from waste tea leaves. *Int. Biodeterior. Biodegrad.* **1998**, *41*, 75–84. [CrossRef]

45. Guntekin, E.; Karakus, B. Feasibility of using eggplant (Solanum melongena) stalks in the production of experimental particleboard. *Ind. Crops Prod.* **2008**, *27*, 354–358. [CrossRef]

46. Suleiman, I.Y.; Aigbodion, V.S.; Shuaibu, L.; Shangalo, M.; Workshop, M.E.; Umaru, W.; Polytechnic, F. Development of eco-friendly particleboard composites using rice husk particles and gum arabic. *J. Mater. Sci. Eng. Adv. Technol.* **2013**, *7*, 75–91.

47. Çöpür, Y.; Güler, C.; Akgül, M.; Taşçıoğlu, C. Some chemical properties of hazelnut husk and its suitability for particleboard production. *Build. Environ.* **2007**, *42*, 2568–2572. [CrossRef]

48. Nemli, G.; Demirel, S.; Gümüşkaya, E.; Aslan, M.; Acar, C. Feasibility of incorporating waste grass clippings (Lolium perenne L.) in particleboard composites. *Waste Manag.* **2009**, *29*, 1129–1131. [CrossRef]

49. Wong, E.-D.; Yang, P.; Zhang, M.; Wang, Q.; Nakao, T.; Li, K.-F.; Kawai, S. Analysis of the effects of density profile on the bending properties of particleboard using finite element method (FEM). *Holz als Roh- und Werkst.* **2003**, *61*, 66–72. [CrossRef]

50. Schulte, M.; Frühwald, A. Some investigations concerning density profile, internal bond and relating failure position of particleboard. *Holz als Roh- und Werkst.* **1996**, 289–294. [CrossRef]

applied sciences

MDPI

Article

Bonding of Selected Hardwoods with PVAc Adhesive

Ján Iždinský [1,*], Ladislav Reinprecht [1], Ján Sedliačik [2], Jozef Kúdela [3] and Viera Kučerová [4]

[1] Department of Wood Technology, Faculty of Wood Science and Technology, Technical University in Zvolen, T. G. Masaryka 24, 960 01 Zvolen, Slovakia; reinprecht@tuzvo.sk
[2] Department of Furniture and Wood Products, Faculty of Wood Science and Technology, Technical University in Zvolen, T. G. Masaryka 24, 960 01 Zvolen, Slovakia; sedliacik@tuzvo.sk
[3] Department of Wood Science, Faculty of Wood Science and Technology, Technical University in Zvolen, T. G. Masaryka 24, 960 01 Zvolen, Slovakia; kudela@tuzvo.sk
[4] Department of Chemistry and Chemical Technology, Faculty of Wood Science and Technology, Technical University in Zvolen, T. G. Masaryka 24, 960 01 Zvolen, Slovakia; viera.kucerova@tuzvo.sk
* Correspondence: jan.izdinsky@tuzvo.sk; Tel.: +421-455-206-385

Abstract: The bonding of wood with assembly adhesives is crucial for manufacturing wood composites, such as solid wood panels, glulam, furniture parts, and sport and musical instruments. This work investigates 13 hardwoods—bangkirai, beech, black locust, bubinga, ipé, iroko, maçaranduba, meranti, oak, palisander, sapelli, wengé and zebrano—and analyzes the impact of their selected structural and physical characteristics (e.g., the density, cold water extract, pH value, roughness, and wettability) on the adhesion strength with the polyvinyl acetate (PVAc) adhesive Multibond SK8. The adhesion strength of the bonded hardwoods, determined by the standard EN 205, ranged in the dry state from 9.5 MPa to 17.2 MPa, from 0.6 MPa to 2.6 MPa in the wet state, and from 8.5 MPa to 19.2 MPa in the reconditioned state. The adhesion strength in the dry state of the bonded hardwoods was not influenced by their cold water extracts, pH values, or roughness parallel with the grain. On the contrary, the adhesion strength was significantly with positive tendency influenced by their higher densities, lower roughness parameters perpendicular to the grain, and lower water contact angles.

Keywords: hardwoods; extractives; pH value; roughness; wettability; PVAc adhesive; adhesion strength

check for
updates

Citation: Iždinský, J.; Reinprecht, L.; Sedliačik, J.; Kúdela, J.; Kučerová, V. Bonding of Selected Hardwoods with PVAc Adhesive. *Appl. Sci.* **2021**, *11*, 67. https://dx.doi.org/10.3390/app11010067

Received: 25 November 2020
Accepted: 18 December 2020
Published: 23 December 2020

Publisher's Note: MDPI stays neutral with regard to jurisdictional claims in published maps and institutional affiliations.

1. Introduction

The strength and stability of glued joints are the priority properties of all construction and decorative composites based on metals, wood, glass, plastics, and also other traditional and modern materials. This also applies to glued solid wood products for industrial, building, and transport structures, furniture, musical instruments, sports equipment, and other uses. With regards to the glued wood products, not only is the initial strength of glued joints important, but also the stability of the joints during indoor and mainly outdoor exposures, causing one-off or cyclical changes in wood moisture and temperature.

The most essential parameters influencing the overall bonding quality of wood products include the following: (a) the wood's species, density, chemical and anatomical structure, physical and strength characteristics, surface machining determining the surface roughness, grain orientation, moisture content, and pre-treatment with biocides or other additives, (b) the adhesive's chemical structure, weight solid, viscosity, surface tension, and mechanism of hardening, and (c) the bonding technology's pressure, time, and temperature [1–13].

The low density, high permeability, and high surface roughness of the individual wood species are basic factors that play an important role in terms of better adhesive penetration depth, usually in connection with a positive impact on the bonding quality. However, Aicher et al. [8] found that the adhesion strength of bonded wood is not always most prominently connected with its density. When compared, experiments of several researchers who tested bonded hardwoods with density (ρ) in a range of 300–1000 kg/m^3 showed that the greater values of the adhesion strength were not in all cases found in specimens prepared from denser species (adhesion = 4.095 + 0.014 ρ/MPa/; R^2 = only 0.25), but almost always in those prepared from species characterized by a higher shear strength (adhesion = 0.628 + 0.912 τ/MPa/; R^2 = 0.88). Shida and Hiziroglu [14] examined these tendencies and found out that the adhesion strength of bonded woods was greater in the denser karamatsu species than in the less dense sugi species. Similar tendencies for nine European wood species observed Konnerth [15]. On the contrary, Alamsyah et al. [16] demonstrated a higher adhesion strength in bonded specimens made from the less dense *Paraserianthes falcataria* tropical wood than from the denser *Acacia mangium*.

Water-based adhesives, to which water dispersed adhesives also belong (e.g., polyvinyl acetate (PVAc) adhesives [17]), may reach an adhesion optimum when the water has totally penetrated into the wood substrate [18]. PVAc adhesives are commonly used in the wood industry for general assembly applications, film overlay and high-pressure lamination, edge gluing, wood veneer, and edge bonding [17]. They are safe, non-toxic, non-combustible, easy cleanable, without pollution, cure at room temperature, colorless, transparent and tough after curing, and give a high adhesion strength to bonded wood elements.

Özçifçi and Yapici [19] determined that there was a greater adhesion strength for beech and Scotch pine woods bonded with PVAc adhesive along the tangential direction than the radial one. Burdurlu et al. [20] obtained similar results for Calabrian pine wood bonded with PVAc and polyurethane (PUR) adhesives and recommended performing the bonding process on the tangential surfaces with higher pressures. However, in spite of these results, it is well known that the penetration depth of liquid adhesives within the wood structure is influenced not only by the wood's anatomical direction, density, moisture, and final permeability, but also by the physical and chemical characteristics of the adhesive and the technological bonding conditions, such as pressure, temperature, and time. For example, Sernek et al. [1] found better penetration of the water-based urea-formaldehyde (UF) adhesive into beech wood in the tangential direction at pressure application, while no significant difference between penetrations in the tangential and radial directions occurred when pressure was not applied.

The roughness of wood surfaces depends, first of all, on the wood anatomy and the mode of its machining [19,21–23]. Recognition and quantification of the surface roughness are important from the viewpoint of wood bonding and surface treatment, as the wood surface morphology significantly influences the wood wetting with film-forming materials and the adhesion of these materials to the wood substrate. The circular rotary saw usually causes a higher surface roughness of woods, in comparison with their planning or sanding [22,23]. Shida and Hiziroglu [14] inspected four Japanese wood species—sugi, hinoki, hiba, and karamatsu—and determined that their adhesion strengths with the PVAc adhesive achieved greater values if their surfaces were pre-finished with 80-grit sandpaper, compared to those surfaces pre-finished with finer 120- and 240-grit sizes. Burdurlu et al. [20] found out that a greater roughness of Calabrian pine wood surfaces was caused by their machining in this order, from most to least influential: sawing with a circular ripsaw, sanding, and planning. The shear strengths of specimens bonded with the PVAc adhesive were better for sanded or sawed wood surfaces compared with planned ones. Hiziroglu et al. [5,24] also documented the increased roughness of wood surfaces resulting from using sandpapers with lower grit sizes and the better wood surface adhesion with the PVAc adhesive. However, the experiments of Özçifçi and Yapici [19] showed that higher adhesion with various adhesive types had smoother planed wood surfaces than those prepared by band or circular sawing.

The high polarity and good wettability of wood surfaces is given mainly by the presence of hydroxyl, carbonyl, and carboxyl groups in the lignin-polysaccharide matrix of the cell walls. This results in the formation of strong physical bonds with various polar adhesives. Wood surfaces with higher polarities are more wettable with water-based adhesives [16,25]. The consequence is a higher penetration of the adhesives through the lumens of cell elements on the wood surface. However, the penetration rate can partly be limited by the formation of Van der Waals interactions, dipolar interactions, and hydrogen bonds of polar adhesives with the lignin-polysaccharide matrix of wood cell walls [26].

The wettability of wood, stability of the adhesive systems, and quality of the final adhesion can negatively or positively be influenced by wood extractives and also by preservatives or other excipients added to the wood [27–30]. Polar and nonpolar extractives play a major role in wood bonding processes, as they can contribute to or determine the relevant bonding properties of wood, such as acidity (pH value), wettability (contact angle, surface free energy), or even permeability (clogging of lumens by crystals). Extractives of a high acidity accelerate the curing of acid curing urea-formaldehyde (UF) and melamine-urea-formaldehyde (MUF) resins, decelerate bonding with alkaline hardening phenol-formaldehyde (PF) resins, or degrade PUR adhesives [9,11,31]. Starch and monomeric sugars, which belong to the primary polar water-soluble extractives present in all wood species, have a negative effect on the bonding of wood with cement, MUF, and PF adhesives [11,31]. Secondary extractives specifically occur in various hardwood and softwood species. These extractives, which are typically situated in the heart zones of some European and several tropical hardwood species, contain either various polar polyphenols with a hydrophilic character, such as flavonoids, tannins, sterols, flobafenes, rubrenolide, rubrynolide, and quinones or coumarins, as well as various nonpolar or semi-polar waxes, fats, and oils with a hydrophobic characteristic [32–35]. Studies of tropical woods bonded together with a PVAc adhesive showed that the extractive content of the wood species had an adverse effect on the bonding quality [36]. In addition, extractives have pronounced inhibitory or supportive effects on the wood sorption capacity, which is reflected in the swelling and shrinkage coefficients associated with moisture changes. The consequence is a stress state at the wood–adhesive–wood interface, impairing the adhesion of the glued joint [37]. Generally, the impact of wood extractives manifests itself more significantly when the bond line is exposed to multiple negative factors. For example, the loss of the adhesion strength in regard to polar extractives is more apparent in the wet state of bonded woods [11].

In the case of wood bonding with adhesives, it is necessary to consider wood's rheological properties that might induce additional retardation of the process of wood surface wetting with the gluing substance on its own. The impact of rheological performance can be, to a considerable extent, mitigated with the aid of mechanical and physical forces applied during adhesive application, during the pressing of the wood at bonding, as well as at exposure of the bonded wood to climatic changes.

The issue of the bonding of European and tropical hardwoods for construction and furniture purposes has been addressed by several researchers, evaluating the factors influencing the shear strength and delamination of glued joints in particular [7,8,38–40]. In summary, they proved that several European and tropical hardwood species meet the requirements for the adhesion strength of glued joints set by the relevant standards and are potentially suitable for glued furniture and construction products.

The aim of this work was to analyze the impact of the selected structural and physical characteristics of hardwoods (e.g., the density, cold water extract, pH value, roughness, and wettability) on their adhesion strength with a water-based PVAc adhesive in dry, wet, and reconditioned states.

2. Materials and Methods

2.1. Hardwoods

2.1.1. Wood Species

The heart zones or central zones of 10 tropical and 3 European wood species were used for the experiment (Figure 1). Their names are usually defined by the standard EN 13556 [41]: bangkirai (*Shorea obtusa* Wall.; Sh. Spp.), European beech (*Fagus sylvatica* L.), black locust (*Robinia pseudoacacia* L.), bubinga (*Guibourtia demeusii* (Harms) J. Léon.), ipé

(*Tabebuia serratifolia* (Vahl) Nicholson), iroko (*Milicia excelsa* (Welw.) C. C. Berg), maçaranduba (*Manilkara bidentata* A. Chev.), dark red meranti (*Shorea curtisii* Dyer ex. King), European oak (*Quercus robur* L.), Santos palisander (*Machaerium scleroxylon* Tul.), sapelli (*Entandrophragma cylindricum* Sprague), wengé (*Millettia laurentii* De Wild.), and zebrano (*Microberlinia brazzavillensis* A. Chev.). The experimental wood material was obtained from the trading company JAF Holz, Ltd. (Špačince, Slovak Republic) in the form of naturally dried boards, having a moisture content of 13% ± 2.5%. Test samples with dimensions of 80 mm × 20 mm × 5 mm (longitudinal × radial × tangential) were prepared with a circular ripsaw (Freud Pro LP30M 026P) having these parameters: a diameter of 255 mm, a cutting thickness of 2.8 mm, a sawblade body thickness of 1.8 mm, a number of teeth of 40, and a maximum rotation speed of 7800 rpm. The wood samples were of a high quality (i.e., without bio-damages, knots, or other inhomogeneities), and before other technological operations, they were conditioned at a temperature of 20 °C ± 2 °C and a relative air humidity of 50% ± 5%, achieving an equilibrium moisture content of 8% ± 2%.

Figure 1. Hardwoods used in the experiment, numbered (from 1 to 13) according to the increasing values of the adhesion strength in the dry state demonstrated in Section 3.2: (1) bangkirai, (10) beech, (4) black locust, (13) bubinga, (7) ipé, (2) iroko, (11) maçaranduba, (3) meranti, (8) oak, (12) palisander, (9) sapelli, (5) wengé, and (6) zebrano.

2.1.2. Characteristics of Hardwoods: Density, Cold Water Extract, pH Value, Roughness, and Wettability

The density ρ of hardwoods was determined in accordance with the standard EN 323 [42].

The cold water extract from hardwoods was obtained in accordance with the standard ASTM D1110 [43], followed by measurement of the pH values of these extracts with a pH meter 7110 (WTW, Wellheim, Germany).

The wettability of the hardwood surfaces was associated with determining the contact angle with a redistilled water drop with a volume of 0.0018 mL up to its complete soaking into the wood substrate, using a goniometer Krüss DSA30 Standard (Krüss, Hamburg, Germany). The course of the water drop profile evolving parallel to the wood grain, from first contact up to the complete soaking, was recorded with a camera, its scanning frequency set in accordance with the wetting interval. The initial contact angle θ_0 was evaluated at the beginning of the wetting process, meaning at the moment of first contact between the water drop and the wood substrate. The drop's contact angle, from the moment of reversion from the acceding contact angle into the receding one, was considered the equilibrium contact angle θ_e. From the values of the contact angles θ_0 and θ_e, the abstract contact angle θ_w, corresponding to an ideal smooth surface, was calculated by the method of Liptáková and Kúdela [44].

Three parameters of wood roughness, the Ra (arithmetic mean deviation), Rz (arithmetic mean of the heights and depressions of the profile at the basic length), and R_{Sm} (mean distance between the valleys), were inspected on the radial surfaces of samples parallel with and perpendicular to the grain, using the Surfcom 130A surface roughness measuring instrument (Carl Zeiss, Jena, Germany) in accordance with the standard EN ISO 4287 [45]. A total measured length for one replicate was 12.5 mm, and a basic length for one analysis was 2.5 mm.

2.2. Adhesive

A bonding of hardwoods was performed with the water resistant, one-component crosslinking polyvinyl acetate (PVAc) adhesive Multibond SK8 (Franklin International, Columbus, Ohio USA). This adhesive is characterized by the following basic technical parameters: weight solids 48.7–52.3%, pH 2.4–3.5, viscosity approximately 4000 mPa.s, and specific gravidity 1.1 g cm^{-3}.

2.3. Wood Bonding

For the adhesion strength, each individual specimen was prepared by bonding two samples (80 mm × 20 mm × 5 mm) of the same wood species. The PVAc adhesive was applied on the contacting surfaces of both wood pieces in an amount of 120 g ± 10 g per square meter. The bonding process was performed in the press for 60 min at a pressure of 1.2 MPa and a temperature of 20 °C ± 2 °C. The conditioning of the bonded hardwood specimens lasted 7 days at a temperature of 20 °C ± 2 °C and a relative air humidity of 50% ± 5%.

2.4. Adhesion Strength: Tensile Shear Strength

The adhesion strength test—the tensile shear strength of lap joints—of bonded hardwoods was performed by the standard EN 205 [46] in the dry state, wet state, and reconditioned state and evaluated in accordance with the criteria of the standard EN 204 [47] for water-resistant adhesives belonging to the D3 class (Figure 2).

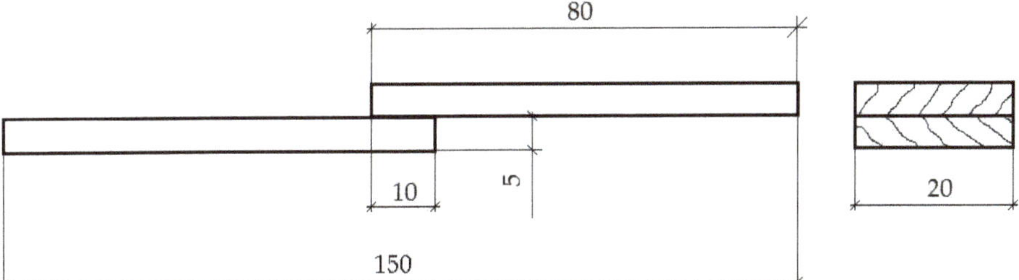

Figure 2. Adhesion test for bonded hardwood specimens according to the standard EN 205 [46]. Dimensions are in mm.

2.5. Statistical Analyses

The statistical software STATISTICA 12 was used to analyze the gathered data. Descriptive statistics deals with the basic statistical characteristics of studied properties: arithmetic mean and standard deviation. Differences of the adhesion strength in the dry state of the bonded hardwoods were analyzed by the Duncan test. The simple linear correlation analyses together with the coefficient of determination R^2 and the significance level parameter p were used as method of inductive statistics to evaluate the measured data.

3. Results and Discussion

3.1. Density, Cold Water Extract, pH Value, Roughness, and Wettability of Hardwoods

The selected structural and physical characteristics of 13 hardwoods, theoretically important from the point of view of their bonding with adhesives, are present in Tables 1–3.

The density ρ of 13 hardwood species, determined at a moisture content of 8% ± 2%, ranged from 636 kg/m^3 for meranti to 1105 kg/m^3 for maçaranduba (Table 1). The cold water extract of the hardwoods, obtained from their sawdust by the standard ASTM D1110 [43], ranged from lower values of 0.9–1.6% for zebrano, wengé, meranti, and sapelli to higher values of 4.2–4.6% for palisander, oak, and black locust (Table 1). The pH of individual hardwood species ranged from a neutral acidic value of 5.8 for palisander and beech to more acidic values in the scope of 3.4–3.9 for meranti, bangkirai, oak, and ipé (Table 1).

The type and amount of extractives in wood varies from species to species, and for the same wood species this can also be influenced by the geographical origin, climate conditions, tree age, and part of the tree from which a sample originates [48]. Several studies have been carried out on the extractives of tropical woods. For example, Wanschura et al. [49] described the benefits of extractives present in tropical woods for their surface treatments. Kilic and Niemz [50], in the structures of 12 tropical wood species, found very low amounts of lipophilics (0.05–0.38 mg/g); the constituent consisted mainly of fatty acids, while the hydrophilics were composed of phenolic acids, flavonoids, sterols, stilbenes, and a lignan. Jankowska et al. [30] determined that there were large differences in quantity of the hot water soluble extractives in the European and tropical hardwood species, which have been researched by us as well (e.g., 2.92% in light red meranti, 4.24% in beech, 4.33% in wengé, 6.07% in sapelli, 6.47% in iroko, 11.21% in oak, and 12.63% in ipé). The values of the cold water extracts of the same species determined in our experiment were evidently lower, from 1.53% in wengé to 4.33% in oak (Table 1). This difference was probably caused by the fact that hot water dissolves not only polar extractives, which also easily dissolve in cold water (e.g., tannins, gums, sugars, and coloring matter), but also starches. The value of the cold water extract for meranti wood was comparable with that found by Yamamoto and Hong [51]; however, those for the wengé and zebrano woods were smaller compared with the values obtained by [50].

Similar pH values for some of the same tropical hardwood species (Table 1) were reported by Yamamoto and Hong [51], Torelli and Čufar [52], and Ikenyiri et al. [53].

Table 1. Densities, cold water extracts and pH values of the hardwoods.

Wood Species	Scientific Name	Density (kg/m^3)		Cold Water Extract (%)	pH
		EN 350 [54]	Obtain		
Bangkirai	*Shorea obtusa*	700-930-1150	834	2.70	3.82
Beech	*Fagus sylvatica*	690-710-750	705	2.13	5.79
Black locust	*Robinia pseudoacacia*	720-740-800	726	4.56	4.65
Bubinga	*Guibourtia demeusii*	700-830-910	887	3.15	4.12
Ipé	*Tabebuia serratifolia*	900-1050-1150 [1]	957	2.08	3.94
Iroko	*Milicia excels*	630-650-670	641	3.51	5.60
Maçaranduba	*Manilkara bidentate*	1000-1100-1150 [1]	1105	3.66	4.60
Meranti	*Shorea curtisii*	600-680-730	636	1.55	3.38
Oak	*Quercus robur*	670-710-760	779	4.33	3.83
Palisander	*Machaerium scleroxylon*	700-900-1000 [1]	818	4.21	5.84
Sapelli	*Entandrophragma cylindricum*	640-650-700	693	1.62	5.23
Wengé	*Millettia laurentii*	780-830-900	881	1.53	4.32
Zebrano	*Microberlinia brazzavillensis*	700-770-850 [1]	777	0.91	5.62

Notes: Mean values of density are from six measurements, and the cold water extract and pH value are from three measurements. [1] By Wagenführ [55].

The roughness parameters Ra and Rz determined for 13 hardwoods exhibited, on average, 48.2% and 79.1% higher values, respectively, measured perpendicularly to the grains (on average: Ra = 10.4 µm, Rz = 85.6 µm) than those determined parallel with the grains (on average: Ra = 7.0 µm, Rz = 47.8 µm). Generally, these represented differences from 8% to 114%, according to the relevant wood species (Table 2). However, the parameter R_{Sm} was approximately comparable in both measured directions of the wood (Table 2).

Table 2. Roughness parameters Ra, Rz, and R_{Sm} of the hardwood surfaces.

Wood Species	Roughness Parallel with Grain (µm)			Roughness Perpendicular to Grain (µm)		
	Ra	Rz	R_{Sm}	Ra	Rz	R_{Sm}
Bangkirai	10.4 (3.0)	68.4 (18.5)	698.4 (193.6)	17.1 (3.6)	131.9 (19.7)	560.5 (89.5)
Beech	6.0 (1.8)	41.2 (12.6)	673.9 (142.9)	8.2 (1.9)	63.7 (12.8)	408.5 (56.7)
Black locust	5.8 (2.7)	40.5 (19.3)	572.4 (121.8)	9.6 (2.7)	86.7 (20.7)	565.5 (149.7)
Bubinga	6.5 (3.6)	44.9 (22.8)	591.4 (150.1)	7.0 (3.1)	69.3 (22.2)	688.5 (233.6)
Ipé	7.9 (4.6)	51.5 (25.8)	737.1 (197.5)	9.3 (3.8)	77.2 (20.5)	572.4 (105.9)
Iroko	7.1 (1.8)	51.6 (13.7)	580.5 (134.8)	10.0 (3.2)	82.5 (24.7)	661.1 (256.5)
Maçaranduba	3.6 (1.5)	26.7 (8.7)	555.1 (104.7)	4.2 (0.9)	46.8 (13.5)	643.9 (213.3)
Meranti	9.9 (5.1)	60.8 (27.3)	655.0 (179.8)	17.6 (4.6)	126.6 (21.3)	760.2 (197.7)
Oak	9.4 (6.4)	59.5 (35.2)	574.5 (195.2)	16.8 (5.1)	123.4 (33.7)	534.5 (119.5)
Palisander	4.7 (1.3)	34.3 (10.1)	679.2 (239.1)	5.9 (2.1)	57.6 (14.7)	555.7 (138.8)
Sapelli	7.3 (2.8)	52.8 (19.4)	565.1 (109.0)	10.2 (2.3)	87.6 (11.1)	533.8 (93.0)
Wengé	5.4 (3.3)	39.2 (20.0)	545.6 (145.7)	9.8 (2.7)	78.5 (17.8)	712.1 (174.5)
Zebrano	7.1 (3.2)	50.3 (22.4)	588.5 (159.5)	9.5 (3.3)	81.6 (21.7)	622.5 (195.1)

Note: Mean values are determined from 90 values (15 different measuring spots on 6 replicates). Standard deviations are in parentheses.

The surfaces of the densest maçaranduba wood were characterized by the lowest roughness parameters Ra and Rz. On the contrary, the highest roughness parameters were exhibited by the least dense wood, the meranti wood, but at the same time by the bangkirai and oak woods also, which had a medium density (Tables 1 and 2). In these circumstances, an unexpectedly higher surface roughness of the bangkirai and oak, belonging to the ring-porous wood species [55], can be explained by their naturally more porous morphological structure.

The water's potential for surface wetting of the tested hardwood species was assessed based on the water contact angles θ_0, θ_e, and θ_w (Table 3). Iroko had the highest initial and abstract contact angles (θ_0 = 112.5°, θ_w = 78.5°), and black locust had the lowest equilibrium and abstract contact angles (θ_e = 19.7°, θ_w = 25.4°), while bubinga had the lowest initial contact angle (θ_0 = 54.8°) (Table 3).

According to Liptáková et al. [56], the values of the contact angles θ_0 and θ_e depend on the wood surface roughness and chemistry. These authors supposed that the calculated contact angle value θ_w, valid for an ideal smooth surface, depends exclusively on the wood chemistry. Consequently, the different values of the contact angle θ_w in Table 3 indicate species-related differences in the wood surface chemistry. This concerns not only the different types and contents of extractives, but also the chemical composition of the lignin and polysaccharides in tested hardwoods.

The wettability of wood is a substantial parameter which gives basic information on the interaction between the solid wood surface and liquids, such as adhesives and paints (e.g., how easily and efficiently the liquids spread over a solid surface) [57,58]. The smaller water contact angles, which usually result from the rougher and more polar characteristics of wood surfaces, indicate deeper penetration of the water-based adhesive into the wood structure [16,59].

When comparing the roughness parameters and the wettability parameters of 13 hardwoods (Tables 2 and 3), it is evident that there was not always a more apparent connection between these two groups of parameters. For example, the maçaranduba and meranti woods, characterized by totally different values for the density (Table 1) and the parameters of roughness (Table 2), had comparable values for the water contact angles (Table 3).

Table 3. Wettability of hardwood surfaces, characterized by the contact angles.

Wood Species	Contact Angle (°)		
	θ_0	θ_e	θ_w
Bangkirai	87.6 (20.7)	46.1 (14.2)	58.6 (22.3)
Beech	71.8 (9.3)	27.6 (11.6)	32.4 (12.9)
Black locust	80.7 (14.8)	19.7 (10.7)	25.4 (13.8)
Bubinga	54.8 (7.5)	30.5 (17.6)	32.2 (18.7)
Ipé	101.3 (7.5)	46.2 (16.4)	61.8 (21.0)
Iroko	112.5 (10.3)	54.0 (10.7)	78.5 (15.7)
Maçaranduba	94.0 (5.7)	23.5 (10.4)	31.5 (14.5)
Meranti	92.0 (15.1)	32.5 (23.7)	41.1 (25.0)
Oak	80.5 (12.2)	33.6 (14.6)	41.0 (16.7)
Palisander	88.4 (12.7)	29.9 (9.3)	40.0 (14.3)
Sapelli	111.7 (14.0)	32.0 (16.5)	52.2 (26.5)
Wengé	86.8 (8.4)	59.1 (7.6)	68.3 (7.9)
Zebrano	71.4 (9.9)	45.0 (13.0)	50.3 (14.2)

Notes: Mean values are from 30 measurements (5 different measuring spots on 6 replicates). Standard deviations are in parentheses.

3.2. Adhesion of Hardwoods with PVAc Adhesive

The values of the adhesion strength (the tensile shear strength of the lap joints) of hardwood specimens bonded with the PVAc Multibond SK8 adhesive are shown in Table 4 and Figure 3.

Table 4. The adhesion strength (tensile shear strength of lap joints) of hardwood specimens bonded with the PVAc Multibond SK8 adhesive, determined in the dry state, wet state, and reconditioned state by the standard EN 204 [47].

Wood Species	Adhesion-Tensile Shear Strength of Lap Joints (MPa)		
	Dry	Wet	Reconditioned
Bangkirai	9.53 (1.68) -	2.29 (0.54)	11.24 (1.58)
Beech	15.65 (2.73) a	2.64 (0.20)	16.19 (2.56)
Black locust	11.69 (2.65) c	1.58 (0.24)	12.62 (2.04)
Bubinga	17.20 (1.09) a	2.00 (0.17)	19.24 (1.62)
Ipé	13.99 (0.51) a	1.27 (0.08)	12.10 (1.41)
Iroko	10.69 (2.14) d	2.01 (0.10)	10.26 (1.30)
Maçaranduba	15.76 (2.23) a	0.56 (0.19)	8.48 (0.78)
Meranti	11.40 (1.54) d	2.34 (0.24)	10.85 (1.73)
Oak	14.27 (1.47) a	2.33 (0.24)	15.57 (1.97)
Palisander	15.86 (1.18) a	1.26 (0.18)	13.67 (2.86)
Sapelli	15.13 (1.85) a	1.98 (0.29)	14.90 (2.36)
Wengé	11.90 (2.02) c	1.57 (0.22)	13.53 (0.51)
Zebrano	13.85 (2.52) a	1.12 (0.30)	9.80 (2.12)

Note: Mean values are from six values. Standard deviations are in parentheses. The Duncan test, using the indexes (a, b, c, and d), identified the significance level of the higher adhesion strength of bonded wood species in relation to the reference bonded bangkirai wood having the lowest adhesion strength in the dry state (e.g., a: very significantly higher, >99.9%; b: significantly higher, >99%; c: less significantly higher, >95%; and d: insignificantly higher, <95%).

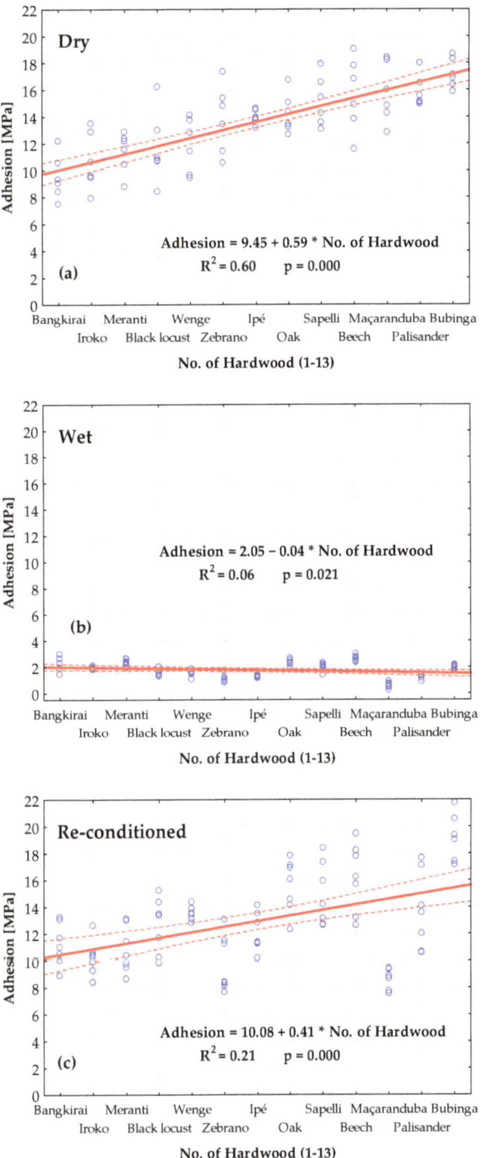

Figure 3. The growth tendency of the adhesion strength from bangkirai (No. 1) to bubinga (No. 13) determined in the dry state of bonded hardwoods (**a**) unequally changed for the individual wood species only in the wet state (**b**), based on the evidently decreased parameter R^2 and reduced significance p (from 99.9% to 99%) of the linear correlation Adhesion = a + b × Number of Hardwood, while this tendency returned in the reconditioned state (**c**).

The Multibond SK8 adhesive was as a good glue type for the bonding of several hardwoods exposed in dry and water-soaked conditions. This type of PVAc adhesive, for all 13 bonded hardwoods, usually secured the minimum adhesion strengths required by the standard EN 204 [47] (i.e., on average, (a) in the dry state of 13.6 MPa, from 9.5 MPa for bangkirai to 17.2 MPa for bubinga (required minimum = 10 MPa), (b) in the wet state of

1.8 MPa, from 0.6 MPa for maçaranduba to 2.6 MPa for beech (required minimum = 2 MPa), and (c) in the reconditioned state of 13.0 MPa, from 8.5 MPa for maçaranduba to 19.2 MPa for bubinga (required minimum = 8 MPa)) (Table 4, Figure 3).

In the summary evaluation, using a linear correlation Adhesion = a + b × Number of Hardwood (for the numbering of hardwoods, see Figure 1), the adhesion strength of 13 bonded hardwoods determined in the dry state (Figure 3a) unequally decreased in the wet state due to 4 days of soaking in water (Figure 3b). This was based on the coefficient of determination R^2 declining from 0.6 to 0.06, and on the change of the significance level parameter p from 0.000 to 0.021. However, due to 7 days of reconditioning of the wet samples in a dry environment, the adhesion strength recovered quite equally to the initial values found in the dry state, which was in the linear correlation confirmed by the increasing of R^2 to 0.21 and with p equal to 0.000 (Figure 3c).

Partially worsened results were achieved in the wet state of the bonded hardwoods, when the adhesion strength decreased in more cases under the criteria value of 2 MPa (i.e., to 0.6 MPa–2.6 MPa) (Table 4, Figure 3), which could be explained, among other factors, with a different penetration of the PVAc adhesive into the individual hardwood species. For example, the lowest adhesion of 0.6 MPa was determined for the maçaranduba wood, the densest species (Table 1) having the lowest roughness (Table 3). On the contrary, the second-highest adhesion of 2.3 MPa was determined for the meranti wood, the least dense species (Table 1) having the highest roughness (Table 3). The highest adhesion of 2.6 MPa was determined for the beech wood (i.e., the species characterized by very good permeability [60,61]). Generally, different microstructures of the wood–adhesive–wood interfaces, when a probable better penetration of adhesives into less dense and more porous woods, as well as into more permeable woods, could be connected with a higher and more water-stable mechanical adhesion of adhesives with wood surfaces.

Results achieved in this work for the bonded beech wood were in accordance with the work of He and Chiozza [62]. For this wood species, bonded it at 23 °C with the PVAC glue Vinavil 2259 L, they determined by the standard EN 205 [46] the adhesion strength in the dry state 15.8 MPa and in the wet state 2.7 MPa.

3.3. Connections between Bonding and Selected Characteristics of Hardwoods

The adhesion strengths of 13 bonded hardwoods, valued in the dry state, were positively influenced—on the 99% or 95% significance level ($p < 0.01$; $p < 0.05$)—by their higher densities ρ, lower roughness parameters Ra and Rz perpendicular to the grain, and lower water contact angles θ_0, θ_e, and θ_w, as was documented by the linear correlations (Adhesion = a + b × Property of wood) (Table 5 and Figure 4a,d–f).

On the contrary, the adhesion strengths of the bonded hardwoods were not significantly influenced by the cold water extract, pH value, or roughness parameters Ra, Rz, and R_{Sm} parallel with the grain ($p > 0.5$) (Table 5 and Figure 4b,c).

Generally, the joint quality of bonded timbers depends on several surface characteristics of the wood, including its wetting capacity. For example, beech and bubinga woods were characterized with good wettability values (Table 3), and this property resulted in a positive impact on their adhesion strength with a PVAc adhesive in the dry state (Table 4 and Figures 3a and 4e,f), as well as in the wet state and reconditioned state (Table 4). However, the wettability may not always be a presuming factor. For example, a good bonding quality in the dry state was also determined for ipé, iroko, sapelli, and wengé woods, whose surfaces had evidently higher water contact angles.

Table 5. Linear correlation analyses between the hardwood characteristics or properties and the adhesion strength, determined for the wood–PVAc adhesive–wood interface in the dry state.

Property of Hardwood	N	R^2	T	p	Adhesion = a + b x Property
Density ρ (kg/m^3)	78	0.086	2.68	0.009	$8.36 + 0.007 \times \rho$
Cold water extract (%)	39	0.004	0.39	0.699	12.94 + 0.16 x extract
pH	39	0.040	1.24	0.224	$10.06 + 0.71 \times$ pH
Roughness parallel with grain (µm)					
Ra	78	0.027	-1.47	0.147	$14.22{-}0.097 \times Ra$
Rz	78	0.015	-1.06	0.292	$14.14{-}0.012 \times Rz$
R_{Sm}	78	0.039	1.75	0.085	$12.43 + 0.002 \times R_{Sm}$
Roughness perpendicular to grain (µm)					
Ra	78	0.071	-2.41	0.018	$14.56{-}0.087 \times Ra$
Rz	78	0.081	-2.59	0.011	$15.05{-}0.016 \times Rz$
R_{Sm}	78	0.006	-0.68	0.498	$14.03{-}0.001 \times R_{Sm}$
Contact angle (°)					
θ_0	78	0.11	-3.03	0.003	$17.52{-}0.044 \times \theta_0$
θ_e	78	0.10	-2.94	0.004	$15.62{-}0.051 \times \theta_e$
θ_w	78	0.13	-3.40	0.001	$15.82{-}0.043 \times \theta_w$

Note: N is the number of samples. However, several times more measurements were performed and by linear correlations analyzed for the roughness parameters and the contact angles, as for each individual sample, the roughness parameters were determined on 15 spots and the contact angles on 5 spots (see Tables 2 and 3).

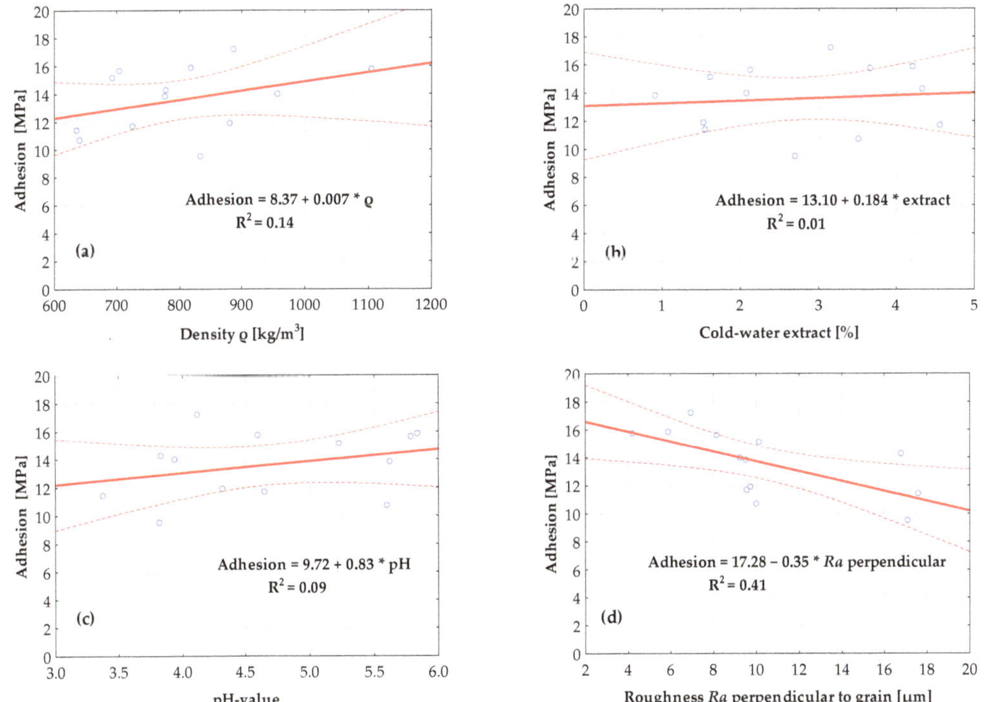

Figure 4. *Cont.*

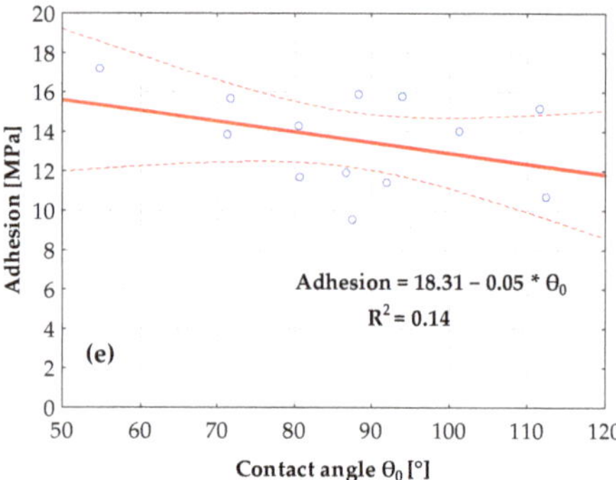

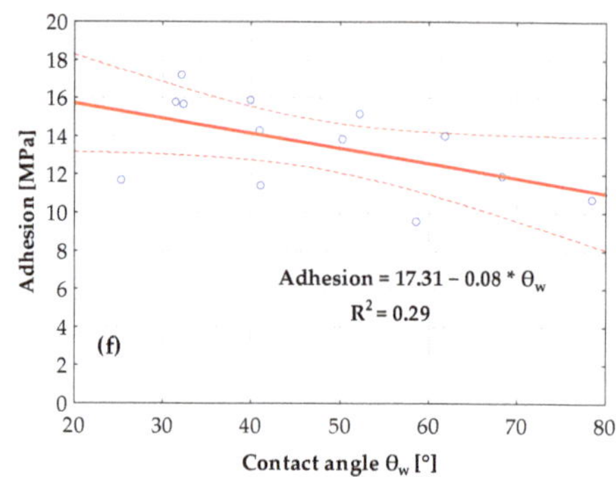

Figure 4. Expression of the linear tendency changes of the mean adhesion strengths of 13 bonded hardwoods, determined in the dry state, in relation to the mean properties of these wood species – (**a**) density, (**b**) cold-water extract, (**c**) pH value, (**d**) roughness parameter *Ra* perpendicular to the grain, (**e**) initial contact angle, (**f**) abstract contact angle. Note: In Table 5, statistical analyses carried out from all individual measurements are present.

4. Conclusions

The adhesion strengths of 13 hardwoods bonded with the PVAc adhesive Multibond SK8 met, in most cases, the requirements of the standard EN 204 [47] (i.e., it ranged from 9.5 MPa to 17.2 MPa (limit 10 MPa) in the dry state, from 0.6 MPa to 2.6 MPa (limit 2 MPa) in the wet state, and from 8.5 MPa to 19.2 MPa (limit 8 MPa) in the reconditioned state).

The water contact angles, expressing the wettability of wood surfaces, were not clearly affected by the other measured structural and physical characteristics of the hardwoods. For example, for the most dense wood, maçaranduba wood (1105 kg/m^3), in comparison with the less dense meranti wood (636 kg/m^3), was determined to have a 136% greater cold water extract, 36% higher pH value, 63.3% lower roughness parameter *Ra* parallel with the grain, and 71.6% lower roughness parameter *Ra* perpendicular to the grain. However, the initial contact angles of these two tropical woods θ_0 were essentially the same, being 94.0° and 92.0°, respectively. This means that the wettability of wood surfaces with water, which usually dominantly affects the adhesion strength between wood surfaces and water-based adhesives or coatings, can simultaneously be dependent on the combination of several other structural parameters of the wood, including its specific molecular structure. This should be analyzed in any following experiments.

The linear correlations indicated that the mutual relations between the adhesion strength values in the dry state of 13 hardwoods bonded with a PVAc adhesive and the cold water extract, pH value, or roughness parameters parallel with the grain were not statistically significant. On the contrary, the adhesion strength values showed negative tendencies on the 95% significance level, influenced by the increased roughness parameters *Ra* and *Rz* perpendicular to the grain, and on the 99% significance level, influenced by the increased water contact angles θ_0, θ_e, and θ_w of the individual hardwoods.

Author Contributions: Conceptualization, J.I., L.R. and J.S.; methodology, J.I., L.R., J.S., J.K. and V.K.; software, J.I. and L.R.; validation, J.I., L.R. and J.K.; formal analysis, J.I., L.R. and J.K.; investigation, J.I., L.R., J.S., J.K. and V.K; resources, J.I., L.R., J.K. and V.K.; data curation, J.I., L.R. and J.K.; writing—original draft preparation J.I., L.R. and J.K.; writing—review and editing, J.I. and L.R; visualization, J.I. and L.R.; supervision, J.I. and L.R.; project administration, L.R. and J.K.; funding acquisition, J.I. and L.R.. All authors have read and agreed to the published version of the manuscript.

Funding: This work was supported by the Slovak Research and Development Agency under the contracts no. APVV-17-0583 and no. APVV-16-0177, and the VEGA project 1/0729/18.

Institutional Review Board Statement: Not applicable.

Informed Consent Statement: Not applicable.

Data Availability Statement: Data sharing is not applicable to this article.

Acknowledgments: The authors would like to thank the Slovak Research and Development Agency under contracts no. APVV-17-0583 and no. APVV-16-0177, and also to the VEGA project 1/0729/18 for funding and financial support. This publication is also the result of the following project implementation: Progressive research of performance properties of wood-based materials and products (LignoPro), ITMS 313011T720 supported by the Operational Programme Integrated Infrastructure (OPII) funded by the ERDF.

Conflicts of Interest: The authors declare no conflict of interest.

References

1. Sernek, M.; Resnik, J.; Kamke, F.A. Penetration of liquid urea-formaldehyde adhesive into beech wood. *Wood Fiber Sci.* **1999**, *31*, 41–48.
2. Kamke, F.A.; Lee, N.J. Adhesive penetration in wood—A review. *Wood Fiber Sci.* **2007**, *39*, 205–220.
3. Mendoza, M.; Hass, P.; Wittel, F.K.; Niemz, P.; Herrmann, H.J. Adhesive penetration of hardwood: A generic penetration model. *Wood Sci. Technol.* **2012**, *41*, 529–549. [CrossRef]
4. Bourreau, D.; Aimene, Y.; Beauchene, J.; Thibaut, B. Feasibility of glued laminated timber beams with tropical hardwoods. *Eur. J. Wood Wood Prod.* **2013**, *71*, 653–662. [CrossRef]
5. Hiziroglu, S.; Zhong, Z.W.; Tan, H.L. Measurement of bonding strength of pine, kapur and meranti wood species as function of their surface quality. *Measurement* **2013**, *46*, 3198–3201. [CrossRef]
6. Stoeckel, F.; Konnerth, J.; Gindl-Altmutter, W. Mechanical properties of adhesives for bonding wood—A review. *Int. J. Adhes. Adhes.* **2013**, *45*, 32–41. [CrossRef]
7. Luedtke, J.; Amen, C.; van Ofen, A.; Lehringer, C. 1C-PUR-bonded hardwoods for engineered wood products: Influence of selected processing parameters. *Eur. J. Wood Wood Prod.* **2015**, *73*, 167–178. [CrossRef]
8. Aicher, S.; Ahmad, Z.; Hirsch, M. Bondline shear strength and wood failure of European and tropical hardwood glulams. *Eur. J. Wood Wood Prod.* **2018**, *76*, 1205–1222. [CrossRef]
9. Bockel, S.; Mayer, I.; Konnerth, J.; Niemz, P.; Swaboda, C.; Beyer, M.; Harling, S.; Weiland, G.; Bieri, N.; Pichelin, F. Influence of wood extractives on two-component polyurethane adhesive for structural hardwood bonding. *J. Adhes.* **2018**, *94*, 829–845. [CrossRef]
10. Bomba, J.; Šedivka, P.; Hýsek, Š.; Fáber, J.; Oberhofnerová, E. Influence of glue line thickness on the strength of joints bonded with PVAc adhesives. *Prod. J.* **2018**, *68*, 120–126. [CrossRef]
11. Bockel, S.; Mayer, I.; Konnerth, J.; Harling, S.; Niemz, P.; Swaboda, C.; Beyer, M.; Bieri, N.; Weiland, G.; Pichelin, F. The role of wood extractives in structural hardwood bonding and their influence on different adhesive systems. *Int. J. Adhes. Adhes.* **2019**, *91*, 43–53. [CrossRef]
12. Petković, G.; Vukoje, M.; Bota, J.; Preprotić, S.P. Enhancement of polyvinyl acetate (PVAc) adhesion performance by SiO_2 and TiO_2 nanoparticles. *Coatings* **2019**, *9*, 707. [CrossRef]
13. Marini, F.; Zikeli, F.; Corona, P.; Vinciguerra, V.; Manetti, M.C.; Portoghesi, L.; Mugnozza, G.S.; Romagnoli, M. Impact of bio-based (tannins) and nano-scale (CNC) additives on bonding properties of synthetic adhesives (PVAc and MUF) using chestnut wood from young coppice stands. *Nanomaterials* **2020**, *10*, 956. [CrossRef] [PubMed]
14. Shida, S.; Hiziroglu, S. Evaluation of shear strength of Japanese wood species as a function of surface roughness. *For. Prod. J.* **2010**, *60*, 400–404. [CrossRef]
15. Konnerth, J. Survey of selected adhesive bonding properties of nine European softwood and hardwood species. *Eur. J. Wood Wood Prod.* **2016**, *74*, 809–819. [CrossRef]
16. Alamsyah, E.M.; Nan, L.C.; Taki, M.Y.K.; Yoshida, H. Bondability of tropical fast-growing tree species I: Indonesian wood species. *J. Wood Sci.* **2007**, *53*, 40–46. [CrossRef]
17. Ülker, O. Wood adhesives and bonding theory. In *Adhesives–Application and Properties*, 1st ed.; Rudawska, A., Ed.; IntechOpen: London, UK, 2016; pp. 271–288. [CrossRef]
18. Boehme, C.; Hora, G. Water absorption and contact angle measurement of native European, North American and tropical wood species to predict gluing properties. *Holzforschung* **1996**, *50*, 269–276. [CrossRef]
19. Özçifçi, A.; Yapici, F. Effects of machining method and grain orientation on the bonding strength of some wood species. *J. Mater. Process. Technol.* **2008**, *202*, 353–358. [CrossRef]
20. Burdurlu, E.; Kilic, Y.; Elibol, G.C.; Kilic, M. The shear strength of Calabrian pine (*Pinus brutia* Ten.) bonded with polyurethane and polyvinyl acetate adhesives. *J. Appl. Polym. Sci.* **2006**, *99*, 3050–3061. [CrossRef]
21. Kilic, M.; Hiziroglu, S.; Burdurlu, E. Effect of machining on surface roughness of wood. *Build. Environ.* **2006**, *41*, 1074–1078. [CrossRef]
22. Gurau, L. Analyses of roughness of sanded oak and beech surface. *ProLigno* **2013**, *9*, 741–750.
23. Kúdela, J.; Mrenica, L.; Javorek, Ľ. The influence of milling and sanding on wood surface morphology. *Acta Fac. Xylologiae Zvolen* **2018**, *60*, 71–83. [CrossRef]

24. Hiziroglu, S.; Zhong, Z.W.; Ong, W.K. Evaluating of bonding strength of pine, oak and nyatoh wood species related to their surface roughness. *Measurement* **2014**, *49*, 397–400. [CrossRef]
25. Petrič, M.; Oven, P. Determination of wettability of wood and its significance in wood science and technology: A critical review. *Rev. Adhes. Adhes.* **2015**, *3*, 121–187.
26. Frihart, C.R. Wood structure and adhesive bond strength. In *Characterization of the Cellulosic Cell Wall*; Stokke, D.S., Groom, L.H., Eds.; USDA Forest Service, Blackwell Publishing: Grad Lake, CO, USA, 2006; pp. 241–253.
27. Hse, C.Y.; Kuo, M. Influence of extractives an wood gluing and finishing–A review. *Prod. J.* **1988**, *38*, 52–56.
28. Prayitno, T.A.; Widyorini, R.; Lukmandaru, G. The adhesion properties of wood preserved with natural preservatives. *Wood Res.* **2016**, *61*, 197–204.
29. Bhatt, S.; Tripathi, S.; Khali, D.P. Performance evaluation of boric and silicic acid treatment in plywood by shear strength. *Indian For.* **2017**, *143*, 38–42.
30. Jankowska, A.; Boruszewski, P.; Drożdżek, M.; Rebkowski, B.; Kaczmarczyk, A.; Skowronska, A. The role of extractives and wood anatomy in the wettability and free surface energy of hardwoods. *BioResources* **2018**, *13*, 3082–3097. [CrossRef]
31. Roffael, E. Significance of wood extractives for wood bonding. *Appl. Microbiol. Biotechnol.* **2016**, *100*, 1589–1596. [CrossRef]
32. Waliszewska, B.; Zborowska, M.; Prądzyński, W.; Robaszyńska, M. Chemical composition of selected species of exotic trees. In *Wood Structure and Properties '06*; Kurjatko, S., Kúdela, J., Lagaňa, R., Eds.; Arbora Publishers: Zvolen, Slovakia, 2006; pp. 171–174.
33. Hernández, R.E. Swelling properties of hardwoods as affected by their extraneous substances, wood density, and interlocked grain. *Wood Fiber Sci.* **2007**, *39*, 146–158.
34. Rodrigues, A.M.S.; Theodoro, P.N.E.T.; Eparvier, V.; Basset, C.; Silva, M.R.R.; Beauchêne, J.; Espíndola, L.S.; Stien, D. Search for antifungal compounds from the wood of durable tropical trees. *J. Nat. Prod.* **2010**, *73*, 1706–1707. [CrossRef] [PubMed]
35. Valette, N.; Perrot, T.; Sormani, R.; Gelhaye, E.; Morel-Rouhier, M. Antifungal activities of wood extractives. *Fungal Biol. Rev.* **2017**, *31*, 113–123. [CrossRef]
36. Sakuna, T.; Moredo, C.C. Bending of selected tropical woods–Effect of extractives and related properties. In Proceedings of the. Adhesive Technology and Bonded Tropical Wood Products, Taipei, Taiwan, 25–28 May 1993; pp. 166–189.
37. Knorz, M.; Niemz, P.; van de Kuilen, J.W. Measurement of moisture related strain in bonded ash depending on adhesive type and glueline thickness. *Holzforschung* **2016**, *70*, 145–155. [CrossRef]
38. Schmidt, M.; Glos, P.; Wegener, G. Gluing of European beech wood for load bearing timber structures. *Eur. J. Wood Wood Prod.* **2010**, *68*, 43–57. [CrossRef]
39. Knorz, M.; Schmidt, M.; Torno, S.; van de Kuilen, J.-W. Structural bonding of ash (*Fraxinus excelsior* L.): Resistance to delamination and performance in shearing tests. *Eur. J. Wood Wood Prod.* **2014**, *72*, 297–309. [CrossRef]
40. Morin-Bernard, A.; Blanchet, P.; Dagenais, C.; Achim, A. Use of northern hardwoods in glued-laminated timber: A study of bondline shear strength and resistance to moisture. *Eur. J. Wood Wood Prod.* **2020**, *78*, 891–903. [CrossRef]
41. EN 13556. *Round and Sawn Timber. Nomenclature of Timbers Used in Europe*; European Committee for Standardization: Brussels, Belgium, 2003.
42. EN 323. *Wood-Based Panels-Determination of Density*; European Committee for Standardization: Brussels, Belgium, 1993.
43. ASTM D1110. *Standard Test. Methods for Water Solubility of Wood*; ASTM International: West Conshohocken, PA, USA, 2013.
44. Liptáková, E.; Kúdela, J. Analysis of the wood–wetting process. *Holzforschung* **1994**, *48*, 139–144.
45. EN ISO 4287. *Geometrical Product Specifications (GPS)—Surface Texture: Profile Method—Terms, Definitions and Surface Texture Parameters*; European Committee for Standardization: Brussels, Belgium, 1998.
46. EN 205. *Adhesives. Wood Adhesives for Non-Structural Applications. Determination of Tensile Shear Strength of Lap Joints*; European Committee for Standardization: Brussels, Belgium, 2016.
47. EN 204. *Classification of Thermoplastic Wood Adhesives for Non-structural Applications*; European Committee for Standardization: Brussels, Belgium, 2016.
48. Bougnom, B.P.; Knapp, B.A.; Etoa, F.; Insam, H. Possible use of wood ash and compost for improving acid tropical soils. In *Recycling of Biomass Ashes*; Springer: New York, NY, USA, 2011; pp. 87–105.
49. Wanschura, R.; Windeisen, E.; Richter, K. Analysis of extractives of tropical hardwoods and benefits for the surface treatment. In *Eco-Efficient Resource Wood with Special Focus on Hardwoods, Proceedings of IAWS Plenary Meeting*; Németh, R., Teischinger, A., Schmitt, U., Eds.; University of West Hungary Press: Sopron, Hungary; Vienna, Austria, 2014; pp. 71–72.
50. Kilic, A.; Niemz, P. Extractives in some tropical woods. *Eur. J. Wood Wood Prod.* **2012**, *70*, 79–83. [CrossRef]
51. Yamamoto, K.; Hong, L.T. A laboratory method for predicting the durability of tropical hardwoods. *JARQ* **1994**, *28*, 268–275.
52. Torelli, N.; Čufar, K. Mexican tropical hardwoods. pH-value. *Holz als Roh-und Werkstoff* **1995**, *53*, 133–134. [CrossRef]
53. Ikenyiri, P.N.; Abowei, F.M.N.; Ukpaka, C.P.; Amadi, S.A. Characterization and physicochemical properties of wood sawdust in Niger area, Nigeria. *Chem. Int.* **2019**, *5*, 190–197. [CrossRef]
54. EN 350. *Durability of Wood and Wood-Based Products-Testing and Classification of the Durability to Biological Agents of Wood and Wood-Based Materials*; European Committee for Standardization: Brussels, Belgium, 2016.
55. Wagenführ, R. *Holzatlas*, 6th ed.; Fachbuchverlag: Leipzig, Germany, 2007; p. 816.
56. Liptáková, E.; Kúdela, J.; Bastl, Z.; Spirovová, I. Influence of mechanical surface treatment of wood on the wetting process. *Holzforschung* **1995**, *49*, 369–375. [CrossRef]

57. Martha, R.; Dirna, F.C.; Hasanusi, A.; Rahayu, I.S.; Darmawan, W. Surface free energy of 10 tropical woods species and their acrylic paint wettability. *J. Adhes. Sci. Technol.* **2020**, *34*, 167–177. [CrossRef]

58. Hubbe, M.A.; Gardner, D.J.; Shen, W. Contact angles and wettability of cellulosic surfaces: A review of proposed mechanisms and test strategies. *BioResources* **2015**, *10*, 8657–8749. [CrossRef]

59. Cheng, E.; Sun, X. Effects of wood-surface roughness, adhesive viscosity and processing pressure on adhesion strength of protein adhesive. *J. Adhes. Sci. Technol.* **2006**, *20*, 997–1017. [CrossRef]

60. Siau, J.F. *Flow in Wood*; Syracuse University Press: New York, NY, USA, 1971; p. 131.

61. Kurjatko, S.; Reinprecht, L. *Transport látok v dreve. (Transport of Substances in Wood)*; Vedecké a pedagogické aktuality 7/1993; Technical University: Zvolen, Slovakia, 1993; p. 110. ISBN 80-228-0307-3.

62. He, Z.; Chiozza, F. Adhesive strength of pilot-scale-produced water-washed cottonseed meal in comparison with a synthetic glue for non-structural interior applications. *J. Mater. Sci. Res.* **2017**, *6*, 20–26. [CrossRef]

applied sciences

Article

Prediction of Mechanical Performance of Acetylated MDF at Different Humid Conditions

Sheikh Ali Ahmed [1],* , **Stergios Adamopoulos [2],*** , **Junqiu Li [1]** and **Janka Kovacikova [3]**

[1] Department of Forestry and Wood Technology, Faculty of Technology, Linnaeus University, Georg Lückligs Plats 1, 351 95 Växjö, Sweden; jl223hc@student.lnu.se

[2] Department of Forest Biomaterials and Technology, Division of Wood Science and Technology, Vallvägen 9C-D, 756 51 Uppsala, Sweden

[3] Department of Mechanical Engineering, Faculty of Technology, Linnaeus University, Georg Lückligs Plats 1, 351 95 Växjö, Sweden; janka.kovacikova@lnu.se

* Correspondence: sheikh.ahmed@lnu.se (S.A.A.); stergios.adamopoulos@slu.se (S.A.); Tel.: +46-470-76-7492 (S.A.A.); +46-018-67-2474 (S.A.)

Received: 19 November 2020; Accepted: 2 December 2020; Published: 4 December 2020

Abstract: Change of relative humidity (RH) in surrounding environment can greatly affect the physical and mechanical properties of wood-based panels. Commercially produced acetylated medium density fiberboard (MDF), Medite Tricoya®, was used in this study to predict strength and stiffness under varying humid conditions by separating samples in parallel (//) and perpendicular ($\perp$) to the sanding directions. Thickness swelling, static moduli of elasticity (MOE_{stat}) and rupture (MOR_{stat}), and internal bond (IB) strength were measured at three different humid conditions, i.e., dry (35% RH) and wet (85% RH). Internal bond (IB) strength was also measured after accelerated aging test. A resonance method was used to determine dynamic modulus of elasticity (MOE_{dyn}) at the aforementioned humid conditions. Linear regression and finite element (FE) analyses were used to predict the MDF's static bending behavior. Results showed that dimensional stability, MOE_{stat}, MOR_{stat} and IB strength decreased significantly with an increase in RH. No reduction of IB strength was observed after 426 h of accelerated aging test. A multiple regression model was established using MOE_{dyn} and RH values to predict MOE_{stat} and MOR_{stat}. In both directions (// and $\perp$), highly significant relationships were observed. The predicted and the measured values of MOE_{stat} and MOR_{stat} were satisfactorily related to each other, which indicated that the developed model can be effectively used for evaluating the strength and stiffness of Medite Tricoya® MDF samples at any humid condition. Percent errors of two different simulation techniques (standard and extended FE method) showed highly efficient way of simulating the MDF structures with low fidelity.

Keywords: acetylation; wood fiber; strength; stiffness; internal bonding strength; thickness swelling; regression; finite element analysis

---

1. Introduction

Medium density fiberboard (MDF) is manufactured with wood fibers bonded with water resistant adhesives such as phenol formaldehyde, urea formaldehyde, isocyanate resin, etc. [1,2]. MDF is primarily used in furniture, as a building material and for laminate flooring, since it has good strength and stiffness and it is easy to process. Compared to plywood, MDF panels generally swell more and may not be recovered after drying due to the inherent hygroscopicity of the wood fibers, the residual stresses formed in the fiber mat during hot pressing and some loss of the glue bonds [3]. As a result, when the MDF panel is exposed to any form of water, its constituent wood fibers swell and some of that residual stress is released, resulting in an increase of thickness of the panel. Thickness swelling markedly weakens the product [4], and the mechanical properties that are most directly affected are shear strength and moduli of elasticity and rupture [5]. Exposure to outdoor environment with varying climate conditions can result in dimensional changes and strength loss, and thus MDF panels are generally not recommended for exterior applications. In terms of computational modelling, these phenomena are considered to be the rheological behavior of the material [6].

Several studies have been carried out using various adhesive systems [7–9], post treatments [10], heat treated fibers [11], alternative fibers [12] and recycled adhesives [13] to improve strength and water resistance of MDF panels. In addition, chemical modification was also used to improve the material properties, such as moisture-related properties, durability and weathering resistance [14–16]. The most described chemical modification method for improved dimensional stability of wood particles and fibers to produce panel products is acetylation with acetic anhydride [17,18]. In acetylation process, acetic anhydride substitutes the hydroxyl (-OH) groups in the wood cells with acetyl groups, resulting in decreased hygroscopicity and increased dimensional stability; while reductions in some mechanical properties also occur depending on the extent of temperature and time of the modification reaction [18–20]. Several explanations of the strength loss were given, as for example the type of the adhesive [21], bondability [22], press pressure [23], etc. Higher bondability is required if the panels are used under severe conditions for a long time. Thus, the effect of weathering on the dimensional changes and mechanical properties of acetylated MDF would be beneficial for predicting its long-term service behavior. Mechanical strength and stiffness along and perpendicular to the sanding direction should be also available to achieve better and proper assembly of acetylated MDF panels in outdoor applications.

Fiberboards can be exposed to a range of environmental conditions during service life. Moisture content of MDF, similar to other lignocellulosic materials, changes with the change of surrounding humidity and temperature. Therefore, it is important to know the relationship between moisture content and strength properties of MDF if used as a structural member subjected to these environmental conditions. Use of non-destructive acoustic testing could provide a rapid and reliable measurement of strength properties of MDF panels. Usually, acoustic testing is carried out by using time-of-flight (TOF) and resonance methods [24]. TOF methods use propagation time of a pulse of ultrasound or a stress wave across the material. On the other hand, resonance methods use the free vibration frequency of the material under forced harmonic vibration. Resonance methods provide more information on the elastic properties of materials and are thus considered more reliable than the TOF methods [25]. Previous studies showed a very good to strong relation between dynamic bending properties measured by acoustic tools with the static bending properties of wood panels [26,27]. However, values of dynamic bending properties vary depending on the method used and, most importantly, on the moisture content levels of wood panels. Prediction of static bending properties of acetylated MDF using acoustic techniques under different humid conditions is lacking. Considering that acetylated MDF is more hydrophobic than conventional MDF, fewer internal bond failures and associated changes of internal structure should be expected by repeated swelling and shrinkage. That in turn should lead to more stable static bending behavior. Nevertheless, establishing the relationship between acoustic and static bending properties of acetylated MDF at different humid conditions would ensure reliable and safe predictions of their performance for intended end uses.

Another available method to predict and analyze material behavior of MDF is creating macro scale finite element (FE) models of MDF board's structure. Following classic design procedures for macroscale modelling, the material characteristics obtained from the experiments presented in this paper were used as input values to define the material, and the geometry and boundary conditions of the experiment set up were imitated in the FE model. Here, two analysis approaches that are implemented in a SIMULIA™ Abaqus/CEA (Systèmes®) were used to imitate a static three-point bending test. The first approach was a standard quasi-static stress/displacement procedure to control the time incrementation, and the approach is named T1 in this work [28]. Additionally, a second and more advanced technique, the extended finite element method (XFEM) [29], was used and is named T2. This technique allows us to model discontinuities as an enriched feature and it is an extension to the conventional finite element method [28]. Both techniques are classified as macroscale techniques that are nowadays typically used while designing structures [30,31]. The outcome of both analyses are displacements, reaction forces and maximum principal stresses of the studied FE models. These models should be later optimized to achieve higher fidelity, meaning to create more sophisticated models accounting for the environmental loads as well as mechanical loads, for example, multiscale models [32].

Yet, there is not enough systematic information on the dimensional stability and mechanical properties of acetylated MDF panels under different humid conditions. Establishing correlations between elastic properties measured nondestructively with bending strength and stiffness of acetylated MDF would lead to a quick and reliable means of assessment of the safety margins for different applications. Thus, this study was focused on elucidating the dimensional and static bending properties of acetylated MDF and on evaluating possibilities to predict its bending behavior from acoustic data by using standard statistical and multiscale prediction modelling methods. Dry (35% RH), standard (65% RH) and wet (85% RH) climatic conditions were considered to represent different moisture content situations as well as accelerated weathering.

2. Materials and Methods

2.1. MDF Panels

Commercially produced Medite Tricoya® MDF panels with dimensions of $300 \times 210 \times 18$ mm^3 (length × width × thickness) were used in this study. Formaldehyde free glue is used for the acetylated softwood fibres during the production of Tricoya ® panels. Two different sample sets were prepared, i.e., along (parallel samples symbolized by //) and across (perpendicular samples symbolized by ⊥) the sanding direction. Working samples were prepared according to Table 1 and were stored in three different climatic conditions, i.e., dry (20 °C, 35% RH), standard (20 °C, 65% RH) and wet (20 °C, 85% RH). Five replicates of MDF samples (two different directions and three climatic conditions depending on measured properties) were produced in a total of 80 samples, which were tested for different properties according to EN standards (Table 1).

Table 1. Dimensions (length × width × thickness) of samples used for measuring different physical and mechanical properties of Medite Tricoya® medium density fiberboard (MDF) samples.

Type	Properties Measured	Dimensions [mm^3]	Sample Number	Standard Followed
	Moisture content	$50 \times 50 \times 18$	15	EN 322 [33]
	Density	$50 \times 50 \times 18$	15	EN 323 [34]
Physical properties	Dimensional changes	$300 \times 50 \times 18$	10	EN 318 [35]
	Thickness swelling	$50 \times 50 \times 18$	5	EN 317 [36]
	Accelerated aging	$300 \times 70 \times 18$	5	
Mechanical properties	Internal bonding	$50 \times 50 \times 18$	15	EN 319 [37]
	Three-point bending	$300 \times 36 \times 18$	15	EN 310 [38]

Moisture content and density were measured on samples after conditioning in each climatic condition. Samples were considered to be acclimatized when the differences were smaller than the 0.1% mass of the sample between two weightings within 24 h.

2.2. Experimental

2.2.1. Dimensional Changes

The relative changes in length and thickness of the samples were determined in between two equilibrium conditions. The increases in length and thickness due to swelling were measured from 65% to 85% RH in adsorption (first regime), while the reductions in length and thickness due to shrinkage were measured from 65% to 35% RH in desorption (second regime), according to the standard EN 318 [35]. The samples were exposed to different RH levels until acclimatized at two regimes. The first regime consisted of dimensional changes among consecutive RHs 35%, 65% and 85% at 20 °C constant temperature, whilst the second regime consisted of consecutive RHs in the reverse order, i.e., 85%, 65% and 35%, at 20 °C constant temperature.

Relative expansion and contraction sample's length were calculated using the formulae below:

$$\delta l_{65,85} \text{ (mm/m)} = 1000 \times (l_{85} - l_{65})/l_{65} \tag{1}$$

$$\delta l_{65,35} \text{ (mm/m)} = 1000 \times (l_{35} - l_{65})/l_{65} \tag{2}$$

where $\delta l_{65,85}$ (mm/m) is the relative increase in length due to swelling of sample's length after RH change from 65% to 85%, based on the length l (mm) measured at 65% RH and 85% RH; $\delta l_{65,35}$ (mm/m) is the relative reduction in thickness due to the shrinkage sample's length after RH change from 65% to 35%, based on the length l (mm) measured at 65% RH and 35% RH.

Similar to the calculations of relative change in the sample's length, thickness swelling and shrinkage properties were calculated as follows:

$$\delta t_{65,85} \text{ (\%)} = 100 \times (t_{85} - t_{65})/t_{65} \tag{3}$$

$$\delta t_{65,35} \text{ (\%)} = 100 \times (t_{35} - t_{65})/t_{65} \tag{4}$$

where $\delta t_{65,85}$ (%) is the relative increase in sample´s thickness due to swelling after RH change from 65% to 85%, based on the thickness t (mm) measured at 65% RH and 85% RH; $\delta t_{65,35}$ (%) is the reduction of the sample's thickness due to shrinkage after RH change from 65% to 35%, based on the thickness t (mm) measured at 65% RH and 35% RH.

2.2.2. Thickness Swelling

In this test, conditioned samples at 20 °C and 65% RH were placed in swelling testers (IMAL SW 200, San Damaso, Italy) having water pH of 7 ± 1, and the temperature was controlled to 20 ± 1 °C. Samples were immersed about 25 mm in water and were separated from each other and from the sides of the water bath. After immersion in water for 24 h, the thickness of each test piece was measured by a digital caliper nearest to 0.01 mm. Thickness swelling, G_t (%), was calculated based on the initial thickness t_1 (mm) before and final thickness t_2 (mm) after soaking in water.

$$G_t = 100 \times (t_2 - t_1)/t_1 \tag{5}$$

2.2.3. Accelerated Aging Test

All the edges of the samples were coated with silicone resin, conditioned at 20 °C and 65% RH and were placed in an QUV Accelerated Weathering Tester, QUV/spray (Q-Lab Co., Westlake, NJ, USA). This QUV with AUTOCAL system facilitates testing the external performance of products on their weather ability, light stability or corrosion resistance by simulating sunlight, rain and dew. The test was continued for 426 h and each complete cycle was equal to one week (168 h) following the sequence of condensation at 45 °C for under 24 h, a repeat of UV-radiation 60 °C (0.89 W/m^2/nm) at a wavelength of 340 nm for under 2.5 h and water spray (6–7 L/min) for under 0.5 h. One complete cycle was equal to one week (168 h). Commercial MDF of similar thickness (18 mm) intended for indoor use was used for comparison.

2.2.4. Non-Destructive Testing

An acoustic resonance method was used for measuring the dynamic modulus of elasticity (MOE$_{dyn}$). In this method, a data acquisition logger (PicoScope 4224, Cambridgeshire, UK) connected with the software BING®, version 9.7.2 (Beam Identification by Non-destructive Grading by CIRAD-

French Agricultural Research Centre for International Development, Montpellier, France) that controls, processes data and delivers results. A free-free flexural vibration test set-up was used, and more details about this method can be found in [39]. In flexural vibration, the first four modes of vibration were measured and used for determining the dynamic transversal modulus of elasticity, which represents stiffness under bending stress, i.e., MOE_{dyn}. The test was repeated four times for every sample, two times in each side, and the average was calculated.

2.2.5. Static Bending Test

A three-point bending test was performed to determine the static modulus of elasticity (MOE_{stat}) and modulus of rupture (MOR_{stat}) of the MDF samples following the standard EN 310 [38]. A universal testing machine (Instron 4466, Buckinghamshire, UK) with 10 kN load capacity was used. A static bending test was performed on the same sample used for measuring MOE_{dyn} in nondestructive testing. Uniaxial load was applied on the flat side of the samples. The load was constant (10 mm/m) so that the maximum load was reached within 60 ± 30 s. An increment of load and deflection between 10% and 40% of maximum load was considered for measuring the MOE_{stat} (MPa).

$$MOE_{stat}\ (MPa) = \{l^3(F_2 - F_1)\}/\{4bt^3(a_2 - a_1)\} \tag{6}$$

where l is the span length (mm), b is the width of sample (mm), t is the thickness of sample (mm), F_1 and F_2 are the increment of load at 10% and 40% of maximum load, and a_1 and a_2 are the corresponding deflection at the mid-length of the test pieces due to the load F_1 and F_2, respectively.

Bending strength, MOR_{stat} (MPa), of the test sample was calculated from the maximum load, F_{max} (N), using the equation:

$$MOR_{stat}\ (MPa) = (3F_{max}l)/(2bt^2) \tag{7}$$

2.2.6. Internal Bond Test

Internal bond (IB) or tensile strength perpendicular to the plane of panels was measured following the standard EN 319 [37]. Samples were effectively bonded with a hot-melt glue, and tensile load was applied until rupture using an Instron 4466 universal testing machine (Buckinghamshire, England) with 10 kN load capacity. A loading speed of 8 mm/min was maintained so that the maximum load is reached within 60 ± 30 s. In addition to the conditioned samples at dry (20 °C, 35% RH), standard (20 °C, 65% RH) and wet (20 °C, 85% RH) conditions, IB strength was also measured on samples after accelerated aging. IB or tensile strength (MPa) perpendicular to the plane of MDF test pieces was calculated by following the formula:

$$IB\ strength\ (MPa) = F_{max}/(ab) \tag{8}$$

where F_{max} is the breaking load (N), and a and b are the width (mm) and length (mm) of the test pieces, respectively.

2.2.7. Finite Element Analysis

Geometries of finite element (FE) models were identical to the samples manufactured for three-point bending tests and solved as a 3D problem; thus, full models were considered. Specifically, the beam's part was 300 mm in length, 36 mm in width with a thickness of 18 mm, and the cylindrical pins' parts were 36 mm long with a diameter of 30 mm and placed 26 mm from both ends on a lower side of the beam part and in the middle of the beam's span on an upper side of the beam part. The beam part was created as a deformable body and pins as discrete rigid bodies. The boundary conditions (BCs) and load were applied to the reference points (RP) created on pins' circular areas. Six degrees of freedom (df) were fixed in both bottom pins' reference points for applying BCs and five df were set on the upper pin RP that also serves for applying a load. Namely, the following BCs were assigned: all six df are fastened, thus 1, 2, 3, 4, 5, 6 = 0 for pins at the ends on the lower side of the beam part; for the pin in the middle of the span on the upper side of the beam part movement in the z-axis direction was enabled, only five df are fastened allowing movement in direction 3, thus 1, 2, 4, 5, 6 = 0. The global coordinate system and df notations and plus directions are shown in Figure 1. The load applied to the beam was a concentrated nodal force of value 1200 N, thus, larger than a mean value of the breaking load F_{max} measured in the experiment (Figure 1).

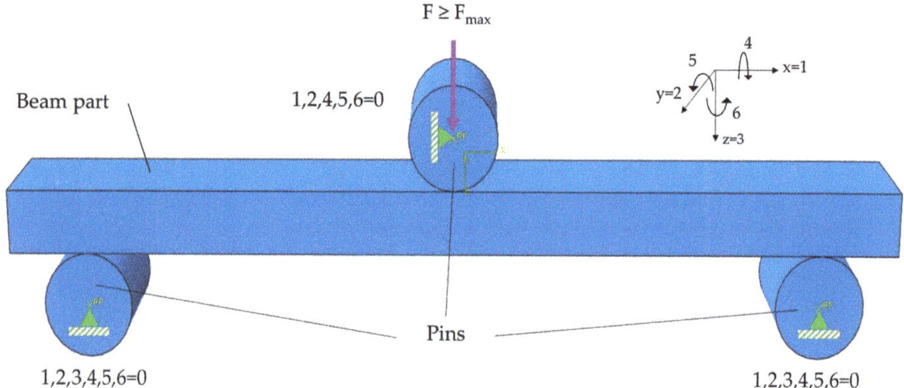

Figure 1. Illustration of parts and boundary conditions assigned in the finite element (FE) model.

3D rigid elements *R3D4* (4-node bilinear quadrilateral) and *R3D3* (3-node triangular facet) were assigned to mesh pins. To create a mesh for the beam, *C3D8R* (8-node linear brick) solid elements with reduced integration were defined.

Normal hard-contact, frictionless tangential behavior, and a general standard surface-to-surface contact, were prescribed to characterize the interaction between pin parts and the beam part. The mechanical properties of the elastic–plastic isotropic material of the studied MDFs had been taken from the experimental data considering the RH of 65% for the standard board perpendicular to bending: thus. Young´s modulus of 2574 MPa, Poisson´s ratio 0.25, and density of 746.2 kg/m^3.

Additionally, to specify a damage initiation, damage criteria for the XFEM model were prescribed to define the constitutive response for cohesive elements: namely, the maximum principal stress damage criterion (MAXPS) of value 34.47 MPa, taken from experiment measures. This means the damage criterion is using traction separation laws and is established as:

$$f = \{\langle \sigma_{max} \rangle / \sigma^0{}_{max}\} \tag{9}$$

where $\sigma^0{}_{max}$ represents the maximum allowable principal stress, and the symbol $\langle \, \rangle$ represents the Macaulay bracket with the usual interpretation. The Macaulay brackets are here to represent that a purely compressive stress does not initiate damage; rather, damage is initiated when the maximum nominal stress ratio reaches a value of one (f = 1) [28]. Additionally, there was no initial crack assigned to the model. Instead, the XFEM crack domain was designated on the whole beam part. For describing a crack geometry and the crack´s growth motion in 3D space, two level sets for a crack is assumed. First set is the crack surface Φ while second, Ψ, is made so the intersection of two level sets that gives the crack front. The nodal value of the function Φ is the signed distance of the node from the crack face and of the function Ψ is the signed distance of the node from an almost-orthogonal surface passing through the crack front [28].

2.2.8. Statistical Analysis

To determine any statistically significant differences between // and $\perp$ samples, mean values were compared by a two-tailed group t-test at 0.05 significance level. In addition, to define relationships between measured parameters (MOE_{stat} and MOR_{stat}, MOR_{stat} and IB), linear regression analyses were performed.

Prediction models for MOE_{stat} and MOR_{stat} by using MOE_{dyn} and RH as input variables were built from multiple regression analysis using the following equation:

$$MOE_{stat} \text{ or } MOR_{stat} \text{ (predicted)} = MOE_{dyn} \times b_{MOEdyn} + RH \times b_{RH} + C \tag{10}$$

where b is coefficient and C is the intercept.

All regressions (linear, multiple) were performed at 95% confidence level using the Microsoft Excel 365 program (Microsoft, Redmond, WA, USA). In addition, the ANOVA is used to check the adequacy of the regression model developed.

3. Results and Discussion

3.1. Physical Properties

Moisture content, density and thickness swelling of commercially produced Medite Tricoya® MDF samples at three different RH levels are presented in Table 2. As expected, the EMC and density of samples increased with an increase in RH levels. Medite Tricoya® samples had 48% lower equilibrium moisture content when compared with commercial indoor MDF samples of similar thickness (Li et al., unpublished data) at 85% RH. Similar results can be found in a previous work [17]. As the acetylation process is a single site reaction, which means that one acetyl group is attached to one hydroxyl group resulting in a reduction of moisture absorption sites in wood polymers [40], acetylated MDF panels have low EMC even at high humidity level. Only 7.1% thickness swelling (after soaking in water for 24 h) was observed, meaning that acetylated panels do not swell severely when they are exposed to the water. These results showed that acetylation plays a significant role in the thickness swelling of the Medite Tricoya® samples. However, the extent of the reduced thickness swelling of acetylated MDF depends on the weight percent gain levels by acetylation process [2].

Table 2. Physical properties of Medite Tricoya® MDF samples at different RH levels. Values in parenthesis are the standard deviations.

Properties	35% RH	65% RH	85% RH
EMC (%)	4.6 (0.14)	7.6 (0.14)	7.9 (0.12)
Density (kg/m^3)	729.2 (9.59)	739.6 (1.37)	742.3 (7.51)
Thickness swelling (%) *		7.1 (0.36)	

* Thickness swelling is measured after immersion in water for 24 h at 20 °C.

Concerning the results of dimensional changes, the acetylation caused the MDF panels to absorb little moisture during the conditioning and are thus dimensionally stable (Table 3). The linear expansion of sample length and thickness swelling values obtained in adsorption conditions (relative humidity change from 65% to 85%) were higher than those values obtained in desorption conditions (relative humidity change from 65% to 35%). The amount of water held by wood fibers at a given temperature and RH depends on the direction from which equilibrium is approached. The moisture adsorbed at high relative humidity exposure is not entirely released when re-drying by lowering the relative humidity levels, and this phenomenon was well observed for wood-based panels in other studies [41,42]. As a result, acetylation had a major effect on the dimensional stability of Medite Tricoya® samples. In addition, the density of panels can also adversely affect the dimensional stability [41]. Average linear expansion and retraction of Medite Tricoya® samples were, respectively, 23% and 57% lower than those of standard samples (Li et al., unpublished data) and for thickness changes, those differences were 67% and 45% (comparison was done from Li et al., unpublished data). However, no significant differences of relative shrinkage or swelling in thickness and length were observed in the two principle directions (// and $\perp$ direction of sanding). An exception was seen for the relative swelling in length when RH increased from 65% to 85% where swelling was significantly higher in perpendicular samples. However, the reason for that is not quite clear.

Table 3. Relative changes in thickness and length of Medite Tricoya® MDF sample at different relative humidity (RH) levels. Values in parenthesis are the standard deviations.

Properties	Direction	Relative Change	
		$\delta_{65,35}$	$\delta_{65,85}$
Thickness (%)	//	−1.12 (0.08)	1.44 (0.07)
	$\perp$	−1.17 (0.07)	1.36 (0.03)
t-value		−0.905 [NS]	−2.140 [NS]
Length (mm/m)	//	−0.34 (0.03)	0.58 (0.08)
	$\perp$	−0.44 (0.09)	0.68 (0.03)
t-value		−1.819 [NS]	3.624 *

//: parallel sample to the sanding direction; $\perp$: perpendicular sample to the sanding direction; $\delta_{65,35}$: samples conditioned from 65% to 35% RH; $\delta_{65,85}$: samples conditioned from 65% to 85% RH. * Significant at the 0.05 level as determined by two-sample t-test. NS, non-significant.

3.2. Mechanical Properties

Table 4 shows MOE_{stat}, MOR_{stat} and IB properties of parallel and perpendicular samples from Medite Tricoya® MDF conditioned at dry, standard and wet conditions. With the increase of RH from 35 to 65%, the average MOE_{stat} and MOR_{stat} reduction of Medite Tricoya® samples was 8% and 10%, respectively. When the RH was further raised from 65% to 85%, reduction of those properties was respectively 37% and 20%. Higher humidity environment had a detrimental effect on the MOE_{stat} and MOR_{stat} and IB strength values for both sample types.

Table 4. Static bending properties and internal bond (IB) strength of Medite Tricoya® MDF samples. Values in parentheses are the standard deviations.

Direction	Dry, 35% RH			Standard, 65% RH			Wet, 85% RH		
	IB Strength [MPa]	MOE_{stat} [MPa]	MOR_{stat} [MPa]	IB Strength [MPa]	MOE_{stat} [MPa]	MOR_{stat} [MPa]	IB Strength [MPa]	MOE_{stat} [MPa]	MOR_{stat} [MPa]
//	0.99 (±0.08)	2938 (±43.38)	38.64 (±0.65)	0.79 (±0.10)	2670 (±61.28)	33.95 (±0.81)	0.57 (±0.07)	1656 (±46.94)	26.80 (±0.97)
$\perp$		2759 (±66.47)	35.73 (±0.99)		2557 (±76.30)	33.25 (±0.49)		1628 (±46.40)	26.77 (±0.63)
t-value		4.516 *	4.906 *		2.306 *	1.460 [NS]		0.841 [NS]	0.044 [NS]

//: parallel sample to the sanding direction; $\perp$: perpendicular sample to the sanding direction. * Significant at the 0.05 level as determine by two-sample T test. NS, non-significant.

Mechanical properties like MOE_{stat} and MOR_{stat} measure the elastic behavior and resistance to bending, respectively, and are important properties when MDF is placed under load. Aforementioned properties determine largely the applicability of MDF as a structural component in furniture or other constructions. Those properties also depend on the sample properties, i.e., sanding direction (parallel or perpendicular), density, moisture content and type of MDF [2,5,43]. Dimensional stability is a good indication of the acetylation effect. However, moisture content had a great influence on the strength properties of both types of MDF samples. Significant reductions in MOE_{stat}, MOR_{stat} and IB strength values were observed with an increase in RH (Table 4). These findings are in agreement with the results found in previous studies [2,43]. The decrease in such properties at increasing RH level can be attributed to the separation of fibers resulting from the thickness swell of the panel materials [44]. Differences in the static bending properties were also found between parallel and perpendicular samples, and especially at the dry condition these differences were statistically significant. In general, MOE_{stat} and MOR_{stat} values were higher in parallel samples compared to perpendicular samples (see Table 4). Lower bending strength and stiffness values of perpendicular samples in MDF and other wood-based panels have been reported previously [26].

Figure 2 shows the linear regressions that determine how well the MOR_{stat} are related with MOE_{stat} values in parallel and perpendicular samples at different humid conditions. Statistical analysis showed strong positive and significant relationships ($p < 0.05$) for both sample directions (parallel and perpendicular) at varying humidity levels. Strong correlation between bending strength and modulus of elasticity is known for wood-based panels [45]. When linear regression analyses were performed by separating humid condition, very poor and insignificant relationships were observed. In the parallel samples, the coefficients of determination (R^2) were 0.08 and 0.39 at 65% and 85% RH, respectively; whilst in the perpendicular samples, those values were, respectively, 0.20 and 0.25. However, in a dry condition (35% RH), significant relationships ($p < 0.05$) were observed, and R^2 values were 0.85 and 0.88 in parallel and perpendicular samples, respectively. Previous findings also showed higher correlation coefficients at lower moisture content levels for wood-based panels [46,47].

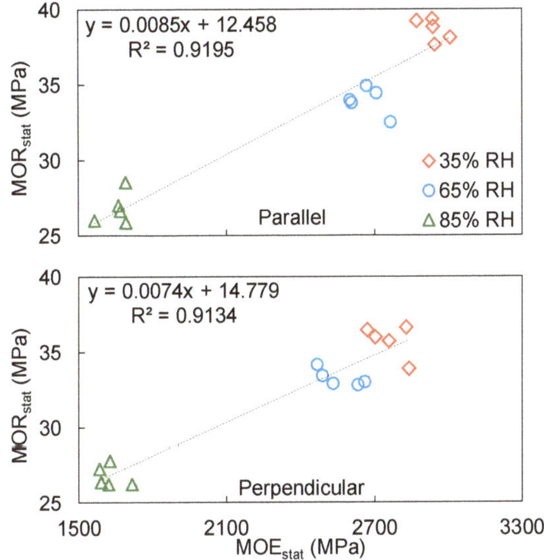

Figure 2. Linear relationship between MOE_{stat} and MOR_{stat} of the Medite Tricoya® medium density fiberboard (MDF) samples in two different directions conditioned at three different climatic conditions.

A positive and significant relationship ($p < 0.05$) between MOR_{stat} and IB was also observed (Figure 3). As shown in Table 4, Medite Tricoya® samples had the highest IB strength at 35% RH. With the increase of RH from 35 to 65%, the average IB strength reduction of Medite Tricoya® samples was 20%. When the RH was further raised from 65% to 85%, reduction of IB strength was 27%.

UV radiation causes photochemical degradation mainly in lignin polymers of the cell walls. UV light in combination with water plays a major role in such type of weathering. When lignin is degraded, water washes away degraded products and subsequently loosens the surface fibers to erode. However, acetylation reduces the loss of surface lignin and thus the erosion caused by accelerated weathering [48]. When compared with commercial MDF samples (indoor use), Medite Tricoya® samples performed much better in terms of the extent of being weathered by moisture and UV light. Thickness swelling of the test pieces from indoor MDF was almost twice as much as Medite Tricoya® (Figure 4). After the accelerated aging test, fibers in the indoor samples were separated and were possible to shred by finger rub. That implied that no residual strength was left. Medite Tricoya® samples showed better resistance against thickness swelling. This could be attributed to the fiber–fiber bonding efficacy of the resin to retain bonding in a very hydrophobic fiber network. After accelerated aging and conditioning (20 °C, 65% RH), the IB strength of the Medite Tricoya® samples was found to be 0.79 ± 0.06 MPa. This result showed that Medite Tricoya® samples made from acetylated wood fibers were able to retain the initial IB strength of non-weathered samples. Previous results also showed that a higher residual strength was observed in acetylated MDF after a cyclic test [19]. Such retention of strength can also be attributed to other parameters than the acetylation of fibers, such as the adhesive used [49].

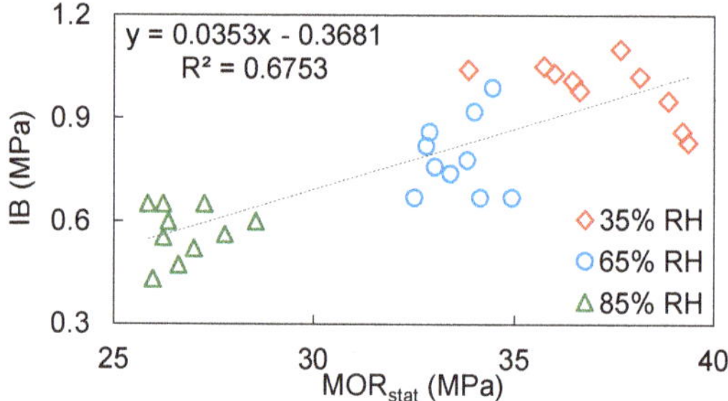

Figure 3. Linear relationship between MOR$_{stat}$ and internal bond (IB) strength of the Medite Tricoya® MDF samples conditioned at three different humidity.

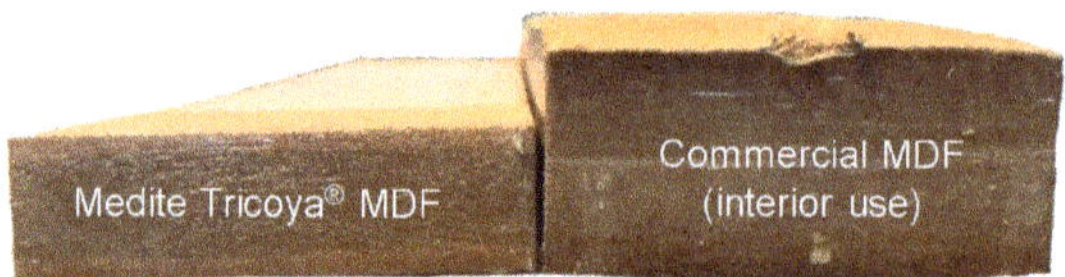

Figure 4. Medite Tricoya® and commercial standard MDF (for interior use) samples after accelerated aging test. Note the double thickness in the commercial MDF sample.

Acoustic resonance measurements were used to determine the MOE$_{dyn}$ (Figure 5). These values represent the mean stiffness, whilst MOE$_{stat}$ represents the local stiffness of the samples at the highly stressed areas of a specific test set-up [50]. MOE$_{dyn}$ was found to differ significantly between // and ⊥ samples only for the dry condition, i.e., 35% RH. As expected, the increase of RH from dry/standard (35 and 65% RH) to wet (85% RH) conditions resulted in a considerable decrease in MOE$_{dyn}$ in both directions.Resonance frequency of MDF samples decreased with the increase in moisture content and thus affected the MOE$_{dyn}$. This is because at a dryer state, molecular chains in the amorphous regions of the wood cell wall are distorted with the presence of microvoids between the molecular chains, resulting in lower internal friction, resulting in higher MOE$_{dyn}$. On the contrary, when moisture content increases, water molecules are embedded in the microvoids and rearrange the distorted molecular chains in the amorphous region. If the moisture content increases further, water acts as a plasticizer and decreases the cohesive forces between molecules, resulting in a higher internal friction and leading to the decrease in MOE$_{dyn}$ [51]. With the increase of RH from 35 to 65%, the average MOE$_{dyn}$ reduction of Medite Tricoya® samples was negligible, i.e., 1%. When the RH was further raised from 65% to 85%, the reduction of MOE$_{dyn}$ was 26%. In addition, MOE$_{dyn}$ values in the perpendicular samples were found lower than the corresponding values in the parallel samples. However, those differences were found to be statistically insignificant. A previous study by Han et al. [26] showed that reduction of stress wave velocity along the perpendicular direction compared to the parallel direction for wood-based panels (plywood, oriented strand board and particleboard) depends on the anisotropic properties of the products. Smaller differences between the two directions in MDF samples implies a uniform product. However, as expected, MOE_{dyn} values were higher than the static values approximately by 40%. Wood-based panels contain a significant number of glued interfaces, and such a difference is generally explained by the different rates of stresses applied in the dynamic and static tests [52]. The differences between the MOE_{dyn} and MOE_{stat} values were higher at a higher RH. It was 33%, 37% and 49% higher at 35%, 65% and 85% RH conditions, respectively. This is because moisture affects the stress-wave properties for wood-based panels, as they can swell considerably during moisture uptake. The swelling often leads to bond failures and to changes of their internal structure, and as a result, to a decrease of stress wave velocity with the increase in panel moisture content [26].

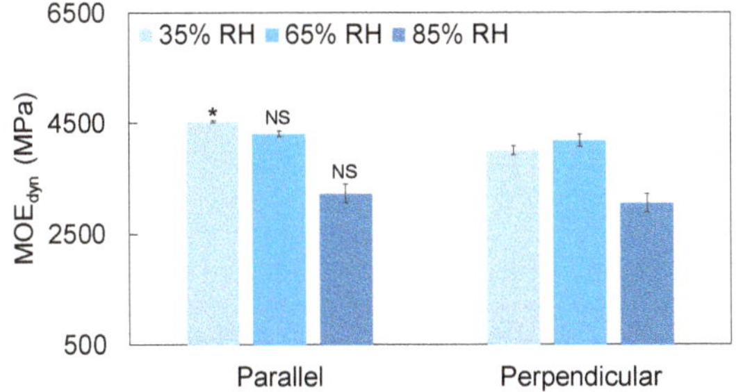

Figure 5. Dynamic modulus of elasticity (MOE_{dyn}) of Medite Tricoya® MDF samples at different relative humidity levels measured by the acoustic resonance method. Error bars represent 95% confidence intervals for the means. * Significant differences between parallel and perpendicular samples as determined by two-sample t-test at the 0.05 level. NS, non-significant.

Acoustic tools have been used quite successfully to predict MOE_{stat} and MOR_{stat} [26,27,53]. Table 5 shows the multiple regression model summary results. Overall, linear and significant relationships ($p < 0.05$) of MOE_{stat} and MOR_{stat} with MOE_{dyn} were observed at different humid conditions. It was noted that R^2 values were slightly higher between MOE_{dyn} and MOE_{stat} than those between MOE_{dyn} and MOR_{stat}.

Table 5. Model summary of regression statistics for MOE_{stat} and MOR_{stat} prediction.

Parameter	MOE_{stat}		MOR_{stat}	
	Parallel	Perpendicular	Parallel	Perpendicular
R^2	0.987	0.969	0.966	0.925
Adjusted R^2	0.970	0.964	0.960	0.913
Standard error	99.484	97.717	1.020	1.175
Intercept	−412.519	883.65	26.822	28.071
RH coefficient	−5.101	−11.624	−0.140	−0.120
MOE_{dyn} coefficient	0.784	0.573	0.004	0.003
F	227.009	187.664	169.889	74.404
Significance F *	0.000	0.000	0.000	0.000

* Significant at the 0.05 level.

The comparison between the experimental and predicted values is shown in Figure 6. The result indicates that the predicted values are very close to the experimental values. F-values for all models (MOE_{stat} and MOR_{stat} in parallel and perpendicular samples) were highly significant at the 0.05 level (Table 5).

Statistical analysis showed that MOE_{dyn} and RH can be used as predictors of MOE_{stat} and MOR_{stat} of MDF samples. The results also indicated that MOE_{stat} and MOR_{stat} have a positive relationship with MOE_{dyn}.

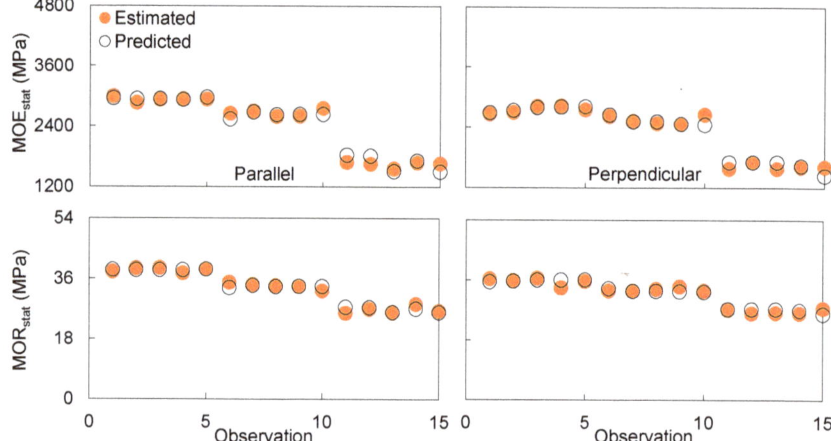

Figure 6. Estimated and predicted values of MOE$_{stat}$ and MOR$_{stat}$ for Medite Tricoya® MDF samples conditioned at three different climatic conditions.

3.3. Finite Element Analysis

Magnitudes of reaction forces and the deflections under these reactions forces in the z-axis direction, as well as values of maximal principal stress for T1, T2 procedures and experimental values, are presented in Table 6. Reaction forces are equal to half of maximal loading force.

Table 6. Magnitudes of displacements, reaction forces, and maximal principal stress for first approach (T1) and second approach (T2) finite element (FE) procedure, and experimental values.

Procedure	Displacement [mm]	Reaction Force [N]	Maximal Principal Stress [MPa]
T1-quasi- static stress /displacement	8.3	599	30.98
T2-extended finite element method	8.9	641	35.60
Experimental value	7.6	563	35.89

Finally, it is necessary to add that the values for the T2 model show the maximal values at the time point when the jump in the damage dissipation energy occurs (Figure 7): thus, in this particular case, at the time increment 0.4623 of 1 with index number 30. At this time, the increment of the crack growth is initiated and the crack propagates [28], i.e., we are at beginning of a fracture mechanism.

The crack occurrence is visualized in Figure 8 using the PHILSM function, i.e., a singled distance function that describes the crack surface [28].

When we compare results obtained from the two computational models using different simulations techniques T1 (quasi-static stress/displacement) and T2 (extended finite element) with the experimental values, we see percental errors of 8.6% and 14.3%. As expected, these percent errors are relatively high. This demonstrates that the used computational methods do not provide reliable results, and thus, can be considered to have a low fidelity. Despite this fact, these techniques are today used as the main techniques in standard design procedures because they are highly efficient [29,31]. To be sure, the errors can be minimized by optimizing the input parameters that imitate the real nature of the studied MDF material; however, that requires more necessary testing, like, for example, creep testing and fracture testing, to obtain those data. This means spending more time to set up experiments, more wasted material for creating samples, as well as time to collect and analyze data sets.

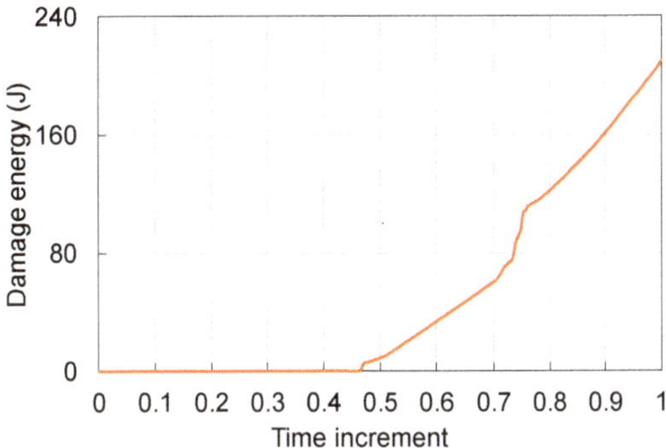

Figure 7. Damage dissipation energy for the extended finite element method model (T2).

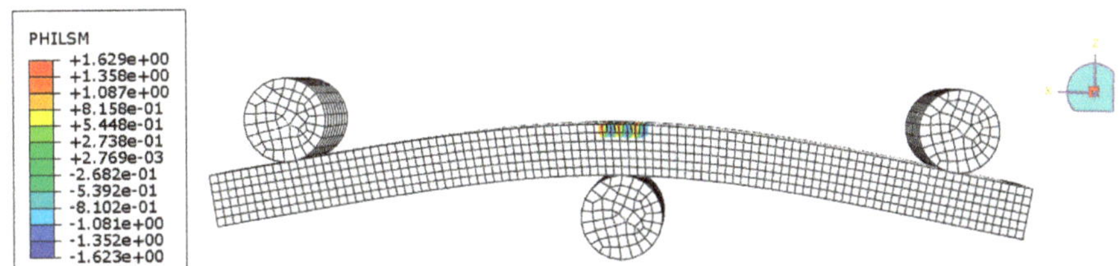

Figure 8. Singled distance function (PHILSM) for T2 model.

A better solution here is to use the help of micromechanical modelling approaches to specify material behaviour on a constituent level that is summarized in [32,54,55], and that will provide more reliable results compared with experimental ones and reduce the need for new sets of tests. More precisely, the employment of concurrent material and structural design applying multiscale models comes in since viscoelastic creep damage models and rheological behavior of MDF material are considered [56]. Here, we mean the time-dependent mechanical repose of MDF on a constant load considering the influence material´s density as well as fiber orientation, and the existence of voids and interfaces in the material microstructure. More importantly, the macroscale computational techniques used in this work do not take into consideration the very significant dependence of mechanical properties on the moisture and temperature. This, as demonstrated in this work, should be included in the computations to precisely predict mechanical material characteristics such as strength, toughness and elasticity as well as to predict the damage mechanism of the panel product.

4. Conclusions

The results obtained in this study showed that, due to the lower hygroscopicity of Medite Tricoya® samples, they absorbed less moisture and became more dimensionally stable even at the highest humidity condition. This MDF type can also retain its IB strength after an accelerated aging test. However, IB strength, MOE_{stat} and MOR_{stat} were reduced from dry to humid conditions. At the highest humid condition (85% RH), strength and stiffness values did not differ significantly between parallel and perpendicular samples. In addition, multiple regression models were developed from MOE_{dyn} and RH to predict the strength and stiffness of Medite Tricoya® MDF. In both parallel and perpendicular directions, highly significant relationships were observed. Developed models could predict the MOE_{stat} and MOR_{stat} values of Medite Tricoya® MDF samples at any humid conditions, which produced an excellent fit to the measured values. This experimental outcome could ensure reliable and safe predictions of Medite's Tricoya® MDF strength and stiffness properties for intended end uses.

This study also showed that employing macroscale computational modelling approaches that are nowadays broadly used in engineering practice, such as quasi-static stress/displacement (T1) and extended finite element (T2) techniques, are not sufficient to obtain reliable results for MDF. These modelling methods are highly efficient and fast, but low in fidelity. Therefore, in future work, it is necessary to employ more precise multiscale models to secure more efficient material and structure design approaches. Additionally, multiscale models will reduce the necessity to set up new testing whenever we need to change the components proportion or component material in composite material.

Author Contributions: Conceptualization, S.A.A. and S.A.; methodology, S.A.A. and S.A.; software, S.A.A.; validation, S.A.A.; formal analysis, S.A.A. and J.K.; investigation, J.L.; data curation, J.L.; writing—original draft preparation, S.A.A. and J.K.; writing—review and editing, S.A.; visualization, S.A.A. and J.K.; supervision, S.A. All authors have read and agreed to the published version of the manuscript.

Funding: This research received no external funding.

Conflicts of Interest: The authors declare no conflict of interest.

References

1. Bianchi, S.; Thömen, H.; Junginger, S.; Pichelin, F. Medium density boards made of groundwood fibres: An analysis of their mechanical and physical properties. *Eur. J. Wood Wood Prod.* **2018**, *77*, 71–77. [CrossRef]
2. Ayrilmis, N.; Winandy, J.E. Effects of Post Heat-Treatment on Surface Characteristics and Adhesive Bonding Performance of Medium Density Fiberboard. *Mater. Manuf. Process.* **2009**, *24*, 594–599. [CrossRef]
3. Palardy, R.D.; Haataja, B.A.; Shaler, S.M.; Williams, A.D.; Laufenberg, T.L. Pressing of wood composite panels at moderate temperature and high moisture content. *For. Prod. J.* **1989**, *39*, 27–32.
4. Li, X.; Li, Y.; Zhong, Z.; Wang, D.; Ratto, J.A.; Sheng, K.; Sun, X.S. Mechanical and water soaking properties of medium density fiberboard with wood fiber and soybean protein adhesive. *Bioresour. Technol.* **2009**, *100*, 3556–3562. [CrossRef] [PubMed]
5. Mohebby, B.; Gorbani-Kokandeh, M.; Soltani, M. Springback in acetylated wood based composites. *Constr. Build. Mater.* **2009**, *23*, 3103–3106. [CrossRef]
6. Findley, W.N.; Lai, J.S.; Onaran, K. *Creep and Relaxation of Nonlinear Viscoelastic Materials*; North-Holland Publishing Company: New York, NY, USA, 1976.
7. Ji, X.; Li, B.; Yuan, B.; Guo, M. Preparation and characterizations of a chitosan-based medium-density fiberboard adhesive with high bonding strength and water resistance. *Carbohydr. Polym.* **2017**, *176*, 273–280. [CrossRef] [PubMed]
8. Gao, S.; Liu, Y.; Wang, C.; Chu, F.; Xu, F.; Zhang, D. Synthesis of Lignin-Based Polyacid Catalyst and Its Utilization to Improve Water Resistance of Urea-formaldehyde Resins. *Polymers* **2020**, *12*, 175. [CrossRef]
9. Mamiński, M.Ł.; Trzepałka, A.; Auriga, R.; H'Ng, P.S.; Chin, K.L. Physical and mechanical properties of thin high density fiberboard bonded with 1,3-dimethylol-4,5-dihydroxyethyleneurea (DMDHEU). *J. Adhes.* **2020**, *96*, 679–690. [CrossRef]
10. Oliveira, S.L.; Freire, T.P.; Mendes, R.F. The Effect of Post-Heat Treatment in MDF Panels. *Mater. Res.* **2017**, *20*, 183–190. [CrossRef]
11. Garcia, R.A.; Cloutier, A.; Riedl, B. Dimensional stability of MDF panels produced from heat-treated fibres. *Holzforschung* **2006**, *60*, 278–284. [CrossRef]
12. Lee, T.C.; Mohd Pu'ad, N.A.S.; Selimin, M.A.; Manap, N.; Abdullah, H.Z.; Idris, M.I. An overview on development of environmental friendly medium density fibreboard. *Mater. Today Proc.* **2020**, *29*, 52–57. [CrossRef]
13. Dazmiri, M.K.; Kiamahalleh, M.V.; Kaiamahalleh, M.V.; Mansouri, H.R.; Moazami, V. Revealing the impacts of recycled urea-formaldehyde wastes on the physical-mechanical properties of MDF. *Eur. J. Wood Wood Prod.* **2018**, *77*, 293–299. [CrossRef]
14. Papadopoulos, A.N.; Gkaraveli, A. Dimensional stabilization and strength of particleboard by chemical modification with propionic anhydride. *Holz Roh Werkst.* **2003**, *61*, 142–144. [CrossRef]
15. Kajita, H.; Imamura, Y. Improvement of physical and biological properties of particleboards by impregnation with phenolic resin. *Wood Sci. Technol.* **1991**, *26*, 63–70. [CrossRef]

16. Nasir, M.; Gupta, A.; Beg, M.D.H.; Chua, G.K.; Asim, M. Laccase application in medium density fibreboard to prepare a bio-composite. *RSC Adv.* **2014**, *4*, 11520–11527. [CrossRef]

17. Rowell, R.M.; Youngquist, J.A.; Rowell, J.S.; Hyatt, J.A. Dimensional stability of aspen fiberboard made from acetylated fiber. *Wood Fiber Sci.* **1991**, *23*, 558–566.

18. Mai, C.; Direske, M.; Varel, D.; Weber, A. Light medium-density fibreboards (MDFs): Does acetylation improve the physico-mechanical properties? *Eur. J. Wood Wood Prod.* **2016**, *75*, 739–745. [CrossRef]

19. Gomez-Bueso, J.; Westin, M.; Torgilsson, R.; Olesen, P.O.; Simonson, R. Composites made from acetylated lignocellulosic fibers of different origin—Part I. Properties of dry-formed fiberboards. *Holz Roh Werkst.* **2000**, *58*, 9–14. [CrossRef]

20. Mahlberg, R.; Paajanen, L.; Nurmi, A.; Kivisto, A.; Koskela, K.; Rowell, R.M. Effect of chemical modification of wood on the mechanical and adhesion properties of wood fiber/polypropylene fiber and polypropylene/veneer composites. *Eur. J. Wood Wood Prod.* **2001**, *59*, 319–326. [CrossRef]

21. Vick, C.B.; Krzysik, A.; Wood, J.E., Jr. Acetylated, isocyanate-bonded flakeboards after accelerated aging: Dimensional stability and mechanical properties. *Holz Roh Werkst.* **1991**, *49*, 221–228. [CrossRef]

22. Korai, H. Effects of low bondability of acetylated fibers on mechanical properties and dimensional stability of fiberboard. *J. Wood Sci.* **2001**, *47*, 430–436. [CrossRef]

23. Ghorbani, M.; Bavaneghi, F. Effect of Press Cycle Time on Application Behavior of Board Made from Chemically Modified Particles. *Drv. Ind.* **2016**, *67*, 25–31. [CrossRef]

24. Legg, M.; Bradley, S. Measurement of stiffness of standing trees and felled logs using acoustics: A review. *J. Acoust. Soc. Am.* **2016**, *139*, 588–604. [CrossRef]

25. Wang, X. Acoustic measurements on trees and logs: A review and analysis. *Wood Sci. Technol.* **2013**, *47*, 965–975. [CrossRef]

26. Han, G.; Wu, Q.; Wang, X. Stress-wave velocity of wood-based panels: Effect of moisture, product type, and material direction. *For. Prod. J.* **2006**, *56*, 28–33.

27. Guan, C.; Guan, C.; Zhou, L.; Wang, X. Dynamic determination of modulus of elasticity of full-size wood composite panels using a vibration method. *Constr. Build. Mater.* **2015**, *100*, 201–206. [CrossRef]

28. Dassault Systèmes. *SIMULIA User Assistance/Analysis Procedure*; Dassault Systèmes: Vélizy-Villacoublay, France, 2019.

29. Belytschko, T.; Black, T. Elastic crack growth in finite elements with minimal remeshing. *Int. J. Numer. Methods Eng.* **1999**, *45*, 601–620. [CrossRef]

30. Barbero, E. *Finite Element Analysis of Composite Materials Using Abaqus™*; CRC Press: Boca Raton, FL, USA, 2013; p. 444.

31. Barbero, E. *Introduction to Composite Materials Design*, 3rd ed.; CRC Press: Boca Raton, FL, USA, 2017; p. 534.

32. Aboudi, J.; Arnold, S.M.; Bednarcyk, B.A. *Micromechanics of Composite Materials: A Generalized Multiscale Analysis Approach*, 1st ed.; Butterworth-Heinemann Ltd.: Oxford, UK, 2013.

33. European Committee for Standardization. *Wood-Based Panels—Determination of Moisture Content*; EN 322; European Committee for Standardization: Brussels, Belgium, 1993.

34. European Committee for Standardization. *Wood-Based Panels—Determination of Density*; EN 323; European Committee for Standardization: Brussels, Belgium, 1993.

35. European Committee for Standardization. *Wood-Based Panels—Determination of Dimensional Changes Associated with Changes in Relative Humidity*; EN 318; European Committee for Standardization: Brussels, Belgium, 2002.

36. European Committee for Standardization. *Particleboards and Fiberboards—Determination of Swelling in Thickness after Immersion in Water*; EN 317; European Committee for Standardization: Brussels, Belgium, 1993.

37. European Committee for Standardization. *Particleboards and Fibreboards—Determination of Tensile Strength Perpendicular to the Plane of the Board*; EN 319; European Committee for Standardization: Brussels, Belgium, 1993.

38. European Committee for Standardization. *Wood-Based Panels—Determination of Modulus of Elasticity in Bending and of Bending Strength*; EN 310; European Committee for Standardization: Brussels, Belgium, 1993.

39. Ahmed, S.A.; Adamopoulos, S. Acoustic properties of modified wood under different humid conditions and their relevance for musical instruments. *Appl. Acoust.* **2018**, *140*, 92–99. [CrossRef]

40. Rowell, R.M.; Ibach, R.E.; McSweeny, J.; Nilsson, T. Understanding decay resistance, dimensional stability and strength changes in heat-treated and acetylated wood. *Wood Mater. Sci. Eng.* **2009**, *4*, 14–22. [CrossRef]

41. Ayrilmis, N. Effect of panel density on dimensional stability of medium and high density fiberboards. *J. Mater. Sci.* **2007**, *42*, 8551–8557. [CrossRef]

42. Ganev, S. Modeling of the Hygromechanical Warping of Medium Density Fiberboard. Ph.D. Thesis, Forestry Faculty, University of Laval, Québec, QC, Canada, 2002.

43. Bekhta, P.; Niemz, P. Effect of relative humidity on some physical and mechanical properties of different types of fibreboard. *Eur. J. Wood Wood Prod.* **2009**, *67*, 339–342. [CrossRef]

44. Pritchard, J.; Ansell, M.P.; Thompson, R.J.H.; Bonfield, P.W. Effect of two relative humidity environments on the performance properties of MDF, OSB and chipboard. *Wood Sci. Technol.* **2001**, *35*, 395–403. [CrossRef]

45. McNatt, J.D.; Wellwood, R.W.; Bach, L. Relationships between small-specimen and large panel bending tests on structural wood-based panels. *For. Prod. J.* **1990**, *40*, 10–16.

46. Wu, Q.; Suchsland, O. Effect of moisture on the flexural properties of commercial oriented strandboards. *Wood Fiber Sci.* **1997**, *29*, 47–57.

47. Halligan, A.F.; Schniewind, P. Prediction of particleboard mechanical properties at various moisture contents. *Wood Sci. Technol.* **1974**, *8*, 68–78.

48. Feist, W.C.; Rowell, R.M.; Ellis, W.D. Moisture sorption and accelerated weathering of acetylated and methacrylated aspen. *Wood Fiber Sci.* **1991**, *23*, 128–136.

49. Kojima, Y.; Suzuki, S. Evaluating the durability of wood-based panels using internal bond strength results from accelerated aging treatments. *J. Wood Sci.* **2011**, *57*, 7–13. [CrossRef]

50. Nocetti, M.; Brancheriau, L.; Bacher, M.; Brunetti, M.; Crivellaro, A. Relationship between local and global modulus of elasticity in bending and its consequence on structural timber grading. *Eur. J. Wood Wood Prod.* **2013**, *71*, 297–308. [CrossRef]

51. Akitsu, H.; Norimoto, M.; Morooka, T.; Rowell, R.M. Effect of humidity on vibrational properties of chemically modified wood. *Wood Fiber Sci.* **1993**, *25*, 250–260.

52. Bos, F.; Casagrande, S. On-line non-destructive evaluation and control of wood-based panels by vibration analysis. *J. Sound Vib.* **2003**, *268*, 403–412. [CrossRef]

53. Guan, C.; Liu, J.; Zhang, H.; Wang, X.; Zhou, L. Evaluation of modulus of elasticity and modulus of rupture of full-size wood composite panels supported on two nodal-lines using a vibration technique. *Constr. Build. Mater.* **2019**, *218*, 64–72. [CrossRef]

54. Souza, F.V.; Castro, L.S.; Camara, S.L.; Allen, D.H. Finite-element modeling of damage evolution in heterogeneous viscoelastic composites with evolving cracks by using a two-way coupled multiscale model. *Mech. Compos. Mater.* **2011**, *47*, 95–108. [CrossRef]

55. Šliseris, J.; Andrä, H.; Kabel, M.; Dix, B.; Plinke, B.; Wirjadi, O.; Frolovs, G. Numerical prediction of the stiffness and strength of medium density fiberboards. *Mech. Mater.* **2014**, *79*, 73–84. [CrossRef]

56. Huč, S.; Hozjan, T.; Svensson, S. Rheological behavior of wood in stress relaxation under compression. *Wood Sci. Technol.* **2018**, *52*, 793–808. [CrossRef]

Article

Structural Application of Eco-Friendly Composites from Recycled Wood Fibres Bonded with Magnesium Lignosulfonate

Petar Antov [1,*] , Vassil Jivkov [2], Viktor Savov [1], Ralitsa Simeonova [2] and Nikolay Yavorov [3]

[1] Department of Mechanical Wood Technology, Faculty of Forest Industry, University of Forestry, 1797 Sofia, Bulgaria; victor_savov@ltu.bg

[2] Department of Interior and Furniture Design, Faculty of Forest Industry, University of Forestry, 1797 Sofia, Bulgaria; v_jivkov@ltu.bg (V.J.); r_simeonova@ltu.bg (R.S.)

[3] Department of Pulp, Paper and Printing Arts, University of Chemical Technology and Metallurgy, 1797 Sofia, Bulgaria; yavorof@uctm.edu

* Correspondence: p.antov@ltu.bg

Received: 14 October 2020; Accepted: 24 October 2020; Published: 26 October 2020

Abstract: The pulp and paper industry generates substantial amounts of solid waste and wastewater, which contain waste fibres. The potential of using these recycled wood fibres for producing eco-friendly composites that were bonded with a formaldehyde-free adhesive (magnesium lignosulfonate) and their use in structural applications was evaluated in this study. Fibreboards were produced in the laboratory with a density of 720 $kg \cdot m^{-3}$ and 15% magnesium lignosulfonate gluing content, based on the dry fibres. The mechanical properties (bending strength, modulus of elasticity and internal bond strength), physical properties (thickness swelling and water absorption) and formaldehyde content were determined and compared with the European Standards requirements for wood-based panels. In general, the laboratory-produced panels demonstrated acceptable mechanical properties, such as bending strength (18.5 $N \cdot mm^{-2}$) and modulus of elasticity (2225 $N \cdot mm^{-2}$), which were higher than the minimum requirements for type P2 particleboards and equal to the requirements for MDF panels. The moisture properties, i.e., thickness swelling (24 h) and water absorption (24 h) significantly deteriorated. The free formaldehyde content of the laboratory-produced composites (1.1 mg/100 g) reached the super E0 grade ($\leq$1.5 mg/100 g), which allowed for their classification as eco-friendly, low-emission wood-based composites. The L-type corner joints, made from the developed composites, demonstrated significantly lower bending capacity (from 2.5 to 6.5 times) compared to the same joints made from MDF panels. Nevertheless, the new eco-friendly composites can be efficiently utilised as a structural material in non-load-bearing applications.

Keywords: wood composites; recycled fibres; bioadhesives; magnesium lignosulfonate; corner joints; bending strength capacity

1. Introduction

The depletion of fossil fuels and the increasing concern about the negative effects of global warming on the environment and human health are the main driving forces for the development of a bio-economy [1]. In this respect, the resource efficiency optimisation and the valorisation of lignocellulosic biomass in different high-value products from renewable sources as an alternative to their petroleum-based equivalents are one of the key objectives for implementing the circular economy principles in the wood-based panel industry [2–4]. Significant amounts of non-hazardous solid waste and sludge are generated annually at pulp and paper facilities worldwide, which require further utilisation [5–8]. This solid and liquid discharge contains residual fibres and represents a potential feedstock that can be utilised in the production of new panels [9–12].

Traditional thermosetting adhesives that are used for the production of wood-based composites, such as phenol-formaldehyde, urea-formaldehyde (UF), melamine-urea-formaldehyde and melamine-formaldehyde resins, are derived from the by-products of petroleum processing [13–15]. Currently, about 95% of the total wood adhesives that are used for manufacturing engineered wood composites are based on formaldehyde [16]. UF resins are the most predominant type, accounting for almost 85% of the total worldwide, followed by melamine at 10% and phenolics at 5% [16–18]. These adhesives have been extensively used in the production of wood-based panels because of their exceptional adhesion properties and water resistance, low curing temperatures, chemical versatility and cost-effectiveness [13,15,19–24]. The main disadvantage of these adhesives is the release of volatile organic compounds and formaldehyde from the finished engineered wood panels, especially in indoor applications [25,26], causing adverse human health effects, such as skin and respiratory tract irritation, skin sensitisation, nausea, genotoxicity and sinonasal cancer [27–30]. Formaldehyde emission from engineered wood panels is dependent on endogenic factors, e.g., wood species, the adhesive used, the resin content, the technological conditions employed and the type of hot press [13,16,31,32], and exogenic factors, which are related to the ambient conditions, processing and ageing of wood-based panels [32–35]. Thus, the growing social concern about hazardous formaldehyde emissions from engineered wood panels, along with the increased environmental consciousness related to the sustainability of the final products, have been the main factors for changing the industrial and scientific interest from the traditional petroleum-based synthetic resins to the development of renewable, bio-based and less toxic adhesives for manufacturing eco-friendly wood composites by partially or completely replacing formaldehyde in their compositions [36–48]. Different biomass sources, such as lignin [37,49,50], soy proteins [51,52], starch [53,54] and tannins [55,56], have been used as feedstocks for the development of biobased adhesives.

Lignin is a polyaromatic macromolecule and the second most abundant natural biopolymer in nature, preceded only by cellulose [57,58]. It is also generated as waste or a by-product of the pulp and paper industry, with an annual global production of approximately 50–75 million tons [59]. The main categories of lignin are lignosulfonates obtained from the sulfite pulping, kraft lignin from the sulfate process (Kraft pulping) and organosolv lignin from the production of bioethanol [60,61]. Currently, only 10% of technical lignin is further re-used at an industrial scale, while the rest is mainly burnt for energy, used to recover chemicals or disposed of as a waste [60,62]. Thus, the valorisation of lignin as an abundant and renewable component in the production of value-added products, including wood adhesives [13,42,63–65], could be an efficient way to achieve sustainable resource management. Lignin-based adhesives represent an eco-friendly alternative to the traditional petroleum-based resins as a promising strategy for integrating bio-refineries in the wood-based panel sector. Due to its phenolic hydroxyl groups, lignin is similar to phenol and may be used to partially replace phenol in the composition of phenol formaldehyde adhesives [66–69]. However, further research and chemical modification, such as methylolation and phenolation, is required to increase its lower reactivity to formaldehyde, especially in applications where a fast curing time is needed [37,42].

The production of lignosulfonates, which are the salts of lignin sulfonic acid, dominates the global technical lignin market, where the annual production is approximately 800,000–1.1 million tons of solids [70,71]. They are obtained as by-products from the production of wood pulp by the sulphite lignin processing and are characterised by a very high molecular weight (10,000–40,000 Da) and a high content of ash and sulfur of approximately 4.0–8.0% and 3.5–8.0%, respectively [72]. The current industrial and scientific interest in incorporating lignosulfonates in adhesive formulations for the production of wood composites is due to their potential environmental and health advantages, i.e., the reduction of harmful formaldehyde emissions.

The aim of this study was to evaluate the potential of using novel eco-friendly fibreboards from recycled wood fibres that were bonded with magnesium lignosulfonate in structural applications, such as furniture, interior structures and claddings. For this purpose, the bending strength capacity of L-type corner joints, constructed from the developed composites using four different connection techniques for assembling, i.e., dowels, Confirmat, Minifix and screws, was determined.

2. Materials and Methods

The residual fibre mass, composed of the species Norway spruce (*Picea abies* Karst.) and Scots pine (*Pinus silvestris* L.) and oven-dried to 12% moisture content, was provided by the pulp and paper factory Mondi Stambolyiski EAD. The bulk density of the factory waste fibres was 44.86 kg·m^{-3}. The waste fibres had lengths from 500 to 1000 µm and 7% reduced lignin content.

Magnesium lignosulfonate at 15% gluing content (based on the dry weight of the fibres) was used as a binder. The lignosulfonate additive had the following characteristics: total solids content—51.2%, magnesium content—6%, reduced sugars—7% and sulfate content—2%. European beech (*Fagus sylvatica* L.) veneers, made using the centric peeling process with an average thickness of 1.13 mm and a moisture content after drying and conditioning of approximately 7%, were supplied by the factory Welde Bulgaria AD, and used for veneering the laboratory-produced fibreboards.

Under laboratory conditions, the fibreboards were produced with a 16 mm thickness and a 720 kg·m^{-3} target density. Residual fibres were mixed with the magnesium lignosulfonate in a high-speed laboratory adhesive mixer at 850 min^{-1}. The hot pressing process was carried out using a single opening hydraulic press (PMC ST 100, Italy). The press temperature was 210 °C. The following four-stage pressing regime was used: in the first stage, the pressure applied was 4.5 MPa for 1 min; in the second stage, the pressure was steadily decreased to 2.23 MPa for 3 min; in the third stage, the pressure was decreased to 0.74 MPa for 10 min. The fourth pressing stage was performed at a pressure of 1.78 MPa for 2 min. The fabricated composites, after pressing, were conditioned for 10 days at 20 ± 2 °C and 60–70% relative humidity.

The veneer sheets with dimensions of 400 × 400 mm were cut. After that, the fibreboards were veneered using magnesium lignosulfonate at the content of 80 g·m^{-2} as an adhesive. A hand roller was used to apply the binder to the veneers in order to achieve a uniform adhesive layer. The veneering press factor used was 1 min·mm^{-1} of board thickness; the pressure applied was 0.6 MPa at 200 °C.

The physical and mechanical properties of the fabricated panels (Figure 1) were tested according to the European Standards EN 310, EN 317, EN 322 and EN 323 [73–76]. A precision laboratory balance Kern (Kern & Sohn GmbH, Balingen, Germany) with an accuracy of 0.01 g was used to determine the mass of the test specimens. The dimensions of the test pieces were measured using digital callipers with an accuracy of 0.01 mm. The physical properties (water absorption and thickness swelling) were measured after 24 h of immersion in water. The thickness swelling was assessed using the differences between the initial and final panel thicknesses, and the water absorption was determined using the difference in weight. The mechanical properties of the panels were determined using a universal testing machine Zwick/Roell Z010 (Zwick/Roell GmbH, Ulm, Germany).

Figure 1. Eco-friendly panels from industrial waste fibres bonded with magnesium lignosulfonate and veneered with beech veneers; 720 kg·m^{-3} target density, 18 mm thickness and 15% lignosulfonate content.

The formaldehyde emission of the laboratory-produced panels was tested in the laboratory of Kronospan Bulgaria EOOD (Veliko Tarnovo, Bulgaria) on four test specimens in accordance with the standard perforator method [77].

For the evaluation of the bending strength of the joints made of the veneered eco-friendly composites from recycled wood fibres, L-shape test samples were prepared with the following dimensions: a length $L_1 = L_2 = 106.4$ mm, a width $b = 100$ mm and a thickness $\delta_1 = \delta_2 = 18$ mm, according to the test method [78] and as given in Figure 2.

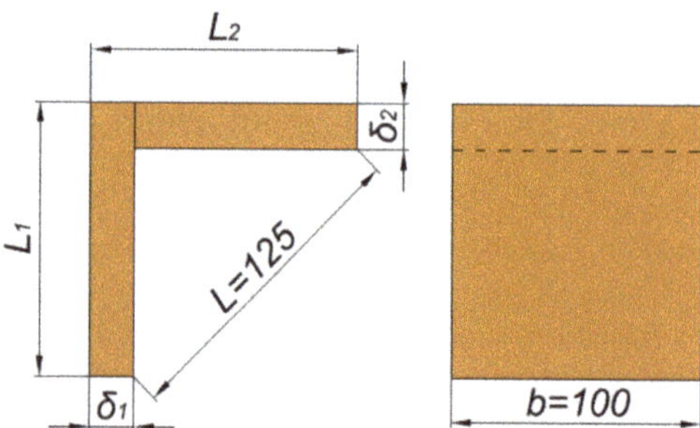

Figure 2. Type and dimensions of the tested samples [78].

Four types of end corner joints were selected, which are among the most used in furniture and interior constructions, where one was fixed with an adhesive and three with dismountable joints. Dowels with the dimensions 8×30 mm and glued with polyvinyl acetate emulsion adhesive (PVA) represented the fixed joints. The following connectors were used in this study as dismountable joints: Minifix with connecting bolts made of steel B 34 for direct screwing (Figure 3a), Confirmat one-piece connector 7×50 mm for screwing in holes Ø 5 mm (Figure 3b) and mounting screws for wood 3.5×50 mm (Figure 3c).

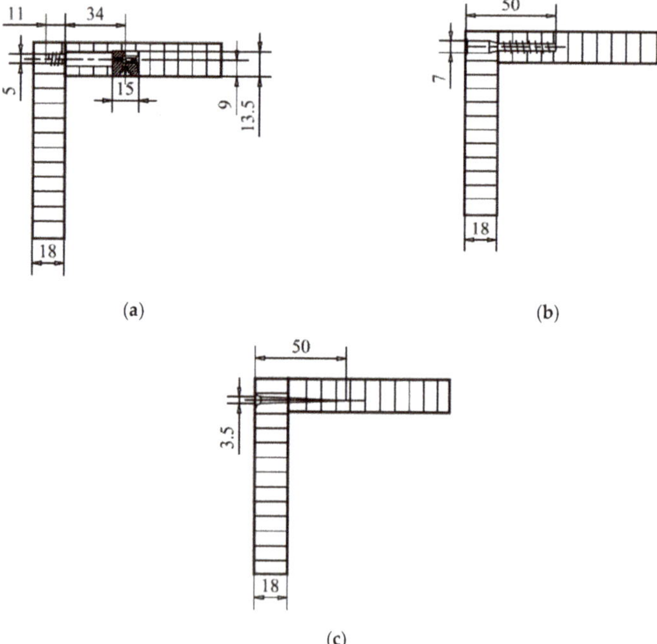

Figure 3. Types of joints: (**a**) Minifix, (**b**) one-piece connector Confirmat ø7 × 50 mm and (**c**) screw for wood ø3.5 × 50 mm.

Seventeen test specimens were made using dowels, Minifix and Confirmat joints, and nine specimens were made with screw joints. All specimens were tested under compression bending loading according to the test scheme presented in Figure 4 [78].

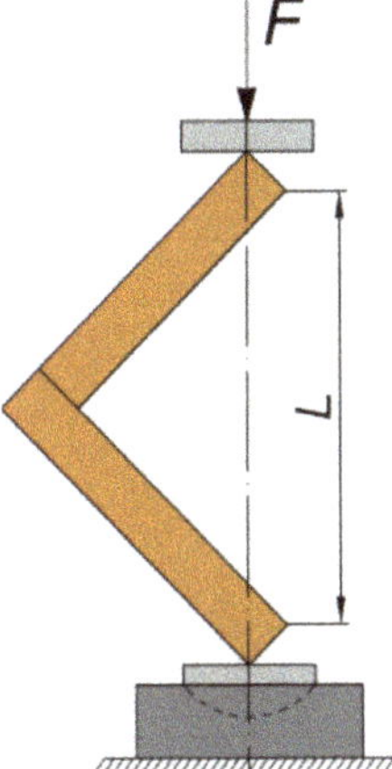

Figure 4. Type of loading of the tested samples [78].

The ultimate bending moment was determined according to the formula:

$$M_{\mathrm{max}} = F_{\mathrm{max}} \cdot l, \tag{1}$$

where:

F_{max} is the maximum value of the loading force (N).

l is arm of bending (m).

The results from a previous study [79] were used to compare the results obtained in the present one, where the same types of joints, made from beech veneered MDF panels with a thickness of 18 mm were tested under compression bending loading.

A statistical analysis of the results was done with XLSTAT (version 2020.2.3, Addinsoft Inc., New York, NY, USA). One-way ANOVA was performed on the results for the bending strength of the L-type corner joints for the analysis of variance at a 95% confidence interval ($p < 0.05$). The statistical differences between the mean values were evaluated using the post hoc Tukey HSD (honestly significant difference) test.

3. Results and Discussion

3.1. Formaldehyde Content

The free formaldehyde content of the fabricated composites, tested in accordance with the standard EN ISO 12460-5, was determined to be 1.1 ± 0.1 mg/100 g, i.e., the super E0 emission grade (≤1.5 mg/100 g) was achieved. This value was remarkably low and can be considered as a zero formaldehyde content [13,77]. Taking into account the fact that natural wood releases low but still measurable amounts of formaldehyde [80], which is formed by its main components and extractives [81–83] at approximately 0.5 to 2 mg/100 g [13,37], our result allowed for defining the produced panels as eco-friendly composites.

3.2. Mechanical and Physical Properties

The results for the mechanical and physical properties of the panels, which consisted of industrial waste fibres bonded with magnesium lignosulfonate, are presented in Table 1. The density of the laboratory-produced panels varied from 681 to 772 kg·m^{-3}, which was rather close to the targeted value. The difference in this main characteristic of the panels was significantly below 5%; thus, it did not have an effect on the mechanical and physical properties.

Table 1. Mechanical and physical properties of the composites produced (MOR: modulus of rupture).

Panel No.	Density ρ, (kg·m^{-3})	Water Absorption (24 h) A, (%)	Thickness Swelling (24 h) Gt, (%)	Bending Strength (MOR) fm, (N·mm^{-2})	Modulus of Elasticity (MOE) Em, (N·mm^{-2})	Internal Bond Strength ft, (N·mm^{-2})
1	772 ± 8.72	150.97 ± 4.87	76.74 ± 2.81	17.0 ± 0.67	1990 ± 53	0.14 ± 0.03
2	736 ± 9.71	175.52 ± 4.25	78.24 ± 2.35	18.9 ± 0.76	2260 ± 61	0.12 ± 0.01
3	681 ± 6.78	184.36 ± 5.73	83.63 ± 2.76	20.6 ± 1.32	2450 ± 72	0.14 ± 0.02
4	757 ± 10.12	144.65 ± 5.91	83.63 ± 2.95	17.0 ± 1.49	1930 ± 49	0.14 ± 0.02
5	743 ± 9.36	180.06 ± 4.74	89.08 ± 3.03	20.6 ± 0.93	2280 ± 67	0.12 ± 0.01
6	751 ± 6.92	163.36 ± 6.01	94.08 ± 3.12	16.6 ± 1.37	2180 ± 56	0.14 ± 0.02
7	737 ± 8.53	178.99 ± 5.38	75.58 ± 2.24	17.8 ± 1.52	2320 ± 42	0.15 ± 0.03

This study followed our previous research work [44], in which the mechanical and physical properties of the fabricated panels, produced from waste fibres and magnesium lignosulfonate as a binder, were investigated in more detail and compared with the European Standard requirements for common engineered wood panels [84,85]. Generally, the laboratory-produced panels demonstrated acceptable mechanical properties in terms of the bending strength (MOR) (18.5 N·mm^{-2}) and modulus of elasticity (MOE) (2225 N·mm^{-2}), which were greater than the lowest requirements for type P2 particleboards and equal to the minimum requirements for MDF panels. The deteriorated dimensional stability, i.e., water absorption (24 h) and thickness swelling (24 h), was the main drawback of the fabricated composites.

3.3. Bending Strength of the L-type Corner Joints Made of the Fabricated Eco-friendly Composites

The results of the bending strength are given in Table 2 and the boxplot graphic in Figure 5. The performance of the joints made of the eco-friendly composites with a Minifix connector was unsatisfactory. In practice, they cannot be used as connecting elements due to the fact that the bolt of Minifix reacts with a limited part of the panel and the pullout strength of this part of the connector is significantly less. The rest of the tested joints showed sufficient bending capacity. The highest bending capacity was demonstrated by the joints with 8 × 30 mm dowels (8.02 N·m), followed by the Confirmat ø7 × 50 mm (6.95 N·m) and screws for wood ø3.5 × 50 mm (4.94 N·m). According to Tukey's test, a significant difference was observed only between the screw joints and the group of dowels and the Confirmat joints. The low strength of the joint with wood screws was due to the inhomogeneity of the eco-friendly panels fabricated from residual wood fibres. In contrast, the corner joints made from MDF panels exhibited an almost equal strength to the joints with the Confirmat and screws due to the homogeneity of the panels. This is proof that the small diameter of the screws had an adverse effect on the strength of the joint. Dowel joints constructed from eco-friendly composites showed almost 2.5 times lower bending capacity compared with the same joints made from MDF panels, while the joints with the Confirmat had a difference of 5 times and the joint with screws had a difference of 6.5 times.

Table 2. Descriptive statistics for the bending strength capacity of L-type corner joints constructed from eco-friendly composites (Eco) and MDF.

Statistic	Dowels Eco	Confirmat Eco	Screw Eco	Minifix MDF	Rafix MDF	Dowels MDF	Screw MDF	Confirmat MDF
No. of observations	17	17	9	15	15	15	15	17
Mean (N·m)	8.02	6.95	4.94	9.75	6.02	19.58	32.06	33.42
Minimum (N·m)	5.45	4.31	4.12	8.13	5.00	17.19	24.06	27.19
Maximum (N·m)	10.40	11.32	6.52	11.25	6.88	21.56	39.38	39.37
Median (N·m)	8.01	6.49	4.56	9.38	6.25	19.69	30.94	33.75
Variance (N·m)	2.07	4.47	0.61	1.11	0.44	1.53	19.97	11.99
St. Dev. (N·m)	1.44	2.12	0.78	1.05	0.66	1.24	4.47	3.46
Covariance	0.18	0.30	0.16	0.11	0.11	0.06	0.14	0.10

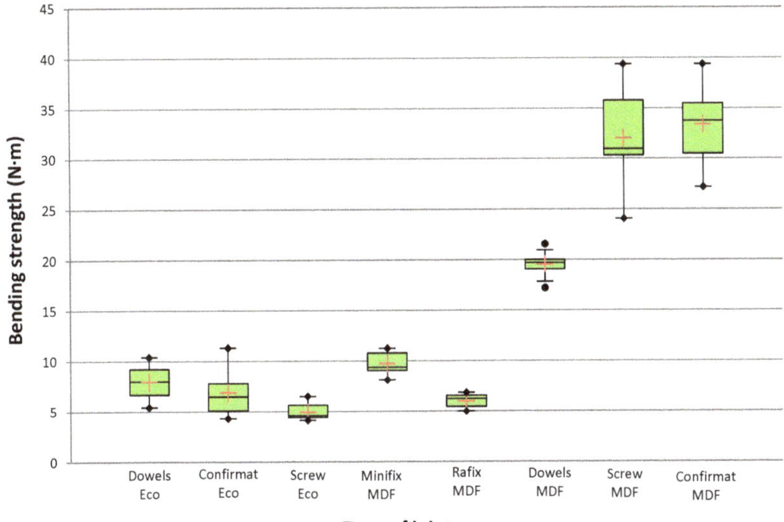

Figure 5. Bending strength under the compression test of L-type corner joints, which were constructed from eco-friendly composites and MDF.

From the statistical analysis of the one-way ANOVA test and the pairwise comparison performed with the Tukey HSD test, a significant difference of $\alpha = 0.05$ at a confidence level of 95% was found in four groups between the obtained bending strength of the L-type end corner joints constructed from eco-friendly composites and MDF panels. The groups are presented in Table 3. All joints made of the eco-friendly composites fell in group "D," together with Rafix in MDF. The dowel joints in the eco-friendly composites were in one group with Minifix in MDF.

Table 3. Tukey honestly significant difference (HSD) analysis of the differences between the groups with a confidence interval of 95% of the bending capacity of joints of two materials of all pairwise comparisons.

Type of Joint	Groups of Homogeneities ($\alpha = 0.05$)			
	A	**B**	**C**	**D**
Confirmat MDF	33.417			
Screw MDF	32.063			
Dowels MDF		19.583		
Minifix MDF			9.750	
Dowels Eco			8.016	8.016
Confirmat Eco				6.951
Rafix MDF				6.021
Screw Eco				4.939

4. Conclusions

Eco-friendly fibreboards with satisfactory mechanical and physical properties according to EN standards were produced from recycled fibres from the pulp and paper industry and bonded with magnesium lignosulfonate at 15% gluing content. The MOR and MOE values of the laboratory-produced composites were quite acceptable in comparison with the standard requirements for type P2 particleboards and almost equal to the minimum requirements for MDF panels [84,85].

The produced composites demonstrated a remarkably low formaldehyde content. The emission values, measured in accordance with the standard perforator method, were $\approx$1.1 mg/100 g, i.e., the composites reached the super E0 emission class, which allowed for their classification as eco-friendly, low-emission composites [77].

The main disadvantage of the laboratory-produced composites was their deteriorated dimensional stability, i.e., water absorption and thickness swelling (24 h). Hence, future investigations should focus on optimising the technological parameters and improving the lignosulfonate formula by adding suitable crosslinkers or catalysts.

The eco-friendly composites produced in this work can be used as a structural material with a limited application only, i.e., when applying minimal loads to the construction or as decorative panels. The Minifix connector was not suitable for joining the eco-friendly laboratory-produced composites. The tested L-type corner joints made of the fabricated eco-friendly composites demonstrated 2.5 to 6.5 times lower bending capacity compared to the same type of joints made of MDF panels with the same connectors. A higher bending capacity of the joints was achieved by improving the homogeneity of the composites and edgebanding of the structural elements.

Author Contributions: Conceptualisation, P.A., V.J. and V.S.; methodology, P.A., V.J. and V.S.; validation, P.A. and V.J.; investigation, V.J., V.S., R.S. and N.Y.; resources, P.A. and V.S.; writing the original draft preparation, P.A. and V.J.; writing—review and editing, P.A. and V.J.; visualisation, P.A., V.J. and V.S.; supervision, P.A. and V.J.; project administration, P.A. All authors have read and agreed to the published version of the manuscript.

Funding: This research was funded by project no. Н С--1002/03.2019 *"Exploitation Properties and Possibilities for Utilisation of Eco-friendly Bio-composite Materials"*, which was carried out at the facilities of the University of Forestry (Sofia, Bulgaria), in cooperation with the University of Chemical Technology and Metallurgy (Sofia, Bulgaria).

Conflicts of Interest: The authors declare no conflict of interest.

References

1. Arias, A.; González-García, S.; González-Rodríguez, S.; Feijoo, G.; Moreira, M.T. Cradle-to-gate Life Cycle Assessment of bio-adhesives for the wood panel industry. A comparison with petrochemical alternatives. *Sci. Total Environ.* **2020**, *738*, 140357. [CrossRef] [PubMed]

2. Vis, M.; Mantau, U.; Allen, B. (Eds.) *Study on the Optimised Cascading Use of Wood, No 394/PP/ENT/RCH/14/7689, Final Report*; Publications Office: Brussels, Belgium, 2016; p. 337.

3. Nitzsche, R.; Budzinski, M.; Gröngröft, A. Techno-economic assessment of a woodbased biorefinery concept for the production of polymer-grade ethylene, organosolv lignin and fuel. *Bioresour. Technol.* **2016**, *200*, 928–939. [CrossRef] [PubMed]

4. Radoykova, T.; Nenkova, S.; Valchev, I. Balck liquor lignin products, isolation and characterization. *J. Chem. Technol. Metall.* **2013**, *48*, 524–529.

5. Bajpai, P. Generation of Waste in Pulp and Paper Mills. In *Management of Pulp and Paper Mill Waste*; Springer: Berlin/Heidelberg, Germany, 2015; ISBN 978-3-319-11788-1.

6. Krigstin, S.; Sain, M. Characterization and potential utilization of recycled paper mill sludge. *Pulp. Paper Can.* **2006**, *107*, 29–32.

7. Abubakr, S.; Smith, A.; Scott, G. Sludge characteristics and disposal alternatives for the pulp and paper industry. Madison. In Proceedings of the 1995 International Environmental Conference, Atlanta, GA, USA, 7–10 May 1995; pp. 269–279.

8. Ochoa de Alda, J.A.G. Feasibility of recycling pulp and pulp and paper sludge in the paper and board industries. *Resour. Conserv. Recycl.* **2008**, *52*, 965–972. [CrossRef]

9. Antov, P.; Savov, V. Possibilities for manufacturing eco-friendly medium density fibreboards from recycled fibres–A review. In Proceedings of the 30th International Conference on Wood Science and Technology-ICWST 2019 "Implementation of Wood Science in Woodworking Sector" and 70th Anniversary of Drvna Industrija Journal, Zagreb, Croatia, 12–13 December 2019; ISBN 978-953-292-062-8.

10. Davis, E.; Shaler, S.M.; Goodell, B. The incorporation of paper deinking sludge into fibreboard. *For. Prod. J.* **2003**, *53*, 46–54.

11. Geng, X.; Deng, J.; Zhang, S.Y. Characteristics of pulp and paper sludge and its utilization for the manufacture of medium density fibreboard. *Wood Fiber. Sci.* **2007**, *39*, 345–351.

12. Migneault, S.; Koubaa, A.; Nadji, H.; Riedl, B.; Zhang, S.Y.; Deng, J. Medium-density fibreboard produced using pulp and paper sludge from different pulping processes. *Wood Fiber. Sci.* **2010**, *42*, 292–303.

13. Mantanis, G.I.; Athanassiadou, E.T.; Barbu, M.C.; Wijnendaele, K. Adhesive systems used in the European particleboard, MDF and OSB industries. *Wood Mater. Sci. Eng.* **2018**, *13*, 104–116. [CrossRef]

14. Youngquist, J.A. Wood-based composites and panel products. In *Wood Handbook: Wood as an Engineering Material*; USDA Forest Service, Forest Products Laboratory: Madison, WI, USA, 1999; pp. 1–31.

15. Frihart, C.R. Wood adhesion and adhesives. In *Handbook of Wood Chemistry and Wood Composites*; Rowell, R.M., Ed.; CRC Press: Boca Raton, FL, USA, 2005; pp. 214–278.

16. Kumar, R.N.; Pizzi, A. Environmental Aspects of Adhesives–Emission of Formaldehyde. In *Adhesives for Wood and Lignocellulosic Materials*; Wiley-Scrivener Publishing: Hoboken, NJ, USA, 2019; pp. 293–312.

17. Byung-Dae, P.; Jae-Woo, K. Dynamic mechanical analysis of urea-formaldehyde resin adhesives with different formaldehyde-to-urea molar ratios. *J. Appl. Polym. Sci.* **2008**, *108*, 2045–2051.

18. Costa, N.; Pereira, J.; Ferra, J.; Cruz, P.; Martins, J.; Magalhães, F.; Mendes, A.; Carvalho, L.H. Scavengers for achieving zero formaldehyde emission of wood based panels. *Wood Sci. Technol.* **2013**, *47*, 1261–1272. [CrossRef]

19. Ružiak, I.; Igaz, R.; Krišťák, L.; Réh, R.; Mitterpach, J.; Očkajová, A.; Kučerka, M. Influence of Urea-formaldehyde Adhesive Modification with Beech Bark on Chosen Properties of Plywood. *BioResources* **2017**, *12*, 3250–3264. [CrossRef]

20. Yang, M.; Rosentrater, K.A. Life cycle assessment and techno-economic analysis of pressure sensitive bio-adhesive production. *Energies* **2019**, *12*, 4502. [CrossRef]

21. Dunky, M. Adhesives in the wood industry. In *Handbook of Adhesive Technology*; Pizzi, A., Mittal, K.L., Eds.; Marcel Dekker: New York, NY, USA, 2003; Chapter 47, pp. 872–941.

22. Jivkov, V.; Simeonova, R.; Marinova, A.; Gradeva, G. Study on the gluing abilities of solid surface composites with different wood based materials and foamed PVC. In Proceedings of the 24th International Scientific Conference Wood Is Good–User Oriented Material, Technology and Design, Zagreb, Croatia, 18 October 2013; pp. 49–55, ISBN 978-953-292-031-4.

23. Solt, P.; Konnerth, J.; Gindl-Altmutter, W.; Kantner, W.; Moser, J.; Mitter, R.; Van Herwijnen, H. Technological performance of formaldehyde-free adhesive alternatives for particleboard industry. *Int. J. Adhes. Adhes.* **2019**, *94*, 99–131. [CrossRef]

24. Zhang, W.; Ma, Y.; Wang, C.; Li, S.; Zhang, M.; Chu, F. Preparation and properties of lignin-phenol-formaldehyde resins based on different biorefinery residues of agricultural biomass. *Ind. Crops Prod.* **2013**, *43*, 326–333. [CrossRef]

25. Tudor, E.M.; Barbu, M.C.; Petutschnigg, A.; Réh, R.; Krišťák, Ľ. Analysis of Larch-Bark Capacity for Formaldehyde Removal in Wood Adhesives. *Int. J. Environ. Res. Public Health* **2020**, *17*, 764. [CrossRef] [PubMed]

26. Tudor, E.M.; Dettendorfer, A.; Kain, G.; Barbu, M.C.; Réh, R.; Krišťák, Ľ. Sound-Absorption Coefficient of Bark-Based Insulation Panels. *Polymers* **2020**, *12*, 1012. [CrossRef]

27. U.S. Consumer Product Safety Commission. *An Update on Formaldehyde (Publication 725)*; U.S. Consumer Product Safety Commission: Bethesda, MD, USA, 2013.

28. Kim, K. Environment-friendly adhesives for surface bonding of wood-based flooring using natural tannin to reduce formaldehyde and TVOC emission. *Bioresour. Technol.* **2009**, *100*, 744–748. [CrossRef]

29. International Agency for Research on Cancer. *IARC Classifies Formaldehyde as Carcinogenic to Humans*; IARC: Lyon, France, 2004.

30. Kelly, T.J. *Determination of Formaldehyde and Toluene Diisocyanate Emissions from Indoor Residential Sources (No. 97-9)*; California Environmental Protection Agency: Sacramento, CA, USA, 1997.

31. Réh, R.; Igaz, R.; Krišťák, L.; Ružiak, I.; Gajtanska, M.; Božíková, M.; Kučerka, M. Functionality of Beech Bark in Adhesive Mixtures Used in Plywood and Its Effect on the Stability Associated with Material Systems. *Materials* **2019**, *12*, 1298. [CrossRef]

32. Antov, P.; Savov, V.; Neykov, N. Reduction of Formaldehyde Emission from Engineered Wood Panels by Formaldehyde Scavengers—A Review. In Proceedings of the 13th International Scientific Conference WoodEMA 2020 and 31st International Scientific Conference ICWST 2020 Sustainability of Forest-based Industries in the Global Economy, Vinkovci, Croatia, 28–30 September 2020; pp. 7–11.

33. Sundin, B.; Risholm-Sundman, M.; Edenholh, K. Emission of formaldehyde and other volatile organic compounds (VOC) from sawdust and lumber, different wood-based panels and other building materials: A comparative study. In Proceedings of the 26th International Particleboard/Composite Materials Symposium, Washington State University, Pullman, DC, USA, 7–9 April 1992.

34. Roffael, E.; Johnsson, B.; Engström, B. On the measurement of formaldehyde release from low-emission wood-based panels using the perforator method. *Wood Sci. Technol.* **2010**, *44*, 369–377. [CrossRef]

35. Kovatchev, G. Influence of the belt type over vibration of the cutting mechanism in woodworking shaper. In Proceedings of the 11th International Science Conference "Chip and Chipless Woodworking Processes", Zvolen, Slovakia, 13–15 September 2018; pp. 105–110.

36. Widyorini, R.; Xu, J.; Umemura, K.; Kawai, S. Manufacture and properties of binderless particleboard from bagasse I: Effects of raw material type, storage methods, and manufacturing process. *J. Wood Sci.* **2005**, *51*, 648–654. [CrossRef]

37. Pizzi, A. Recent developments in eco-efficient bio-based adhesives for wood bonding: Opportunities and issues. *J. Adhes. Sci. Technol.* **2006**, *20*, 829–846. [CrossRef]

38. Kües, U. Wood Production, Wood Technology, and Biotechnological Impacts. Universitätsverlag Göttingen. 2007. Available online: https://univerlag.uni-goettingen.de/handle/3/isbn-978-3-940344-11-3 (accessed on 17 September 2020).

39. Papadopoulou, E. Adhesives from renewable resources for binding wood-based panels. *J. Environ. Prot. Ecol.* **2009**, *10*, 1128–1136.

40. Navarrete, P.; Mansouri, H.R.; Pizzi, A.; Tapin-Lingua, S.; Benjelloun-Mlayah, B.; Pasch, H. Wood panel adhesives from low molecular mass lignin and tannin without synthetic resins. *J. Adhes. Sci. Technol.* **2010**, *24*, 1597–1610. [CrossRef]

41. Nordström, E.; Demircan, D.; Fogelström, L.; Khabbaz, F.; Malmström, E. Green binders for wood adhesives. In *Applied Adhesive Bonding in Science and Technology*; Interhopen Books: London, UK, 2017; pp. 47–71. [CrossRef]

42. Hemmilä, V.; Adamopoulos, S.; Karlsson, O.; Kumar, A. Development of sustainable bio-adhesives for engineered wood panels-A review. *RSC Adv.* **2017**, *7*, 38604–38630. [CrossRef]

43. Pizzi, A.; Papadopoulos, A.; Policardi, F. Wood Composites and Their Polymer Binders. *Polymers* **2020**, *12*, 1115. [CrossRef]

44. Antov, P.; Mantanis, G.I.; Savov, V. Development of wood composites from recycled fibres bonded with magnesium lignosulfonate. *Forests* **2020**, *11*, 613. [CrossRef]

45. Taghiyari, H.R.; Tajvidi, M.; Taghiyari, R.; Mantanis, G.I.; Esmailpour, A.; Hosseinpourpia, R. Nanotechnology for wood quality improvement and protection. In *Nanomaterials for Agriculture and Forestry Applications*; Husen, A., Jawaid, M., Eds.; Elsevier: Amsterdam, The Netherlands, 2020; pp. 469–489.

46. Hosseinpourpia, R.; Adamopoulos, S.; Mai, C.; Taghiyari, H.R. Properties of medium-density fiberboards bonded with dextrin-based wood adhesives. *Wood Res.* **2019**, *64*, 185–194.

47. Antov, P.; Savov, V.; Neykov, N. Sustainable Bio-based Adhesives for Eco-Friendly Wood Composites. A Review. *Wood Res.* **2020**, *65*, 51–62. [CrossRef]

48. Valyova, M.; Ivanova, Y.; Koynov, D. Investigation of free formaldehyde quantity in the production of plywood with modified ureaformaldehyde resin. *Wood Des. Technol.* **2017**, *6*, 72–77.

49. Li, R.J.; Gutierrez, J.; Chung, Y.; Frank, C.W.; Billington, S.L.; Sattely, E.S. A lignin-epoxy resin derived from biomass as an alternative to formaldehyde-based wood adhesives. *Green Chem.* **2018**, *20*, 1459–1466. [CrossRef]

50. Gadhave, R.V.; Srivastava, S.; Mahanwar, P.A.; Gadekar, P.T. Lignin: Renewable Raw Material for Adhesive. *Open J. Polym. Chem.* **2019**, *9*, 27–38. [CrossRef]

51. Pizzi, A. Wood products and green chemistry. *Ann. For. Sci.* **2016**, *73*, 185–203. [CrossRef]

52. Wang, Z.; Zhao, S.; Pang, H.; Zhang, W.; Zhang, S.; Li, J. Developing eco-friendly high-strength soy adhesives with improved ductility through multiphase core–shell hyperbranched polysiloxane. *ACS Sustain. Chem. Eng.* **2019**, *7*, 7784–7794. [CrossRef]

53. Li, Z.; Wang, J.; Li, C.; Gu, Z.; Cheng, L.; Hong, Y. Effects of montmorillonite addition on the performance of starch-based wood adhesive. *Carbohydr. Polym.* **2015**, *115*, 394–400. [CrossRef]

54. Wang, P.; Cheng, L.; Gu, Z.; Li, Z.; Hong, Y. Assessment of starch-based wood adhesive, quality by confocal Raman microscopic detection of reaction homogeneity. *Carbohydr. Polym.* **2015**, *131*, 75–79. [CrossRef]

55. Santos, J.; Antorrena, G.; Freire, M.S.; Pizzi, A.; González-Álvarez, J. Environmentally friendly wood adhesives based on chestnut (*Castanea sativa*) shell tannins. *Eur. J. Wood Wood Prod.* **2017**, *75*, 89–100. [CrossRef]

56. Ndiwe, B.; Pizzi, A.; Tibi, B.; Danwe, R.; Konai, N.; Amirou, S. African tree bark exudate extracts as biohardeners of fully biosourced thermoset tannin adhesives for wood panels. *Ind. Crops Prod.* **2019**, *132*, 253–268. [CrossRef]

57. Lora, J.H.; Glasser, W.G. Recent industrial applications of lignin: A sustainable alternative to nonrenewable materials. *J. Polym. Environ.* **2002**, *10*, 39–48. [CrossRef]

58. Sharma, S.; Kumar, A. (Eds.) *Lignin: Biosynthesis and Transformation for Industrial Applications*; Springer Series on Polymer and Composite Materials, Switzerland AG; Springer Nature: Berlin/Heidelberg, Germany, 2020.

59. Mandlekar, N.; Cayla, A.; Rault, F.; Giraud, S.; Salaün, F.; Malucelli, G.; Guan, J.-P. An overview on the use of lignin and its derivatives in fire retardant polymer systems, lignin-trends and applications, 2018, InTech. Matheus Poletto, IntechOpen. *Lignin Trends Appl.* **2018**. [CrossRef]

60. Bajwa, D.S.; Pourhashem, G.; Ullah, A.H.; Bajwa, S.G. A concise review of current lignin production, applications, products and their environmental impact. *Ind. Crop. Prod.* **2019**, *139*, 111526. [CrossRef]

61. Pizzi, A. Bioadhesives for wood and fibres. *Rev. Adhes. Adhes.* **2013**, *1*, 88–113. [CrossRef]

62. Waldron, K. *Advances in Biorefineries: Biomass and Waste Supply Chain Exploitation*; Woodhead Publishing: London, UK, 2014.

63. Ferdosian, F.; Pan, Z.; Gao, G.; Zhao, B. Bio-based adhesives and evaluation for wood composites application. *Polymers* **2017**, *9*, 70. [CrossRef]

64. Yotov, N.; Valchev, I.; Petrin, S.; Savov, V. Lignosulphonate and waste technical hydrolysis lignin as adhesives for eco-friendly fibreboard. *Bulg. Chem. Commun.* **2017**, *49*, 92–97.

65. Klapiszewski, Ł.; Oliwa, R.; Oleksy, M.; Jesionowski, T. Calcium lignosulfonate as eco-friendly additive of crosslinking fibrous composites with phenol-formaldehyde resin matrix. *Polymers* **2018**, *63*, 102–108. [CrossRef]

66. Jin, Y.; Cheng, X.; Zheng, Z. Preparation and characterization of phenol–formaldehyde adhesives modified with enzymatic hydrolysis lignin. *Bioresour. Technol.* **2010**, *101*, 2046–2048. [CrossRef]

67. Antov, P.; Savov, V.; Mantanis, G.I.; Neykov, N. Medium-density fibreboards bonded with phenol-formaldehyde resin and calcium lignosulfonate as an eco-friendly additive. *Wood Mater. Sci. Eng.* **2020**. [CrossRef]

68. Laurichesse, S.; Avérous, L. Chemical modification of lignins: Towards biobased polymers. *Prog. Polym. Sci.* **2014**, *39*, 1266–1290. [CrossRef]

69. Hemmilä, V.; Adamopoulos, S.; Hosseinpourpia, R.; Sheikh, A.A. Ammonium lignosulfonate adhesives for particleboards with pMDI and furfuryl alcohol as cross-linkers. *Polymers* **2019**, *11*, 1633. [CrossRef]

70. Miller, J.; Faleiros, M.; Pilla, L.; Bodart, A.C. *Lignin: Technology, Applications and Markets, Special Market Analysis Study*; RISI, Inc., Market-Intell LCC: Charlottesville, VA, USA, 2016.

71. Berlin, A.; Balakshin, M. Industrial Lignins: Analysis, Properties, and Applications. In *Bioenergy Research: Advances and Applications*; Chapter 18; Gupta, V.K., Tuohy, M., Kubicek, C., Saddler, J., Xu, F., Eds.; Elsevier: Amsterdam, The Netherlands, 2014; pp. 315–336.

72. Lora, J. Industrial commercial lignins: Sources, properties and applications. In *Monomers, Polymers and Composites from Renewable Resources*; Belgacem, M.N., Gandini, A., Eds.; Elsevier: Amsterdam, The Netherlands, 2008; pp. 225–241.

73. EN 310. *Wood-Based Panels-Determination of Modulus of Elasticity in Bending and of Bending Strength*; European Committee for Standardization: Brussels, Belgium, 1999.

74. EN 317. *Particleboards and Fibreboards-Determination of Swelling in Thickness after Immersion in Water*; European Committee for Standardization: Brussels, Belgium, 1998.

75. EN 322. *Wood-Based Panels-Determination of Moisture Content*; European Committee for Standardization: Brussels, Belgium, 1998.

76. EN 323. *Wood-Based Panels-Determination of Density*; European Committee for Standardization: Brussels, Belgium, 2001.

77. EN ISO 12460-5. *Wood-Based Panels-Determination of Formaldehyde Release–Part 5. Extraction Method (Called the Perforator Method)*; European Committee for Standardization: Brussels, Belgium, 2015.

78. Kyuchukov, G.; Jivkov, V. *Furniture Construction. Structural Elements and Furniture Joints*, 1st ed.; Bismar: Sofia, Bulgaria, 2016; 452p.

79. Jivkov, V.; Grbac, I. Influence of the cyclic loading on bending strength of different end corner joints made of MDF. In Proceedings of the 22nd International Scientific Conference "Wood is good–EU Preaccession Challenges of the Sector", Zagreb, Croatia, 21 October 2011; pp. 59–66.

80. Roffael, E. Volatile organic compounds and formaldehyde in nature, wood and wood based panels. *Holz. Roh. Werkst.* **2006**, *64*, 144–149. [CrossRef]

81. Salem, M.Z.M.; Böhm, M. Understanding of formaldehyde emissions from solid wood: An overview. *BioResources* **2013**, *8*, 4775–4790. [CrossRef]

82. Schäfer, M.; Roffael, E. On the formaldehyde release of wood. *Holz. Roh. Werkst.* **2000**, *58*, 259–264. [CrossRef]

83. Birkeland, M.J.; Lorenz, L.; Wescott, J.M.; Frihart, C.R. Determination of native (wood derived) formaldehyde by the desiccator method in particleboards generated during panel production. *Holzforschung* **2010**, *64*, 429–433. [CrossRef]

84. European Committee for Standardization, EN 312. *Particleboards–Specifications*; European Committee for Standardization: Brussels, Belgium, 2010.

85. European Committee for Standardization, EN 622-5. *Fibreboards-Specifications-Part 5: Requirements for Dry Process Boards*; European Committee for Standardization: Brussels, Belgium, 2010.

Publisher's Note: MDPI stays neutral with regard to jurisdictional claims in published maps and institutional affiliations.

Article

Formaldehyde Emission in Micron-Sized Wollastonite-Treated Plywood Bonded with Soy Flour and Urea-Formaldehyde Resin

Hamid R. Taghiyari [1,*], Seyed Behzad Hosseini [2], Saman Ghahri [2], Mohammad Ghofrani [1] and Antonios N. Papadopoulos [3,*]

[1] Wood Science and Technology Department, Faculty of Materials Engineering & New Technologies, Shahid Rajaee Teacher Training University, Tehran 16788-15811, Iran; ghofrani@sru.ac.ir

[2] Department of Wood and Paper Sciences, Tarbiat Modares University, Tehran 14115-111, Iran; Hosseini_b@modares.ac.ir (S.B.H.); saman.ghahri@modares.ac.ir (S.G.)

[3] Laboratory of Wood Chemistry and Technology, Department of Forestry and Natural Environment, International Hellenic University, GR-661 00 Drama, Greece

* Correspondence: htaghiyari@sru.ac.ir (H.R.T.); antpap@for.ihu.gr (A.N.P.)

Received: 7 September 2020; Accepted: 23 September 2020; Published: 25 September 2020

Abstract: Soy flour was partly substituted for urea-formaldehyde (UF) resin with different content to investigate its effect on formaldehyde emission in three-layer plywood panels. In each square meter of panels, 300 g of resin was used (wet weight basis of resin). Micron-sized wollastonite was added to the resin mixture at 5% and 10% consumption levels (wet weight basis of resin) to determine its potential effects as a reinforcing filler to mitigate the negative effects of addition of soy flour. Results showed a decreasing trend in formaldehyde emission as soy flour content increased to 20%. The highest shear-strength values were observed in panels with 10% and 15% soy flour content. The addition of wollastonite did not have a significant effect on formaldehyde emission, but it decreased the shear strength in soy-treated panels, although the values were still higher than those of control panels. Wollastonite significantly mitigated the negative effects of soy flour on the water absorption and thickness swelling of panels. It was concluded that 10% of soy flour and 5% of wollastonite provided the lowest formaldehyde emission and the most optimum physical and mechanical properties.

Keywords: biobased resins; formaldehyde emission; minerals; wollastonite; wood composite panels

---

1. Introduction

Adhesives play an important role in the efficient utilization of wood resources and in the development and growth of the forest product industry. Adhesive bonding of solid wood and wood particles of various sizes is a key factor for the production of modern, functional wood products, used in a variety of applications. For centuries, wood was bonded using biobased adhesives until synthetic adhesives, mainly thermosetting ones, gradually took over in the 20th century, as they were typically regarded as more effective, cost-efficient [1], and stable for use in humid conditions. Today, the main classes of thermosetting adhesives are amino-based, phenolic, and isocyanate resins. The utilization of these thermosetting adhesives is considered more economical, and reactive adhesives with quick curing behavior are versatile in a range of properties in the cured state. These adhesives have dominated the wood composite industry for many decades. Within this group, urea-formaldehyde (UF) resins are the most important adhesives in terms of quantity. Due to their low-cost raw materials, their rapid curing, their high dry-bond strength, and a colorless glue line, UF-based adhesives are almost exclusively used for producing wood-based materials, such as particleboard or medium-density fiberboards for interior applications [2]. When products are utilized in conditions exhibiting higher humidity, UF resins are usually modified with significantly more expensive compounds such as melamine, phenol, or resorcinol [2]. It has to be mentioned that the final adhesive composition in use depends on the requirements of the wood-based material such as the required strength properties, the expected moisture resistance, the production cost of the finished product, and the desired formaldehyde emission class.

When it comes to formaldehyde emission from wood-based composites, it has to be mentioned that emission can originate from (i) synthetic-free formaldehyde that is not polymerized into the network, which emits during or quickly after panel production, (ii) formaldehyde released due to adhesive hydrolysis, which emits over the lifetime of the panel depending on moisture and temperature, and (iii) biogenic sources [2]. The discussion surrounding formaldehyde started as early as the mid-1960s, as reviewed by Roffael et al. [3], and various stages of reduced emissions were achieved in the 1970s and intensified in the 1980s when the carcinogenicity of formaldehyde in rats and mice after long-term inhalation exposure was reported. The topic of formaldehyde in indoor environments was intensively and controversially discussed by various authors, as reviewed by Salthammer et al. [4]. Driven by the standard requirements specified by the local authorities in, e.g., Europe, Japan, and the United States of America (USA), formaldehyde content and emissions from wood-based composites have been continuously reduced over the last few decades.

During the last few decades, the wood industry made great effort with many innovations in amino resin technology to gradually reduce formaldehyde emissions in order to fulfill the individual product and standard requirements for each type of composite. These can be summarized as follows: (i) low-formaldehyde-content resins, (ii) formaldehyde scavenger additives, (iii) post treatments, and (iv) alternative adhesives [1].

For the first type of approach, the first tendency has been to prepare engineered UF resins of progressively lower molar ratio, at levels much lower than 1:1 [5], which has become rather common in industry today. This is due to the attempt to minimize formaldehyde emissions from wood panels bonded with UF resins. One of the drawbacks of the much lower than 1:1 molar ratio has been identified in an increase in the tendency of the UF resin to form increasingly present crystalline domains upon hardening as a result of hydrogen between linear molecules [6–8]. At higher molar ratios, the hardened resin is amorphous, affording better adhesion and better bonding performance. The overly high crystallinity drawback was very recently solved [9] by blocking the formation of hydrogen bonds using transition metal ion–bentonite nanoclay through in situ intercalation and, thus, converting the crystalline domains of the UF resins to amorphous polymers. Addition of 5% nanoclay to the UF yielded in excess of 50% better adhesion and almost 50% lower formaldehyde emission, thus resulting in a marked improvement in performance with a low level of crystallinity. In the same trend, the potential introduction of a very acidic pH condensation step in the preparation of UF resins, inducing the formation of occasionally considerable amounts of uron (a cyclic intramolecular urea methylene ether) in the UF resins with lower formaldehyde emission, has attracted some research interest [10,11]. This initial work indicated that introduction of such an acid step can lead to UF resins with improved bonding strength, but also higher post-cure formaldehyde emission.

With regard to the second approach, formaldehyde scavengers, capable of capturing formaldehyde either physically or chemically and forming stable products, are added to UF resins or to wood particles before pressing [12,13]. These additives should provide long-term formaldehyde emission reduction, in principle, along the panel's life. Examples used in industry include the addition of urea in an aqueous solution or powder form, organic amines, scavenger resins, sulfites, functionalized paraffin waxes, and porous absorbers such as pozzolan and charcoal [14]. Very good formaldehyde emission reduction is obtainable by adding sodium metabisulfite to the resin, by adding tannin solution to urea-formaldehyde resin, or by using different starch derivatives [15,16].

The third approach involves treatments that are applied after pressing. Currently used methods include panel impregnation with formaldehyde-scavenging species, such as aqueous solutions of ammonia, ammonium salts, or urea. Another option is the creation of diffusional barriers in the panel surfaces that keep formaldehyde confined, by using paints, varnishes, veneers, laminates, or resin-impregnated papers [17,18].

Alternative adhesives involve isocyanate-based adhesives and biobased wood adhesives. The concern about formaldehyde emission vapor levels from UF adhesives has brought isocyanate adhesives to the fore, where formaldehyde emission does not occur as no formaldehyde is added. pMDI (polymeric methylenediphenyl isocyanate) is an excellent adhesive and can be used in markedly smaller proportions than formaldehyde-based adhesives to bind wood composites. Another attractive option is biobased adhesives. The term biobased adhesive has come to be used in a very well specified and narrow sense to only include those materials of natural, nonmineral origin which can be used as such or after small modifications to reproduce the behavior and performance of synthetic resins. Thus, only a limited number of materials can be currently included, at a stretch, in the narrowest sense of this definition. These are tannins, lignin, carbohydrates, unsaturated oils, proteins and protein hydrolysates, dissolved wood, and wood welding by self-adhesion. An excellent review on this topic was recently published by Pizzi et al. [1]

The most common biobased adhesives are protein-based sourced from animal bones and hides, milk (casein), blood, fish skins, and soybeans. Soy protein is obtained from soybean and has been used for centuries as a wood adhesive. In the context of wood composite production, soy protein was added to phenol formaldehyde resin to lower formaldehyde emission, but lower water resistance is an important limitation [19]. Formaldehyde-free wood composites were obtained using an adhesive based on soy flour and glyoxal, a nontoxic, but less reactive, aldehyde [20]. It was reported that the use of soy protein combined with polyamidoamine-epichlorohydrin (PAE) resins yields a strong and water-resistant product that is commercially available for wood composites [21]. Another interesting formaldehyde-free adhesive system, successfully tested in the production of plywood and OSB (Oriented Strand Board) panels, is based on a combination of soy flour, polyethylenimine, maleic anhydride, and sodium hydroxide [19]. Hosseini et al. [22] reported that the partial replacement of urea-formaldehyde adhesive with soy flour, particularly with a substitution rate of 15%, significantly reduced the formaldehyde emission, while it did not significantly influence the shear strength, under both dry and wet conditions. Kawalerczyk et al. [23] applied five types of flours (rye, hemp, coconut, rice, and pumpkin) as fillers with urea-formaldehyde resin in plywood manufacture. It was reported that the type of flour had a major influence on the properties of resin mixture such as gel time, solid content, and viscosity. The use of hemp flour as a filler led to a substantial decrease in free formaldehyde content.

Under this context, the present study was carried out to primarily investigate the effects of partial substitution of soy flour for UF resin and the consequent effects on formaldehyde emission and on some key physical and mechanical properties. In order to minimize possible negative effects on properties due to the use of soy flour, an innovative approach was made in an extra set of boards, to add micron-sized wollastonite. Micron-sized wollastonite is a mineral that was successful in improving the physical and mechanical properties in medium-density fiberboards and particleboards [24–28]. It significantly improved the shear strength of polyvinylacetate resin [29]. It was even effective in improving the fire properties in wood-based composite panels and solid wood [29–35]. Therefore, this study continued with the addition of micron-sized wollastonite to the resin mixture to determine its potential effect on the properties of plywood panels produced with a mixture of UF resin and soy flour.

2. Materials and Methods

2.1. Panel Production

Three-layer plywood panels were produced, using poplar veneer (*Populus deltoides*) with 2.1 mm of thickness. The dimensions of the produced panels were 350 × 350 mm. The target thickness of the panels was 6 mm. For each square meter of panels, 300 g of urea-formaldehyde resin was used (wet weight basis of resin). Glued veneers were hot-pressed for 5 min at 130 °C. The specific pressure of plates was 1.5 MPa, and the total nominal pressure of the plates was 20 MPa. Once produced, 25 mm of each side was trimmed to avoid inconsistent edges (Figure 1). Except for the formaldehyde emission tests, all produced panels were kept in a conditioning chamber (25 ± 3 °C, relative humidity 60–65%) for a week before test specimens were cut to size. The specimens were kept under the same conditions for two more weeks before tests were carried out. The target density of the panels was 0.55 g·m^{-3}.

Figure 1. Panels produced and trimmed, ready to be cut to size for each test according to the relevant standard (**A**); front surface of a trimmed plywood panel (**B**).

2.2. Resin Application

UF resin was purchased from Amol Resin Company (Amol, Iran). The viscosity of the resin was 200–400 cP, with 47 s of gel time, and a density of 1.277 g/cm^3. Defatted soy flour (SF) was purchased from Behpak Company (Behshahr, Iran). SF contained 47% (*w/w*) protein (Ghahri et al. 2016). SF was substituted for UF resin at 5%, 10%, 15%, and 20% (*w/w* dry weight). SF and UF were mixed for 10 min using a magnetic stirrer. Micron-sized wollastonite was prepared by Mehrabadi Machinery Mfg Co. (Tehran, Iran) (Table 1). Wollastonite was mixed with the resin for 10 min at 5% and 10% (*w/w* dry weight basis of resin). Once the resin was prepared, it was applied onto the veneers in less than 1 min. Just before applying the prepared resin on veneers, 1% ammonium chloride was mixed as a hardener (dry weight basis of UF resin). Within 4 min of the application of resin on veneers, the layers were arranged and then hot-pressed.

Table 1. Chemical composition of the wollastonite ingredients on weight basis. (Data from Hassani et al. [35].

Component	Proportion (% *w/w*)
SiO$_2$	46.96
CaO	39.77
Al$_2$O$_3$	3.95
Fe$_2$O$_3$	2.79
TiO$_2$	0.22
K$_2$O	0.04
MgO	1.39
Na$_2$O	0.16
SO$_3$	0.05
Water	4.67

2.3. Measurement of Formaldehyde Emission

Formaldehyde emission (FE) in the present study was measured on the basis of European standard specifications (EN 717-3/Part 3) [36]. From each of the five replicate panels, three specimens were cut for the formaldehyde emission test. Necessary coordination was made so that specimens were tested immediately to minimize sources of error. The dimensions of the FE test specimens were $25 \times 25 \times 6$ mm. Flask type 2 (with a volume of 500 mL) was employed. FE tests were carried out at the temperature of 40 ± 2 °C, for a duration of 180 min. In this test method, FE specimens were hung in vertical position at 40 mm above the distilled water (50 mL) at the bottom of the flask (Figure 2). The flask was then cooled (using a mixture of ice water) for 30 min to ensure absorption of the emitted formaldehyde in the distilled water inside the flask. Acetylacetone spectrophotometric analysis was used to determine the amount of formaldehyde in each flask. The determination of formaldehyde emission was based on the Hantzsch reaction. In this method, aqueous formaldehyde reacted with ammonium ions and acetylacetone to yield diacetyldihydrolutidine (DDL). DDL has a maximum absorption capacity at wavelength of 412 nm. A T60 visible range spectrophotometer was used in the present study, with a fixed 2 nm spectral bandwidth. The absorption amount was then expressed as mg of formaldehyde/kg of dry wood.

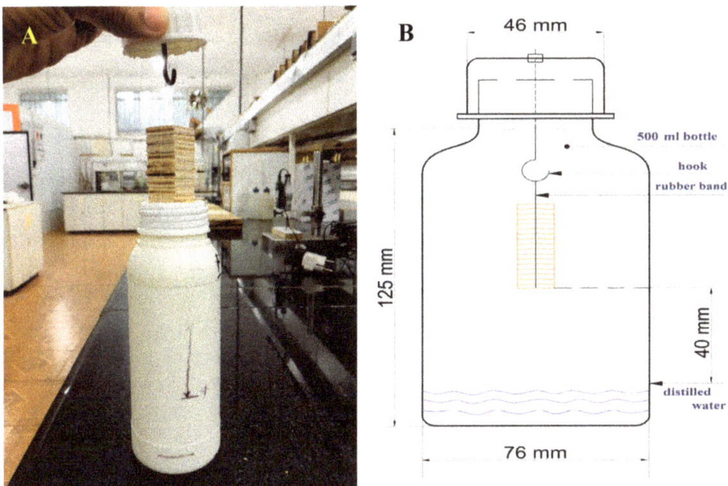

Figure 2. Test apparatus for measurement of formaldehyde gases, type 2 for the flask method (1:500 mL bottle with a top made of polyethylene plastic; 2: hook to hang specimens; 3: rubber band to suspend specimens within the flask; 4: distilled water at the bottom of the flask): (**A**) the actual flask with a set of specimens; (**B**) linear drawing of the flask.

2.4. Shear-Strength Test

Specimens for the shear-strength test were prepared according to European standard specifications EN 314-1: 2004 (Figure 3). Specimens were tested using a universal test machine model STM-20, produced by Santam Engineering Design Co. (Semnan, Iran). Loading speed was 1 mm/min. Shear strength was calculated using Equation (1).

$$S\,S = \frac{F_{\max}}{A} \; (\text{MPa}), \tag{1}$$

where $F_{\max}$ is the maximum failing force, and A is the shear area in the specimen.

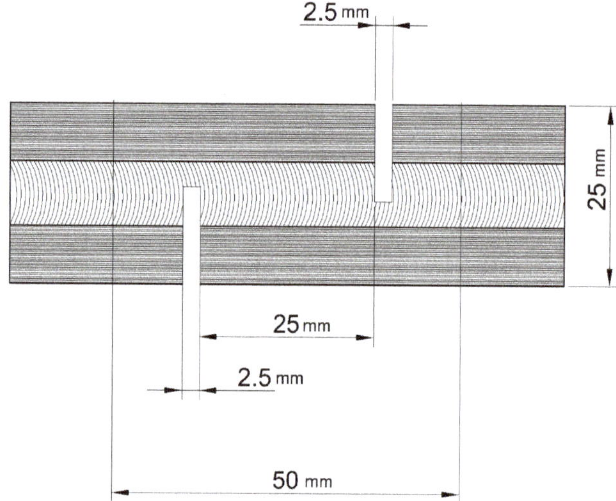

Figure 3. Linear diagram of shear-strength test specimen according to European standard specifications EN 314-1: 2004.

2.5. Delamination Test

Delamination properties of the produced plywood were determined on the basis of voluntary standard specifications for plywood, provided by "The Hardwood Plywood and Veneer Associations (ANSI/HPVA HP-1, 2004). On the basis of the specifications, three specimens (50.8 mm × 127 mm) were cut from each panel. They passed three rounds of soaking/drying cycles. Each cycle comprised 4 h of soaking in water at 24 ± 2 °C, and then a drying period of 19 h with heating at 50 ± 1 °C. According to the specifications, a panel is acceptable for interior use if <5% of specimens delaminate after the first soaking/drying cycle. For exterior applications, delamination should not occur in >15% of the specimens. By definition, delamination is any continuous opening between two layers of plywood that is longer than 5.08 cm (or 2 in), deeper than 0.64 cm (or 0.25 in), and wider than 0.008 cm (or 0.003 in).

2.6. Statistical Analysis

One-way analysis of variance (ANOVA) was carried out in a completely randomized design and experiment with SAS software, version 9.2 (2010) at a 95% level of confidence. Duncan's multiple-range test was then performed to discern similar groupings among treatments for each property. Hierarchical cluster analysis was then performed using SPSS/18 (2010). Clusters included dendrograms by means of Ward's methods using squared Euclidean distance intervals.

3. Results and Discussion

3.1. Preliminary Study

In a preliminary study, 5%, 10%, 15%, and 20% UF resin was replaced with soy flour to gain a better estimation of the most effective combination to decrease formaldehyde emission. Results of the preliminary study demonstrated a clear decreasing trend in formaldehyde emission (FE) as soy flour content increased (Figure 4). This is in line with the data reported in the literature. Hosseini et al. [22] reported that the partial replacement of urea-formaldehyde adhesive with soy flour, particularly with a substitution rate of 15%, significantly reduced the formaldehyde emission. A similar observation was also made by Kawalerczyk et al. [23] who applied five types of flours (rye, hemp, coconut, rice, and pumpkin) as fillers with urea-formaldehyde resin in plywood manufacture. All soy-treated panels showed a significant decrease in formaldehyde emission in comparison to control panels. The highest and lowest formaldehyde emissions were observed in the control panels (130.1 mg/kg) and panels with 20% soy flour (88.6 mg/kg), respectively. No significant difference was observed between formaldehyde emissions in panels produced with 10%, 15%, and 20% soy flour, although the FE trend was a decreasing one as soy flour content increased to 20%. Two factors potentially contributed to this behavior. The first one may have been the great dependence of formaldehyde emission on the resin content. A lower content of formaldehyde in the adhesive mixture causes a lower level of formaldehyde emission from the panel. In fact, a higher substitution of UF resin with soy flour results in a lower formaldehyde emission. Another factor is the reaction of free formaldehyde with amino groups present in the soy flour. Pereira et al. [37] found that the use of soy protein as a natural formaldehyde scavenger in wood particleboard production can contribute to a decrease in the formaldehyde content of particleboard panels, without significantly affecting the properties of the panels

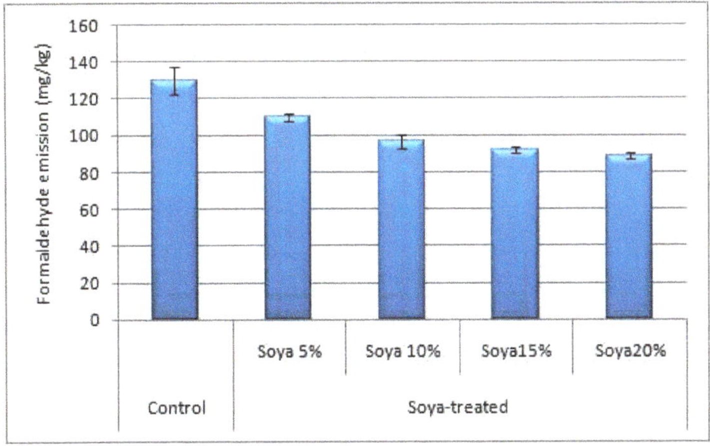

Figure 4. Formaldehyde emission (mg/kg) in three-layer plywood panels produced with urea-formaldehyde resin and 5%, 10%, 15%, and 20% soy flour substitution of UF resin.

The shear strength of all panels produced in the preliminary phase of the study was also measured to find out the effects of the addition of soy flour on at least one mechanical property. Results showed that the shear strength values of all treatments were more than the standard limit of 1 MPa, even the control panels (1.55 MPa). The highest shear strength was observed in panels with 15% soy flour content (2.1 MPa); that is, soy flour resulted in a 34% increase in shear strength. The soy flour content of 5% did not have a significant effect on the shear strength of plywood panels (Figure 5). A further increase in soy flour content to 20% resulted in a decrease in shear strength (1.8 MPa), although it was still higher than that of the control specimens. Results of the delamination tests showed that all panels successfully passed the test (Table 2). This indicated that substitution of soy flour for UF resin (for soy flour contents lower than 20%) cannot significantly affect the delamination property of plywood panels. This finding is in accordance with a previously reported study, in which it was found that the partial replacement of UF adhesive with soy flour, particularly at substitution rates of 10% and 15%, did not significantly affect the shear strength of plywood under both dry and wet conditions [22].

Cluster analysis using the two properties of formaldehyde emissions and shear-strength values in the preliminary phase demonstrated similar clustering of the control panel with panels containing 5% soy flour (Figure 6). This indicated that 5% soy flour is too low to significantly influence the studied properties of plywood panels. The other three panels were distinctly clustered away from control panels. In order to benefit from the maximum shear strength, as well as a satisfactory decrease in formaldehyde emission, the optimal soy flour contents of 10% and 15% were chosen for the next phase in which wollastonite was added to the UF resin along with soy flour.

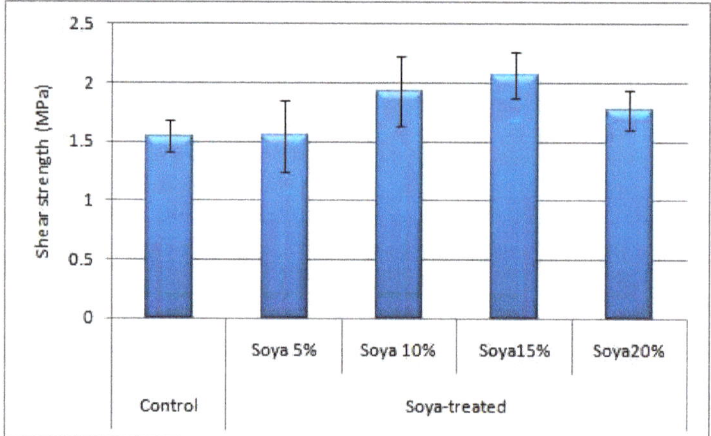

Figure 5. Shear strength (MPa) in three-layer plywood panels produced with urea-formaldehyde resin and 5%, 10%, 15%, and 20% soy flour substitution of UF resin.

Table 2. Delamination test on plywood produced with different binders.

Binder	First Round [1]	Third Round [2]	Result [3]
UF [4] 100%	1/20	1/20	P
UF 95% + SF [5] 5%	0/20	0/20	P
UF 90% + SF 10%	1/20	1/20	P
UF 85% + SF 15%	0/20	0/20	P
UF 80% + SF 20%	0/20	0/20	P

[1] No. of delaminated cases after the first round of the soaking/drying cycle. [2] No. of delaminated cases after the third round of the soaking/drying cycle. [3] P = testing passed, F = testing failed. [4] Urea-formaldehyde resin content. [5] Soy flour content.

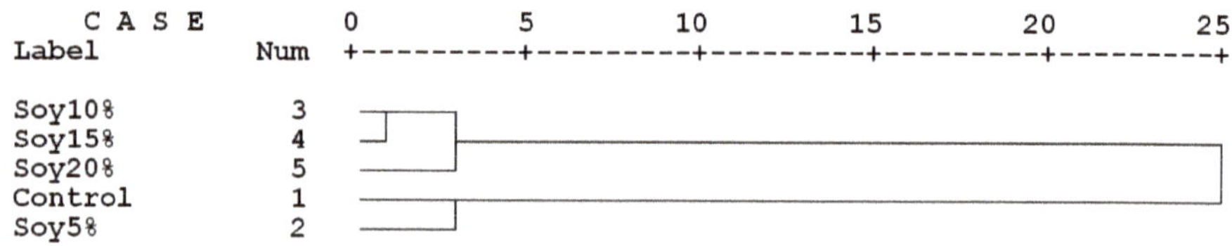

Figure 6. Cluster analysis of five different three-layer plywood panels produced with urea-formaldehyde resin, and 5%, 10%, 15%, and 20% soy flour substitution for UF resin (Soy% = soy flour content).

3.2. Main Phase of the Study

In the second phase of the study, 5% and 10% wollastonite gel (W) was added to panels made with a mixture of UF resin and soy flour (only two optimal SF contents of 10% and 15% were tested in the main phase). Results illustrated that, apart from slight fluctuations which were attributed to the standard deviation among different specimens, the addition of wollastonite had no significant effect on formaldehyde emissions (Figure 7). However, both contents of wollastonite had a decreasing impact on shear strength (Figure 8), although the shear strength in W-added panels was still higher than that of control panels. The decrease in shear strength was attributed to the absorption of part of the resin by wollastonite particles. Results of the delamination tests revealed that all panels passed this test, indicating that the addition of SF or wollastonite did not have a significant impact on the delamination of plywood panels (Table 3).

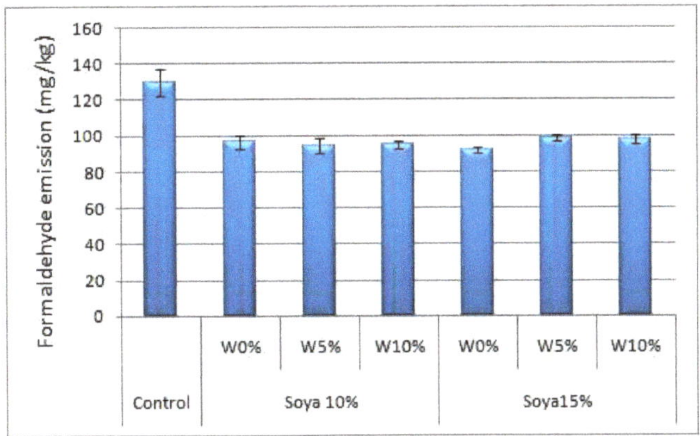

Figure 7. Formaldehyde emission (mg/kg) in three-layer plywood panels (Soy% = soy flour content; W% = wollastonite content).

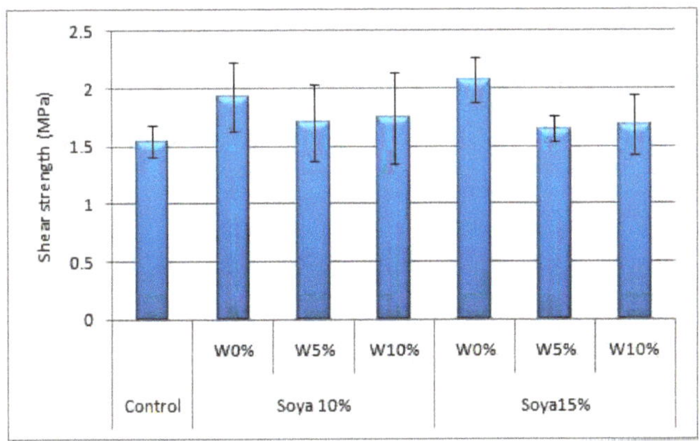

Figure 8. Shear strength (MPa) in three-layer plywood panels (Soy% = soy flour content; W% = wollastonite content).

Substitution of soy flour for UF resin at both levels of 10% and 15% significantly increased the water absorption (WA) and thickness swelling (TS) (Figures 9A and 10A after 2, 24 and 720 h and Figures 9B and 10B at various time intervals). The increases were significant at 2 h, 24 h, and the long-term immersion of 720 h. The increase was attributed to the water hydrophilicity of soy flour [38–41]. Addition of wollastonite (both W contents of 5% and 10%) to the resin mixture resulted in a significant decrease in water absorption, almost reaching the same value as in the control panels. This decrease was attributed to the reinforcing effect of wollastonite in the resin mixture. Similar reinforcing effects were previously reported to improve the shear strength of polyvinyl acetate resin [31] and the fire-retarding property of acrylic–latex paint [42]. With respect to thickness swelling, 5% wollastonite significantly decreased the TS to nearly the same level as in the control panels. However, TS values of panels containing 10% W were still nearly as high as the values of soy-treated panels. That is, a W content of 10% could not mitigate the negative effect of soy flour on thickness swelling. It was concluded that 10% wollastonite was too high, thereby absorbing UF resin rather than acting as reinforcing filler to improve thickness swelling; therefore, a W content of 5% is recommended as an optimum.

Table 3. Delamination test on plywood produced with different binders and with the addition of wollastonite.

Binder	First Round [1]	Third Round [2]	Result [3]
UF [4] 100%	1/20	1/20	P
UF 90% + SF [5] 10%	1/20	1/20	P
UF 90% + SF 10% + W [6] 5%	1/20	2/20	P
UF 90% + SF 10% + W 10%	0/20	0/20	P
UF 85% + SF 15%	0/20	0/20	P
UF 85% + SF 15% + W 5%	0/20	0/20	P
UF 85% + SF 15% + W 10%	0/20	0/20	P

[1] No. of delaminated cases after the first round of the soaking/drying cycle. [2] No. of delaminated cases after the third round of the soaking/drying cycle. [3] P = testing passed, F = testing failed. [4] Urea-formaldehyde resin content. [5] Soy flour content. [6] Wollastonite content.

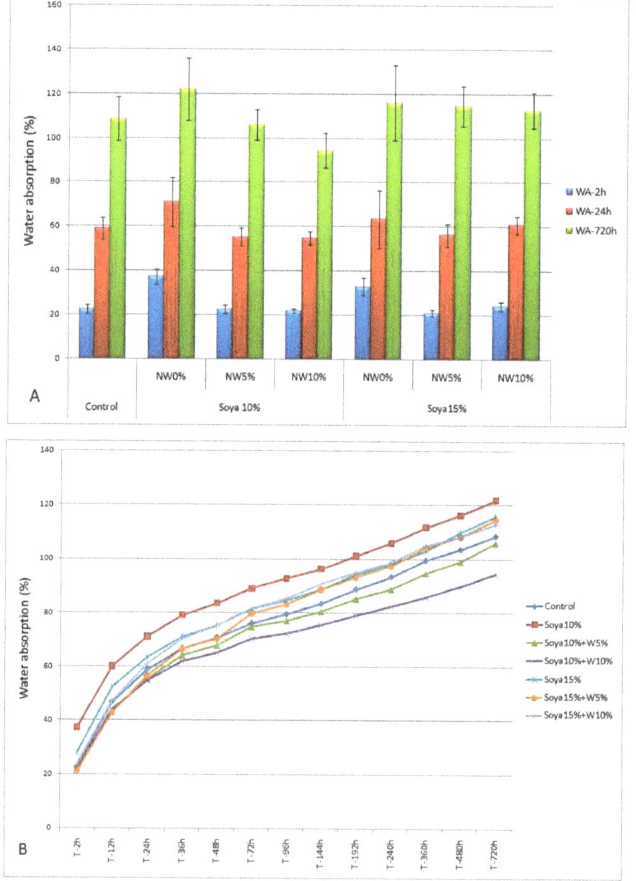

Figure 9. Water absorption (%) in three-layer plywood panels (Soy% = soy flour content; W% = wollastonite content; WA = water absorption).

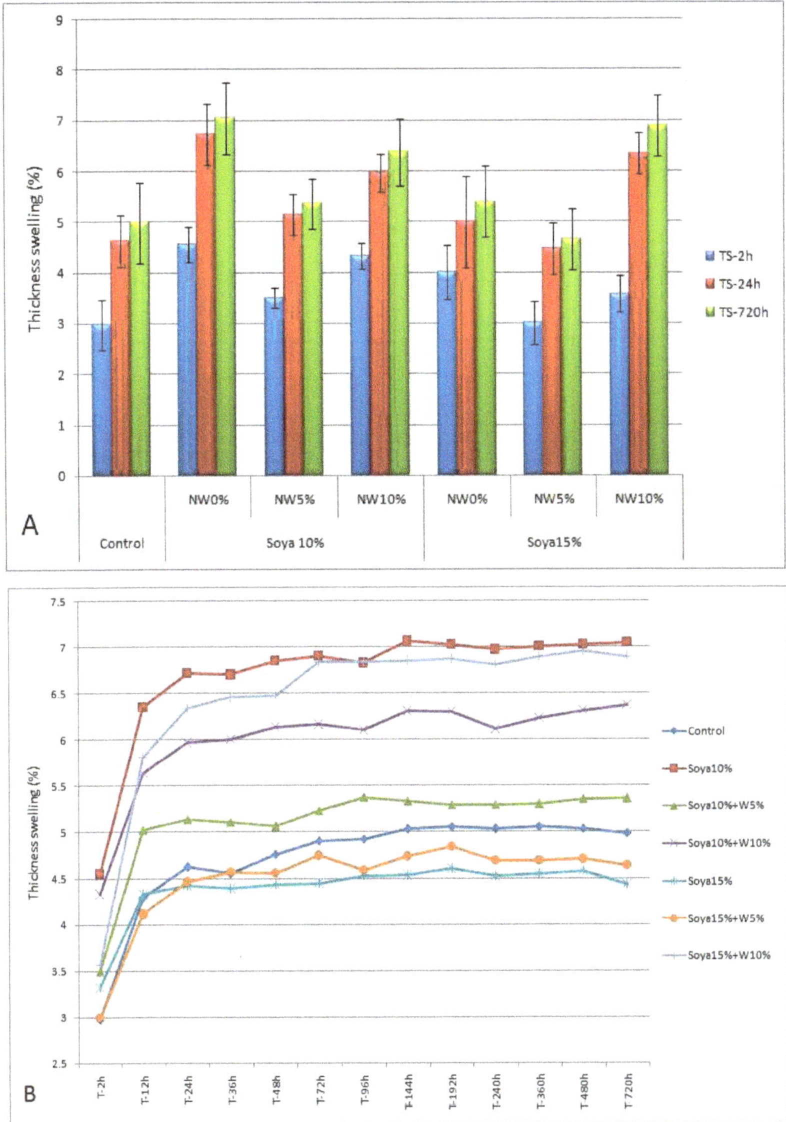

Figure 10. Thickness swelling (%) in three-layer plywood panels (Soy% = soy flour content; W% = wollastonite content; TS = thickness swelling).

The addition of both SF and W to UF resin resulted in a decrease in pH of the resin mixture (Table 4). Gel time was significantly increased as a result of the addition of both SF and W (Table 4); however, the gel times of all mixtures were below the hot-press time of 5 min. Therefore, it is unlikely that alterations in different properties measured in this study can be attributed to the difference in gel time. The addition of SF significantly increased viscosity, while W had a decreasing effect. The decreasing effect of W on viscosity was attributed to the water content of the W gel that was added to the resin mixture.

Table 4. Properties of the 12 resin mixtures used to manufacture three-layer plywood.

Binder	pH	Gel Time (s)	Viscosity (6 rpm) (cP)	Viscosity (10 rpm) (cP)
UF [1] 100%	7.4	75	449.9	455.9
UF 95% + SF [2] 5%	5.4	105	564.9	566.9
UF 90% + SF 10%	5.2	128	824.8	794.8
UF 90% + SF 10% + W [3] 5%	5.6	214	689.9	656.9
UF 90% + SF 10% + W 10%	5.2	230	569.9	551.9
UF 85% + SF 15%	5.6	146	1300	1302
UF 85% + SF 15% + W 5%	5.5	172	1120	1086
UF 85% + SF 15% + W 10%	5.1	158	949.8	902.8
UF 80% + SF 20%	5.6	172	2000	1911

[1] UF = urea-formaldehyde resin; [2] SF = soy flour content; [3] W = wollastonite content

Cluster analysis as a function of the properties measured demonstrated that control panels were remotely clustered away from panels containing soy flour and wollastonite (Figure 11). This showed the significant effect of both soy flour and wollastonite on the overall properties of plywood panels. Panels containing either 5% or 10% wollastonite were also clustered differently from those containing only soy flour, indicating the significant impact of wollastonite. On the basis of the results of each property considered individually and altogether, it was concluded that panels containing 10% soy flour and 5% wollastonite are recommended to achieve the optimum decrease in carcinogenic formaldehyde emission, as well as the optimum physical and mechanical properties. With regard to the promising results of wollastonite as a reinforcing filler in different resins and coatings, further studies can be carried out to investigate the effects of the addition of wollastonite to different wood-based composite panels produced solely using bioresins, such as soy flour.

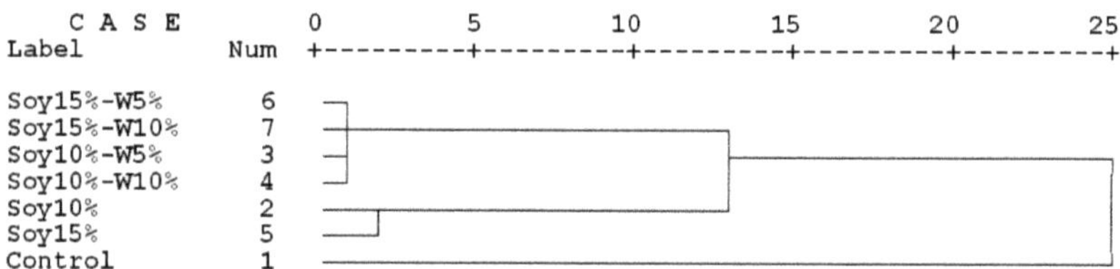

Figure 11. Cluster analysis of seven different three-layer plywood panels produced with urea-formaldehyde resin, and 10% and 15% soy flour substitution for UF resin, plus addition of wollastonite at 5% and 10% (Soy% = soy flour content; W% = wollastonite content).

4. Conclusions

Partial substitution of soy flour for urea-formaldehyde resin has the potential to decrease carcinogenic formaldehyde emission in plywood panels. Shear strength was also improved as soy flour content increased. However, this had a negative effect on the water absorption and thickness swelling of plywood panels. The addition of micron-sized wollastonite mitigated the undesirable increased hydrophilicity in panels caused by soy flour. It was concluded that 10% soy flour and 5% wollastonite provide the lowest formaldehyde emission and the most optimum physical and mechanical properties.

Author Contributions: Methodology, H.R.T. and S.G.; validation, H.R.T., S.G., M.G., and S.B.H.; investigation, H.R.T. and S.B.H.; writing—original draft preparation, H.R.T., and A.N.P.; writing—review and editing, H.R.T., S.B.H., and A.N.P.; visualization, H.R.T. and S.B.H.; supervision, H.R.T. and A.N.P. All authors read and agreed to the published version of the manuscript.

Funding: This research received no external funding.

Acknowledgments: The first author appreciates the constant scientific support of Jack Norton (retired, Horticulture and Forestry Science, Queensland Department of Agriculture and Fisheries, Australia), as well as Alexander von Humboldt Stiftung (Bonn, Germany).

Conflicts of Interest: The authors declare no conflict of interest.

References

1. Pizzi, A.; Papadopoulos, A.N.; Policardi, F. Wood composites and their polymer binders. *Polymers* **2020**, *12*, 1115. [CrossRef]
2. Sopt, P.; Konnerth, J.; Gindl-Altmutter, W.; Katner, W.; Moser, J.; Mitter, R.; van Herwijnen, H. Technological performance of formaldehyde-free adhesive alternatives for particleboard industry. *Int. J. Adhes. Adhes.* **2019**, *94*, 99–131.
3. Roffael, E. Formaldehyde release from wood-based panels—A review. *Holz als Roh Werkst* **1989**, *47*, 41–51. [CrossRef]
4. Salthammer, T.; Mentese, S.; Marutzky, R. Formaldehyde in the indoor environment. *Chem. Rev.* **2010**, *110*, 2536–2572. [CrossRef] [PubMed]
5. Pizzi, A.; Lipschitz, L.; Valenzuela, J. Theory and practice of the preparation of low formaldehyde emission UF adhesives for particleboard. *Holzforschung* **1994**, *48*, 254–261. [CrossRef]
6. Dunker, A.K.; Johns, W.E.; Rammon, R.; Farmer, B.; Johns, S.Y. Slightly Bizarre Protein Chemistry: Urea-Formaldehyde Resin from a Biochemical Perspective. *J. Adhes.* **1986**, *19*, 153–176. [CrossRef]
7. Levendis, D.; Pizzi, A.; Ferg, E.E. The correlation of strength and formaldehyde emission with the crystalline/amorphous structure of UF resins. *Holzforschung* **1992**, *45*, 260–267. [CrossRef]
8. Ferg, E.E.; Pizzi, A.; Levendis, D. A 13 C NMR analysis method for urea-formaldehyde resin strength and formaldehyde emission. *J. Appl. Polym. Sci.* **1993**, *50*, 907–915. [CrossRef]
9. Lubis, M.A.R.; Park, B.D. Enhancing the performance of low molar ratio urea–formaldehyde resin adhesives via in-situ modification with intercalated nanoclay. *J. Adhes.* **2020**, in press. [CrossRef]
10. Hammami, N.; Jarroux, N.; Robitzer, M.; Majdoub, M.; Habas, J.P. Optimized synthesis according to one-step process of a biobased thermoplastic polyacetal derived from isosorbide. *Polymers* **2016**, *8*, 294. [CrossRef]
11. Wang, H.; Cao, M.; Li, T.; Yang, L.; Duan, Z.; Zhou, X.; Du, G. Characterization of the low molar ratio urea–formaldehyde resin with ^{13}C NMR and ESI–MS: Negative effects of the post-added urea on the urea–Fformaldehyde polymers. *Polymers* **2018**, *10*, 602. [CrossRef] [PubMed]
12. Kim, S.; Kim, H.; Kim, H.; Lee, H. Effect of bio-scavengers on the curing behavior and bonding properties of melamine-formaldehyde resins. *Macromol. Mater. Eng.* **2006**, *291*, 1027–1034. [CrossRef]
13. Kim, S. The reduction of indoor air pollutant from wood-based composite by adding pozzolan for building materials. *Constr. Build. Mater.* **2009**, *23*, 2319–2323. [CrossRef]
14. Réh, R.; Igaz, R.; Krišťák, Ľ.; Ružiak, I.; Gajtanska, M.; Božíková, M.; Kučerka, M. Functionality of beech bark in adhesive mixtures used in plywood and its effect on the stability associated with material systems. *Materials* **2019**, *12*, 1298. [CrossRef] [PubMed]
15. Zhang, H.; Zhang, J.; Song, S.; Wu, G.; Pu, J. Modified nanocrystalline cellulose from two kinds of modifiers used for improving formaldehyde emission and bonding strength of urea-formaldehyde resin adhesive. *BioResources* **2011**, *6*, 4430–4438.
16. Kawalerczyk, J.; Siuda, J.; Mirski, R.; Dziurka, D. Hemp flour as a formaldehyde scavenger for melamine-urea-formaldehyde adhesive in plywood production. *BioResources* **2020**, *15*, 4052–4064.
17. Barry, A.; Corneau, D. Effectiveness of barriers to minimize VOC emissions including formaldehyde. *For. Prod. J.* **2006**, *56*, 38–42.
18. Dunky, M. Urea-formaldehyde (UF) adhesive resins for wood. *Int. J. Adhes. Adhes.* **1998**, *18*, 95–107. [CrossRef]
19. Schwarzkopf, M.; Huang, J.; Li, K. A formaldehyde-free soy-based adhesive for making oriented strandboard. *J. Adhes.* **2010**, *86*, 352–364. [CrossRef]
20. Amaral-Labat, G.A.; Pizzi, A.; Gonc, A.R.; Celzard, A.; Rigolet, S.; Rocha, G.J.M.; Eel, D.E.; Paulo, S. Environment-friendly soy flour-based resins without. *J. Appl. Polym. Sci.* **2008**, *108*, 624–632. [CrossRef]
21. Li, K.; Peshkova, S.; Geng, X. Investigation of soy protein-kymene ®adhesive systems for wood composites. *J. Am. Oil Chem. Soc.* **2004**, *81*, 487–491. [CrossRef]
22. Hosseini, S.B.; Assadollahzadeh, M.; Najfai, S.K.; Taherzadeh, J. Partial replacement of urea-formaldehyde adhesive with fungal biomass and soy flour in plywood fabrication. *J. Adhes. Sci. Technol.* **2020**, *34*, 1371–1384. [CrossRef]

23. Kawalerczyk, J.; Dziurka, D.; Mirski, R.; Trocinski, A. Flour fillers with urea formaldehyde resin in plywood. *BioResources* **2019**, *14*, 6727–6735.

24. Taghiyari, H.R.; Karimi, A.; Tahir, P.M.D. Nano-wollastonite in particleboard: Physical and mechanical properties. *BioResources* **2013**, *8*, 5721–5732. [CrossRef]

25. Taghiyari, H.R.; Majidi, R.; Jahangiri, A. Adsorption of nano-wollastonite on cellulose surface: Effects on physical and mechanical properties of medium-density fiberboard (MDF). *CERNE* **2016**, *22*, 215–222. [CrossRef]

26. Taghiyari, H.R.; Mohammad-Panah, B.; Morrell, J.J. Effects of wollastonite on the properties of medium-density fiberboard (MDF) made from wood fibers and camel-thorn. *Maderas Cienc. Tecnol.* **2016**, *18*, 157–166. [CrossRef]

27. Mantanis, G.; Papadopoulos, A.N. Reducing the thickness swelling of wood based panels by applying a nanotechnology compound. *Eur. J. Wood Wood Prod.* **2010**, *68*, 237–239. [CrossRef]

28. Papadopoulos, A.N.; Bikiaris, D.N.; Mitropoulos, A.C.; Kyzas, G.Z. Nanomaterials and chemical modification technologies for enhanced wood properties: A review. *Nanomaterials* **2019**, *9*, 607. [CrossRef]

29. Taghiyari, H.R.; Esmailpour, A.; Adamopoulos, S.; Zereshki, K.; Hosseinpourpia, R. Shear strength of heat-treated solid wood bonded with polyvinyl-acetate reinforced by nanowollastonite. *Wood Res.* **2020**, *65*, 183–194. [CrossRef]

30. Esmailpour, A.; Taghiyari, H.R.; Ghorbanali, M.; Mantanis, G.I. Improving fire retardancy of medium density fiberboard by nano-wollastonite. *Fire Mater.* **2020**, *44*, 1–8. [CrossRef]

31. Papadopoulos, A.N.; Gkaraveli, A. Dimensional stabilisation and strength of particleboard by chemical modification with propionic anhydride. *Eur. J. Wood Wood Prod.* **2003**, *61*, 142–144. [CrossRef]

32. Taghiyari, H.R.; Majidi, R.; Mohseni Armaki, S.M.; Haghighatparast, M. Graphene as reinforcing filler in polyvinyl acetate resin. *Int. J. Adhes. Adhes.* **2020**, in press.

33. Funk, M.; Wimmer, R.; Adamopoulos, S. Diatomaceous earth as an inorganic additive to reduce formaldehyde emissions from particleboards. *Wood Mater. Sci. Eng.* **2015**, *12*, 92–97. [CrossRef]

34. Suganya, S.; Kumar, P.S.; Saravanan, A. Construction of active bio-nanocomposite by inseminated metal nanoparticles onto activated carbon: Probing to antimicrobial activity. *IET Nanobiotechnol.* **2017**, *11*, 746–753. [CrossRef]

35. Hassani, V.; Taghiyari, H.R.; Schmidt, O.; Maleki, S.; Papadopoulos, A.N. Mechanical and physical properties of oriented strand lumber (OSL): The effect of fortification level of nanowollastonite on UF resin. *Polymers* **2019**, *11*, 1884. [CrossRef] [PubMed]

36. EN 717-3. *Wood Based Panels-Determination of Formaldehyde Release-Part 3: Formaldehyde Release by the Flask Method*; Comite Europeen de Normalisation: Brussels, Belgium, 1996.

37. Pereira, F.; Pereira, J.; Paiva, N.; Ferra, J.; Martins, J.; Magalhaes, F.; Carvalho, L. Natural additive for reducing formaldehyde emissions in urea-formaldehyde resins. *J. Renew. Mater.* **2016**, *4*, 41–46. [CrossRef]

38. Allen, A.J.; Marcinko, J.J.; Wagler, T.A.; Sosnowick, A.J. Investigations of the molecular interactions of soy-based adhesives. *For. Prod. J.* **2010**, *60*, 534–540. [CrossRef]

39. Zhu, D.; Damodaran, S. Chemical phosphorylation improves the moisture resistance of soy flour-based wood adhesive. *J. Appl. Polym. Sci.* **2014**, *131*. [CrossRef]

40. Ghahri, S.; Pizzi, A. Improving soy-based adhesives for wood particleboard by tannins addition. *Wood Sci. Technol.* **2017**. [CrossRef]

41. Ghahri, S.; Mohebby, B.; Pizzi, A.; Mirshokraie, A.; Mansouri, H.R. Improving water resistance of soy-based adhesive by vegetable tannin. *J. Polym. Environ.* **2017**. [CrossRef]

42. Esmailpour, A.; Majidi, R.; Taghiyari, H.R.; Ganjkhani, M.; Mohseni Armaki, S.M.; Papadopoulos, A.N. Improving fire retardancy of beech wood by graphene. *Polymers* **2020**, *12*, 303. [CrossRef] [PubMed]

*applied
sciences*

Article

Alien Wood Species as a Resource for Wood-Plastic Composites

Sergej Medved, Daša Krapež Tomec , Angela Balzano and Maks Merela *

Department of Wood Science and Technology, Biotechnical Faculty, University of Ljubljana, Jamnikarjeva 101,
1000 Ljubljana, Slovenia; sergej.medved@bf.uni-lj.si (S.M.); dasa.krapez.tomec@bf.uni-lj.si (D.K.T.);
angela.balzano@bf.uni-lj.si (A.B.)
* Correspondence: maks.merela@bf.uni-lj.si

Abstract: Since invasive alien species are one of the main causes of biodiversity loss in the region and thus of changes in ecosystem services, it is important to find the best possible solution for their removal from nature and the best practice for their usability. The aim of the study was to investigate their properties as components of wood-plastic composites and to investigate the properties of the wood-plastic composites produced. The overall objective was to test the potential of available alien plant species as raw material for the manufacture of products. This would contribute to sustainability and give them a better chance of ending their life cycle. One of the possible solutions on a large scale is to use alien wood species for the production of wood plastic composites (WPC). Five invasive alien hardwood species have been used in combination with polyethylene powder (PE) and maleic anhydride grafted polyethylene (MAPE) to produce various flat pressed WPC boards. Microstructural analyses (confocal laser scanning microscopy and scanning electron microscopy) and mechanical tests (flexural strength, tensile strength) were performed. Furthermore, measurements of density, thickness swelling, water absorption and dimensional stability during heating and cooling were carried out. Comparisons were made between the properties of six WPC boards (five alien wood species and mixed boards). The results showed that the differences between different invasive alien wood species were less obvious in mechanical properties, while the differences in sorption properties and dimensional stability were more significant. The analyses of the WPC structure showed a good penetration of the polymer into the lumens of the wood cells and a fine internal structure without voids. These are crucial conditions to obtain a good, mechanically strong and water-resistant material.

Keywords: alien plants; wood plastic composite; flexural strength; tensile strength; swelling; dimension stability; scanning electron microscopy

Citation: Medved, S.; Tomec, D.K.; Balzano, A.; Merela, M. Alien Wood Species as a Resource for Wood-Plastic Composites. *Appl. Sci.* **2021**, *11*, 44. https://dx.doi.org/10.3390/app 11010044

Received: 27 November 2020
Accepted: 18 December 2020
Published: 23 December 2020

Publisher's Note: MDPI stays neutral with regard to jurisdictional claims in published maps and institutional affiliations.

1. Introduction

The properties of wood-based composites are determined by the components used for their production. This is also demonstrated in the case of wood plastic composites (WPC), where wood can act as a reinforcement or as a filler and in some cases both. Wood and the derived components are an important factor influencing the properties of wood-based panels [1–5]. WPC is basically composed of two main components, namely plastic or polymer and wood, resulting in a material which combines the best properties of both components. Although the wood constituents in WPC are small (usually size class between 0.1 mm and 1.0 mm), and (according to [6,7]) the wood species related differences should be smaller, several authors [8–13] have shown that the wood constituents (in terms of species and size of constituents) influence the properties of WPC. The influence of the wood species used for WPC depends on the size of constituent obtained during breakdown process (particularly in terms of slenderness ratio), its affinity towards polymeric compound and strength of bond between wooden constituent and polymeric matrices. Shebani et al. [14] determined that chemical composition of wood also influences the properties of WPC; namely, they proved that a higher cellulose and lignin content results in better mechanical properties but also in lower moisture resistance when the cellulose content is high. One of the disadvantages of using wood in WPC is the reaction of the wood to UV radiation. The effect of UV radiation (e.g., sunlight) on the wood surface leads to photochemical degradation of wood and thus of WPC. Colour changes (darkening) also occur during the production of WPC. Exposure to elevated temperatures during the pressing process leads to the evaporation of extractives, which darken the wood surface and thus also WPC [10,15].

Based on data from the existing Flora of Slovenia (CCFF) Database, the species *Robinia pseudoacacia* (black locust) is the invasive alien plant species with the potentially most negative impact on biodiversity. Tree of heaven (*Ailanthus altissima*) and boxelder maple (*Acer negundo* L.) can also be classified as invasive alien plants with a high negative impact on biodiversity. It is, therefore, undoubtedly useful to raise awareness of the impact of invasive alien plant species on our environment and to look for the most versatile applications, including their use as WPC components.

Our main objective is to test the suitability of the most widespread invasive alien hardwood species present in Slovenia for the production of WPC and to encourage their removal from native natural ecosystems by transforming them into a source of raw materials that could be processed into useful products. The mechanical properties of these wood species are not well known, and further information about them may encourage the use of these woods in the most appropriate way in new products. Wood anatomical analyses and machining tests have already been carried out on the same wood species, which overall showed good degree of machinability [16,17]. In light of the results obtained so far, we believe that the selected wood species may be suitable for the production of WPC due to its anatomical structure and mechanical properties. To verify their suitability for this application, we carried out classical mechanical tests, such as flexural and tensile strength tests. In addition, we performed microstructural analyses of the surface and internal structure of WPC boards using scanning electron microscopy (SEM).

SEM is a powerful tool for examining the surface and structure of wood. The application of SEM in wood science is well described in the literature [18–22]. Recently SEM has been successfully used to study the morphology and surface evaluation of WPC [13,23]. SEM has been used to evaluate the adhesion between wood and polymer matrix and detect the occurrence of fibre pull outs and voids within the composite [24,25]. It was shown that the evaluation of SEM is consistent with the sorption behaviour and can clearly explain the mechanical properties of WPC. It was observed that an intact composite surface corresponds to a lower moisture transport rate within the matrix [26]. Therefore, an intact and homogeneous material with stronger adhesion of its two components, namely wood and polymer matrix, results in a material with higher mechanical properties. Given the reported advantage of the microscopy method, we used Confocal laser scanning microscopy (CLSM) and Scanning electron microscopy (SEM) for the surface analysis of WPC boards, to evaluate their homogeneity, the quality of adhesion at the interface between the wood fibres and the polymer matrix, and the possible occurrence of voids that could reduce their mechanical properties. We used the results to discuss the mechanical properties as well as the sorption properties of WPC boards considering the observed microscopic features.

2. Materials and Methods

2.1. WPC Boards Preparation

We prepared WPC boards using some of the most widespread invasive alien hardwood species present in Slovenia, namely: boxelder maple (*Acer negundo*), horse chestnut (*Aesculus hippocastanum*), tree of heaven (*Ailanthus altissima*), black locust (*Robinia pseudoacacia*) and honey locust (*Gleditsia triacanthos*). The boards were produced using wood of the aforementioned species and polyethylene (PE) powder Dowlex™ 2631.10UE obtained from local company ROTO-Pavlinjek d.o.o. (MURSKA SOBOTA, Slovenia). The physical properties of the powder used are shown in Table 1.

Table 1. Physical properties of polyethylene powder Dowlex^{TM} 2631.10UE (values were determined by the supplier).

Physical Property	Value
Density	0.935 g·cm^{-3}
Melt index, 190 °C/2.16 kg	7 g/10 min
Melting point	124 °C
Vicat softening point A120	115 °C
Deflection temperature under load HDT B	52 °C

Maleic anhydride grafted polyethylene (MAPE) was used as a coupling agent (donated by Graft Polymere d.o.o., Ljubljana, Slovenia). MAPE was added to increase the affinity and adhesion between wooden constituents and polymeric matrices. MAPE acts as so-called coupling agent or compatibilizing agent.

The polyethylene content was 46.5% and MAPE 3.5%, while 50% of the total mass consisted of wood particles obtained from 5 alien wood species with different densities [17]:

- black locust (*Robinia pseudoacacia*): density 0.778 g·cm^{-3}
- boxelder maple (*Acer negundo*): density 0.560 g·cm^{-3}
- honey locust (*Gleditsia triacanthos*): density 0.705 g·cm^{-3}
- horse chestnut (*Aesculus hippocastanum*): density 0.495 g·cm^{-3}
- tree of heaven (*Ailanthus altissima*): density 0.555 g·cm^{-3}

The 50% wood content was selected based on a report by Leu et al., 2012 [27], which showed that the mechanical properties of WPC increased by up to 50%, while a higher share led to a decrease in mechanical properties.

A two-step decomposition process was used to break down wood into particles (Figure 1).

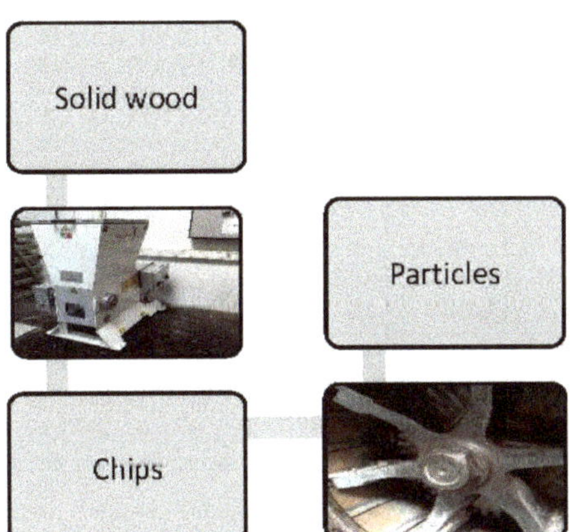

Figure 1. Wood break down process.

The breakdown of solid wood into chips was carried out in a Prodeco M-0 chipper, which has an output screen with openings of 25 mm in diameter. The ring chipper used for production of particles was a Condux CSK 350/N1 ring chipper (the gap between the blade and beating bar was 1.25 mm). After chipping particles were analysed by sieving, whereby 100 g of particles were placed on the top sieve. After 10 min of sieving, the residues on each sieve were weighted. The particles used for the experiment are shown in Figure 2.

As the moisture content of the particles was higher than required for WPC production, the particles were dried at 80 °C for 24 h to achieve a moisture content below 4% (the actual moisture content for board production was between 0.9% and 2.3%). After drying, the particles were mixed by hand with PE powder and MAPE. The mass ratio was 50:46.5:3.5 (wood:PE:MAPE).

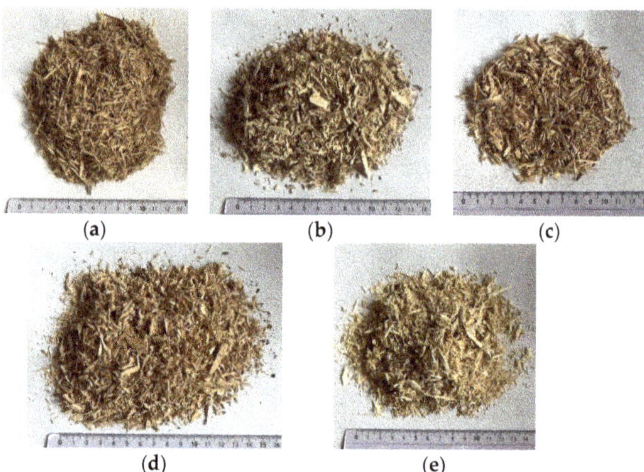

Figure 2. Wood particles used for research; (**a**) black locust, (**b**) boxelder maple, (**c**) honey locust, (**d**) horse chestnut, (**e**) tree of heaven.

The prepared mixture was hand formed into a frame measuring 300×300 mm^2, which was placed on a steel plate. The target thickness was 4 mm, the target density 0.9 g·cm^{-3}.

Wood-PE mat was flat pressed at 180 °C for 10 min at a specific pressure of 3 MPa. After 10 min the boards were transferred to the cold press. The specific pressure during cold pressing was the same as during hot pressing (3 MPa), while the pressing temperature was set at 25 °C (equal to room temperature). The cooling process also hardened the PE. The process for preparing the WPC is shown in Figure 3.

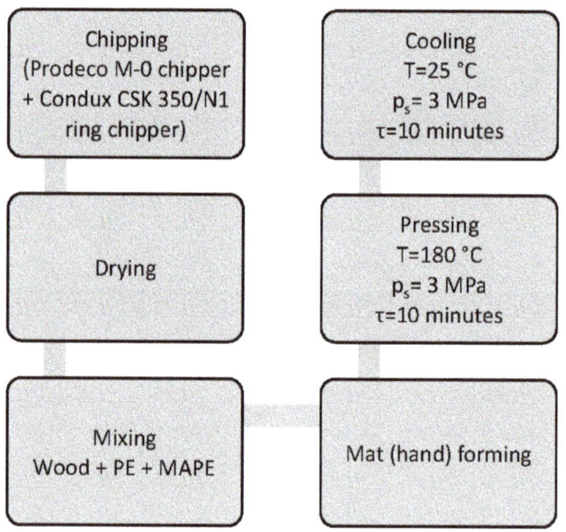

Figure 3. Schematic layout of wood plastic composites (WPC) preparation.

Six sets of WPC boards were prepared from different wood species as shown in Table 2:

Table 2. WPC board types regarding wood species used.

Series	Wood Species Used
A	Boxelder maple (*Acer negundo*)
B	Horse chestnut (*Aesculus hippocastanum*)
C	Tree of heaven (*Ailanthus altissima*)
D	Black locust (*Robinia pseudoacacia*)
E	Honey locust (*Gleditsia triacanthos*)
F	Mixture

In the mixture, 20% of each wood species was used. After climatization period, boards were cut for testing:

- thickness and density (EN 323): 50×50 mm^2, 6 samples [28]
- flexural strength (EN ISO 178): 80×50 mm^2, 10 samples [29]
- tensile strength (EN ISO 527-1): 165×13 mm^2, 10 samples [30]
- thickness swelling and water absorption (EN 317): 50×50 mm^2, 10 samples [31]
- dimensional stability due heating and cooling: 100×25 mm^2, 2×5 samples

2.2. Physical and Mechanical Properties Testing

Flexural strength was determined by a three-point bending test on the testing machine Zwick Roell Z005 testing machine. Since particles were evenly distributed over the width and length of board and no difference in fibre orientation was expected, only one direction was tested. The span distance was 64 mm, while the loading speed was set to 2 mm·min^{-1}. Maximum force, deformation at maximum force, flexural strength and modulus of elasticity were determined.

The tensile strength was determined on Zwick Roell Z005 testing machine, also in one direction only. The loading speed was set to 5 mm·min^{-1}. Maximum force, deformation at maximum force and tensile strength were determined.

Thickness swelling and water absorption were determined by immersion of samples in water. The immersion time was 2 and 24 h. Thickness swelling (TS) in % and water absorption (WA) in % were calculated by Equations (1) and (2):

$$TS_y = \frac{t_2 - t_1}{t_1} \times 100 \tag{1}$$

$$WA_y = \frac{m_2 - m_1}{m_1} \times 100 \tag{2}$$

where y represents immersion time, t sample thickness in mm, m mass of samples in g, while 1 denotes the thickness or mass before and 2 the thickness or mass after 2 h or 24 h of immersion.

The dimensional stability of samples was determined by exposing one set of 5 samples to a temperature of $-25\,°C$ and one set of 5 samples to a temperature of $+65\,°C$. The exposure time was 60 ± 1 min. The dimensional stability (δ_x) in % was calculated by Equation (3)

$$\delta_x = \frac{x_2 - x_1}{x_1} \times 100 \tag{3}$$

where x represents length or thickness respectively, while 1 denotes the dimension before and 2 the dimension after 60 ± 1 min exposure. All results were evaluated using Statistica software by ANOVA and LSD test at $\alpha = 0.05$.

2.3. WPC Structural Analyses

To evaluate WPC surface and internal structure, sub-samples of boards were prepared and observed using a Confocal Laser Scanning Microscope (CSLM) and a Scanning Electron Microscope (SEM). For the structural analyses we used WPC boards made of a mixed material (different wood species). Before the observation, the sub-samples were cut on their cross-section surface with a blade on a sliding microtome (Leica SM2000, Nussloch, Germany) to obtain a flat and smooth surface and then dried at room temperature (T = 22 °C and RH = 65%) [16]. To obtain a panoramic view of the sub-sample and to inspect its entire surface, it was placed on the stage of Confocal Laser Scanning Microscope (CLSM) Olympus LEXT OLS5000 (Olympus Corporation Tokyo 163-0914, Tokyo, Japan) and observed with the optical system using the MPLFLN10xLEXT objective (numerical aperture 0.3, working distance 10.4 mm). Images of the entire surface area were obtained by combining several images at different focus positions, which were recorded in real time using the stitching function by moving the stage. SEM was used to investigate the quality of adhesion at the interface between wood fibres and the polymer matrix and to detect possible voids. Before SEM observations, samples were mounted on stubs with a conductive carbon adhesive tape and coated with an Au/Pd sputter coater (Q150R ES Coating System; Quorum technologies, Laughton, UK) for 30 s with a constant current of 20 mA. The SEM micrographs were then recorded in a high vacuum with 5 kV voltage and with a large field detector (LFD) in a FEI Quanta 250 SEM microscope (FEI Company, Hillsboro, OR, USA) at 9.3 mm working distance and at 100×, 250×, 500× and 2500× magnification.

2.4. Pilot Production of 3D Composites Based on the Proposed Methodology

Wood residues, which arise from the primary processing of wood and the production of wood products, were firstly chipped in a mill and secondly in a knife ring chipper (as presented in Figure 1). Subsequently, the obtained particles were additionally ground with the Retsch SM2000 rotary wood mill (Retsch, Haan, Germany) with a 1 mm sieve.

The particles were then dried at 80 °C for 24 h to achieve a moisture content of less than 4%. Polyethylene (PE) DowlexTM powder, maleic anhydride grafted polyethylene (MAPE) and wood particles obtained from 5 different invasive alien wood species were used in a ratio of 46.5:3.5:50 (PE:MAPE:wood).The mixture was formed by hand into a 3D mould, which was primarily sprayed with a non-stick agent (Silicone H1 spray, Panolin, Madetswil, Switzerland) and then pressed in a hot press at 180 °C and a specific pressure of 3 MPa for 10 min. The mould was then transferred to the cold press with the same pressing parameters, which differ only in temperature, namely 25 °C.

The 3D WPC product was then removed from the mould and, where necessary, edge milling and sealing was carried out.

3. Results and Discussion

Most of particles (65–70%) used in experiment was size class 1.5 and lower, as classified by screening (particles that fell through sieve with opening 2.0 mm), while minority of particles (30–35%) was size class 2 mm and higher (Table 3).

Table 3. WPC board types regarding wood species used.

Size Class	A Boxelder Maple	B Horse Chestnut	C Tree of Heaven	D Black Locust	E Honey Locust
Dust	9%	6%	6%	7%	7%
0.237	11%	10%	13%	12%	14%
0.6	12%	12%	13%	13%	14%
1.0	9%	9%	9%	10%	9%
1.27	9%	9%	9%	8%	8%
1.5	18%	19%	17%	17%	16%
2.0	27%	30%	26%	27%	27%
4.0	5%	5%	6%	5%	7%

Although particles were prepared under same conditions, there are differences between them related to the particle size class (share of residue on the sieve).

3.1. Physical and Mechanical Properties of WPC Boards

The properties of WPC depend on the polymer type used and the type of wood species used for its production. Since polymer was the same the differences between WPC bords (Figure 4) are caused by wood species used through their structure (chemical and anatomical), generated particles, their compressibility, interaction with polymer and through their mechanical properties (Tables 4 and 5).

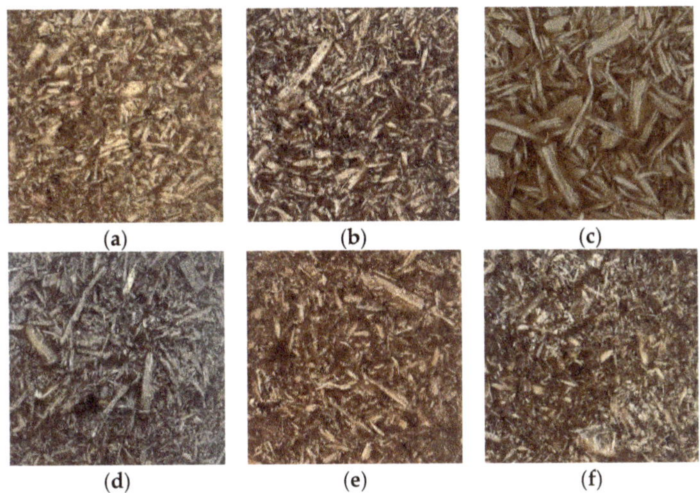

Figure 4. Produced WPC; (**a**) boxelder maple, (**b**) horse chestnut, (**c**) tree of heaven, (**d**) black locust, (**e**) honey locust, (**f**) mixture. Samples are 50×50 mm^2.

Table 4. Physical and strength properties of flat pressed WPC (letters in bracket denote same homogeneous group determined by LSD test at $\alpha = 0.05$).

Series	Thickness	Density	Flexural Strength	Modulus of Elasticity	Tensile Strength
	mm	g·cm^{-3}	MPa	MPa	MPa
A	4.23	0.901 (a)	15.04 (a,b)	759 (a,b)	6.46 (a)
B	4.36	0.929 (c)	15.14 (a,b)	876 (a)	7.03 (a,b,c)
C	4.33	0.912 (b,c)	15.21 (a,b)	812 (a)	7.29 (b,c)
D	4.15	0.956 (d)	15.60 (b)	721 (a,b)	7.10 (a,b,c)
E	4.07	0.961 (d)	14.57 (a,b)	594 (b,c)	6.48 (a,b)
F	3.80	0.889 (a)	13.69 (a,b)	550 (c)	7.52 (c)

Table 5. Sorption properties and dimensional stability of flat pressed WPC (letters in bracket denote same homogeneous group determined by LSD test at $\alpha = 0.05$).

Series	TS2	WA2	TS24	WA24	Length		Thickness	
					$\delta + 65$	$\delta - 25$	$\delta + 65$	$\delta - 25$
	%	%	%	%	%	%	%	%
A	3.39 (b)	4.12 (b)	7.01 (c)	41.69 (a)	−0.05	0.13	−2.18	−1.23
B	1.42 (c)	3.83 (a)	3.24 (a)	42.72 (a,b)	−0.02	0.10	−0.53	−1.32
C	2.51 (a,b)	4.12 (b)	5.02 (b)	43.15 (a,b)	−0.07	0.05	−0.51	−0.61
D	2.24 (a,c)	2.06 (a)	3.91 (a,b)	38.72 (c)	−0.04	0.02	−0.74	−0.74
E	2.35 (a,b,c)	2.23 (a)	3.51 (a)	42.32 (a)	−0.07	0.05	−1.09	−2.27
F	2.82 (a,b)	4.81 (b)	4.92 (a,b)	45.15 (b)	0.14	−0.09	−2.07	−2.65

The PE matrix (or the matrix of WPC in general) is responsible for the load transfer between constituents and moisture resistance, while wood is responsible for density, strength and stiffness. The result (properties of WPC) should be the combination of the best properties of the components. The impact of the wood species can already be seen in the WPC density (Table 4 and Figure 5), where the densities of WPC boards made of higher density wood species differ from those of lower density and mixture. With regard to the density of the wood species, we can divide the material into two different categories, namely wood species below 0.6 g·cm^{-3}, and those with higher density. From this perspective, we can observe an interesting relationship between the density of the wood itself and the density of the WPC (Figure 5).

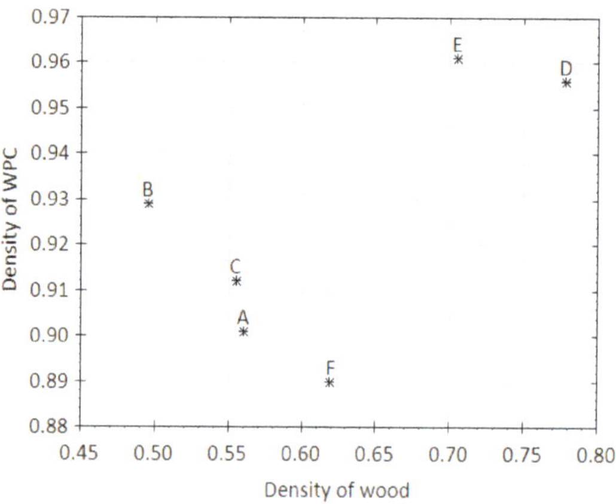

Figure 5. Density of WPC with respect to wood species used (A—boxelder maple; B—horse chestnut; C—tree of heaven; D—black locust; E—honey locust; F—mixture).

The highest increase in density was at use of the horse chestnut (compaction ratio (Ratio between WPC density and density of wood species) 1.88), while the lowest was in black locust (compaction ratio 1.23), which was also expected. Wood species with low density are indeed more compressible at the same condition as those with higher density. Furthermore, we can observe from Figure 5 that an increase in density results in a decrease in WPC. Again, we can notice two different set namely behaviour of wood species with a density below and those above 0.6 g·cm^{-3}. The decrease in density is more pronounced for boards made from wood species with low density, while differences are smaller for boards made from wood species with density above 0.7 g·cm^{-3}. Such behaviour could be related to compressibility of wood or to penetration of polymer matrix into cell lumens. Tangential diameter of vessel lumina are: 50–100 µm in *Acer negundo*, ≤50–100 µm in *Aesculus hippocastanum*, ≥200 µm in *Ailanthus altissima*, 100–≥200 µm in *Robinia pseudoacacia* (with common tyloses) and 100–≥200 µm in *Gleditsia triacanthos* [32]. The polymer compound penetrated easily into larger lumens at more dense wood species, while at wood species with lower density the compression of cell wall occurred prior to mobilization of polymeric compound into cell lumens. For low density wood species, the increase in density is due to the compression of the cell walls, while for high density wood species the penetration of PE in lumens lead to a higher density (although the compression ratio is lower).

The differences between alien wood species are less obvious in terms of flexural (Figure 6; *p* value 0.40) and tensile strength (Figure 7; *p* value 0.06), while the differences in modulus of elasticity (Figure 8; *p* value 0.00), sorption properties (Figures 9 and 10; *p* value 0.01 respectively 0.00) and dimensional stability (Table 4) are more significant (at $\alpha = 0.05$).

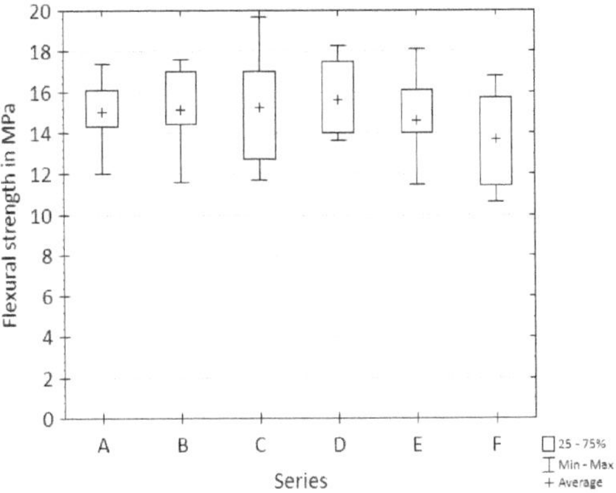

Figure 6. Flexural strength of WPC with respect to wood species used (A—boxelder maple; B—horse chestnut; C—tree of heaven; D—black locust; E—honey locust; F—mixture).

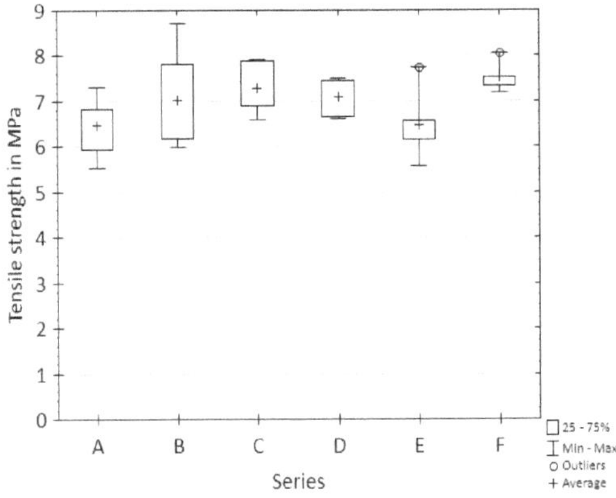

Figure 7. Tensile strength of WPC with respect to wood species used (A—boxelder maple; B—horse chestnut; C—tree of heaven; D—black locust; E—honey locust; F—mixture).

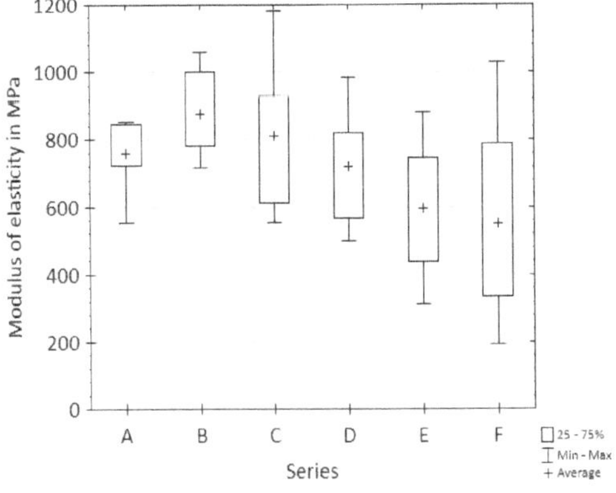

Figure 8. Modulus of elasticity of WPC with respect to wood species used (A—boxelder maple; B—horse chestnut; C—tree of heaven; D—black locust; E—honey locust; F—mixture).

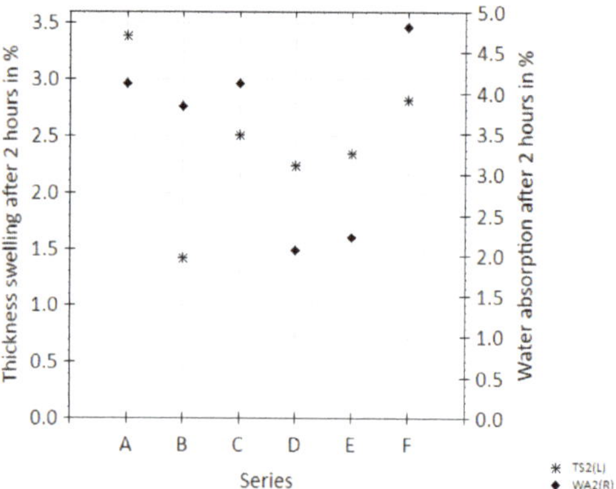

Figure 9. Thickness swelling and water absorption after 2 h immersion (A—boxelder maple; B—horse chestnut; C—tree of heaven; D—black locust; E—honey locust; F—mixture).

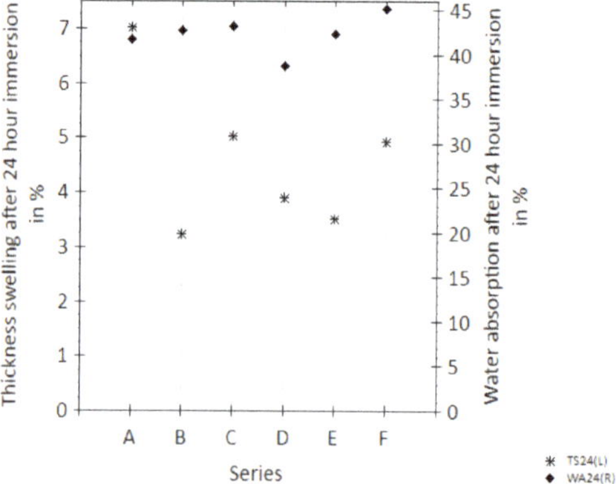

Figure 10. Thickness swelling and water absorption after 24 h immersion (A—boxelder maple; B—horse chestnut; C—tree of heaven; D—black locust; E—honey locust; F—mixture).

The WPC strengths (flexural and tensile) are related to the combined effect of particle size (Figure 2, Table 4) and WPC density. Medved et al. [33] determined size related differences between some alien wood species. They analysed particles of some alien wood species. Black locust and staghorn sumac gave the longest particles, while the shortest were determined at honey locust and tree of heaven. The authors also determined differences in aspect ratio (ratio between particle length and width), tree of heaven gave the particles with the lowest aspect ratio, while black locust the highest. In wood species with lower density the densification (higher compaction ratio) enabled optimal strength values, whereas for WPC made from higher density wood species (D and E series), adequate particle morphology and their mechanical properties are to be considered. In the case of flexural strength, two values stand out, namely in the case of WPC made of black locust (D series) and mixture (F series), one having the highest value and the other the lowest. The high value at black locust (D series) is related to its density and high strength properties, while the low flexural strength of the WPC board made of mixture is the consequence of its low density. An important aspect of WPC strength is related to the size of constituent, its embedding and interaction with the polymer matrix. In order to achieve adequate strength, the fibrous material (in our case wood particles), must be long enough to resist the forces applied to them, especially the shear and tensile forces generated when the fibrous elements are pulled out of the matrix. According to Callister [34], the most likely occurrence of failure is the end of fibrous elements, where the shear stresses are highest and the tensile stresses are lowest. The load is transferred from the matrix through the particle ends through shear, which gradually "moved" to tensile, which was more carried by the particle and less by the matrix. In such a loading behaviour, the aspect ratio of the particles is important, i.e., when the aspect ratio is low, the load transfer overlaps at the ends, so that the strain gradient in the particles does not reach the strain gradient in the PE matrix. [35,36] When the particles are long enough, their ability to withstand the load is much higher, and according to the flexural strength results, black locust particles have reached and exceeded this critical particle length. Although modulus of elasticity and tensile strength should follow the same pattern (related to particle dimensions and density), the results in our experiment do not support this. The highest tensile strength was determined at WPC made of mixture (F series), and the highest modulus of elasticity was found for WPC made of horse chestnut (B series). The lowest modulus of elasticity was determined for WPC made of mixture (F series), and the lowest tensile strength for boxelder maple (A Series).

The differences in the strength properties of WPC could be related to the particle morphology, its densification rate as well as to its interaction with the polymer matrix. A possible reason for the differences could also be the presence of micro- and macro voids in the particles formed during disintegration and drying. Such micro- and macro voids lead to a strength reduction due to a less efficient load transfer from the matrix to the particles. The comparison of the strength properties of WPC made from alien wood species with WPC made from spruce (*Picea abies*) shows similar values. The properties of WPC made from spruce are presented in Table 6. WPC board made of spruce wood was in our laboratory made by same conditions and process as compared boards made of invasive species.

Table 6. Comparison of WPC board properties from spruce wood with properties of WPC from alien wood species.

Property	Spruce	Alien
	%	%
Density in $g \cdot cm^{-3}$	0.946	0.871–0.963
Flexural strength in MPa	15.98	13.69–15.60
Modulus of elasticity in MPa	807	550–876
Tensile strength in MPa	7.18	6.46–7.52

The influence of the type of wood used for WPC was also determined at dimensional stability and moisture resistance.

Although the particles are embedded into PE matrix, water can penetrate the particles to cause the thickness change. However, the degree of change (thickness swelling and water absorption) could be related to the sorption properties of the wood. We assume that the differences are more related to the interaction between wood and PE matrix. In the case of a good interaction, the PE matrix efficiently embedded the wood particles and penetrated into the cell wall lumens, sealing them and thus making them inaccessible to water penetration into the lumen, and the effect (thickness swelling and water uptake) was lower. The amount of water absorbed by wood depends strongly on the number of free hydroxyl groups to which water can attach, and if the bond between PE matrix and wood is compact, then these hydroxyl groups are occupied by another component and therefore unavailable to water. The differences in thickness swelling could also be related to the composite density. Lower swelling was observed for WPC with densities between $0.92 \ g \cdot cm^{-3}$ and $0.96 \ g \cdot cm^{-3}$.

The sensitivity of the particles to moisture, although embedded in the PE matrix, was also determined by dimensional stability test (Table 5). When exposed to a temperature of +65 °C, the size of the WPC decreases in length and thickness (compared to the value before exposure). We hypothesise that this could be related to the shrinkage of the particles, while exposure to lower temperatures (-25 °C) causes an expansion in length that could be related to the expansion of water molecules in the cell wall, in the lumen of the particles, between the particles, and between the particles and the PE matrix. Water is indeed an exception when exposed to low temperatures, namely when water freezes, it can expand by about 9% [37].

3.2. Surface and Internal Structure of WPC Boards

According to the methodology presented above, we investigated the surface and internal structure of WPC boards made of wood species mixture (F series). First Scanning Electron Microscopy (SEM) observations were performed on the unflattened surface of WPC boards. In this case, we could only observe flat homogeneous surface and we could not recognize any wood structure, especially at lower magnification (100×) (Figure 11a). Rarely, in some areas, we could observe some voids on the surface. A detailed analysis (1000× magnification) revealed some wood anatomical structure which were, however, difficult to recognize and could not enable the identification of wood species (Figure 11b).

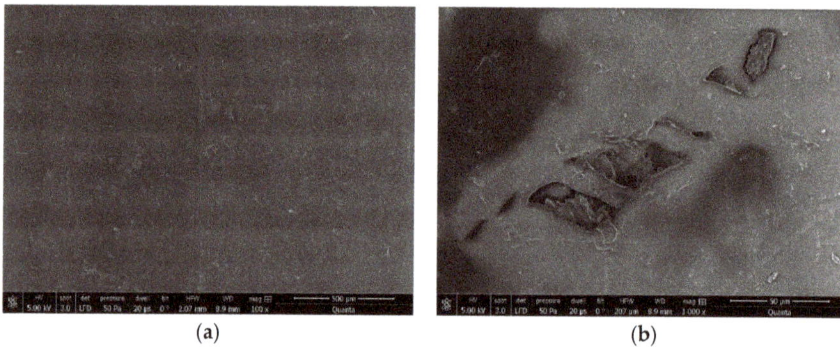

(a)   (b)

Figure 11. Scanning Electron Microscopy observation on the unflattened surface of WPC boards; (**a**) homogeneous surface with no recognizable wood structure, (**b**) higher magnification image revealing some wood structures.

The main objective of the microscopic analysis of the WPC structure was to observe the interaction and adhesion between the polymer matrix and the small wood particles embedded in it. To enhance the observation of the anatomical structures of the wood, WPC cross-section were pre-treated. A first overview of a WPC pre-treated cross-section was made on the region of interest (ROI) by a Confocal Laser Microscope (Figure 12). On Figure 12 we can clearly observe wood particles. We could recognize the polymer matrix between the wood particles and within the wood pores. In the WPC cross-section we could not detect any voids. To observe in detail and with high resolution the interaction between wood particles and polymer we used SEM. Various regions of interest (ROI), which we observed at SEM, are marked A1 to A4 in Figure 12.

In Figure 13 the internal WPC board structure at 100× and 250× magnification is shown (ROI marked on Figure 11). In this case, the structure of wood is clearly identifiable, so we can recognize the arrangement of the vessels as well as the wood fibres and tracheids. According to previous wood structure analysis of IAPS [16,17] we can identify boxelder maple (*Acer negundo*) cross section. This proved that with our SEM methodology it is possible to identify the wood species used for WPC production although the wood particles are small.

A larger magnification (500× and 2500×) in Figure 14 revealed interaction between wood particles and polymer matrix. The embedding of wood components in a polymer matrix, as well as lumens filled with polymer, is clearly visible. In Figure 14b is shown a detailed cross-section structure, in which we can see that all fibres, as well as ray parenchyma cells, are filled with polymer. The absence of considerable voids between wood fibre and matrix indicates good compatibility and good interfacial adhesion.

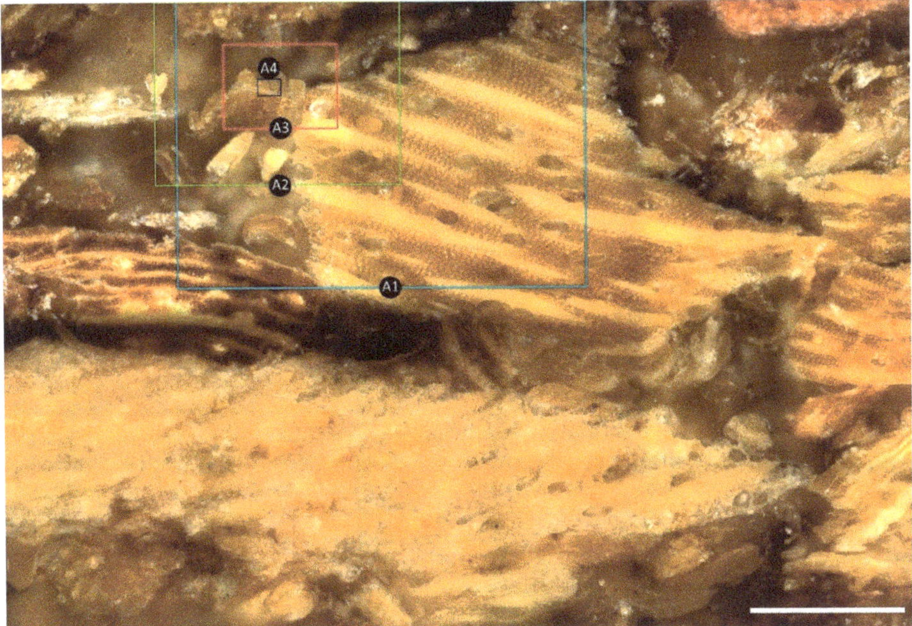

Figure 12. Confocal Scanning Laser image of a cross section WPC board. A1 to A4 are ROI analysed by Scanning Electron Microscopy (scale bar is 500 μm).

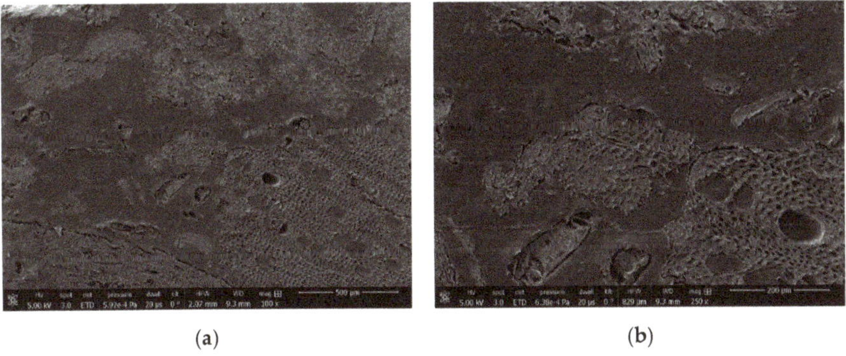

(a)   (b)

Figure 13. Scanning Electron Microscopy image of a cross section WPC board; (a) A1 analysed ROI, (b) A2 analysed ROI.

The detailed structural observation of WPC showed that the WPC production process used (flat pressing) produced a fine and filled structure without large any major voids or fibre pull out, and with good interfacial adhesion, so that the moisture resistance, mechanical properties and thermal stability were relatively high.

The experience gained in the production of flat pressed WPC was used to carry out a pilot production of 3D shaped composites. Following the described procedure, we produced several different 3D-shaped WPC products made from invasive alien woody plants (Figure 15).

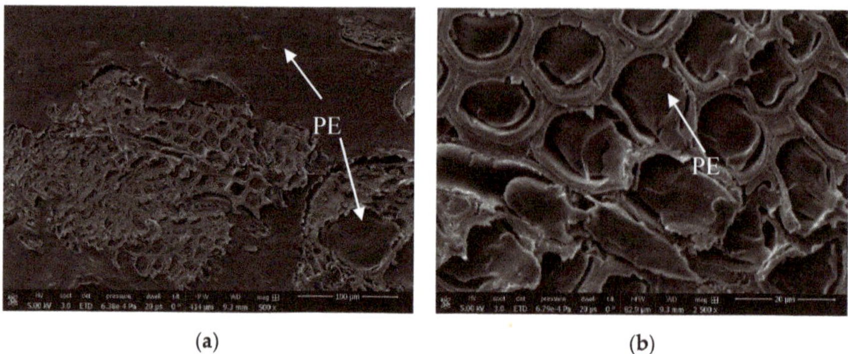

(a)   (b)

Figure 14. Scanning Electron Microscopy image of a cross section WPC board. (**a**) A3 analysed ROI, (**b**) A4 analysed ROI.

Figure 15. Some examples of 3-dimensional shaped WPC products made from invasive alien woody plants.

Based on the results of this research and the newly gained experience, we will further develop the production of wood-based composites from invasive species, analysing the impact of particle morphology differences in order to optimise (increase) the proportion of wood in WPC.

Future studies on replacing polyethylene with polylactic acid (PLA)—A biodegradable, renewable material derived from crops such as corn and sugarcane, would be of great interest. PLA is one of the fastest growing bioplastics in the bio-composites industry due to its good properties such as renewability, biodegradability, biocompatibility, ease of processing and high modulus [38].

4. Conclusions

The present study shows that the differences between the mechanical properties of WPC were less pronounced in all the selected invasive alien wood species studied, while the differences in sorption properties and dimensional stability were more significant. We conclude that good adhesion and complete embedding of the polymer material in the wood cells is crucial to obtain a good, solid, mechanically strong and water resistant WPC material. Invasive alien plant species proved to have a high potential for the production of WPCs. Taking into account the economic indicators, it is currently difficult to demonstrate a high added value of the developed products, but there is the potential to do so in the future. In the processing of wood residues from invasive alien plant species, it is the reuse of harmful invasive alien plants that brings a particular added value to our products, as it contributes (in)directly to the care of the environment and the conservation of biodiversity. That is the greatest contribution of this work.

Author Contributions: Conceptualization, S.M. and M.M.; methodology, S.M., D.K.T., A.B. and M.M.; validation, S.M. and M.M.; experimental analyses, D.K.T. and A.B., formal analysis, S.M. and M.M.; writing—original draft preparation, S.M. and M.M.; writing—review and editing, S.M., D.K.T., A.B. and M.M.; visualization, S.M. and M.M.; supervision, S.M. and M.M.; project administration, M.M.; funding acquisition, M.M. All authors have read and agreed to the published version of the manuscript.

Funding: The research was supported by APPLAUSE (UIA02-228) project, co-financed by the European Regional Development Fund through the Urban Innovative Actions Initiative (www.ljubljana.si/en/applause/), and additionally supported by the Program P4-0015, co-financed by the Slovenian Research Agency.

Acknowledgments: The authors wish to thank Jože Planinšič, Luka Krže and Denis Plavčak (production of 3D WPC) for their immense help with sample preparation.

Conflicts of Interest: The authors declare no conflict of interest.

References

1. Buschbeck, L.; Kehr, E.; Jensen, U. Untersuchungen über die Eignung verschiedener Holzarten und sortimente zur Herstellung von Spanplatten—1. Mitteilung: Rotbuche und Kiefer. *Holztechnologie* **1961**, *2*, 99–110.
2. Buschbeck, L.; Kehr, E.; Jensen, U. Untersuchungen über die Eignung verschiedener Holzarten und sortimente zur Herstellung von Spanplatten—2. Mitteilung: Kiefernreiserholz. *Holztechnologie* **1961**, *2*, 195–201.
3. Kehr, E. Untersuchungen über die Eignung verschiedener Holzarten und sortimente zur Herstellung von Spanplatten—3. Mitteilung: Der Einfluß des Härteranteils auf die eigenschaften von Spanplatten aus Rotbuchen-und Kieferholz. *Holztechnologie* **1962**, *3*, 22–28.
4. Niemz, P. *Physik des Holzes und der Holzwerkstoffe*; Leinfelden—Echterdingen, DRW—Verlag: Tübingen, Germany, 1993; pp. 27–33. ISBN 3-87181-324-9.
5. Schöberl, M. Elastische Rückfederung verdichteter Spänvliese aus Siebfraktionen verschiedener Span– und Holzarten. *Holz Roh Werkst* **2000**, *58*, 46.
6. Marra, A.A. *Technology of Wood Bonding, Principles in Practice*; Van Nostrand Reinhold: New York, NY, USA, 1992; 454p, ISBN 978-0442007973.
7. Dunky, M.; Niemz, P. *Holzwerkstoffe und Leime: Technologie und Einflussfaktoren*; Springer: Berlin, Germany, 2002; 954p, ISBN 978-3-540-42980-7.
8. Berger, M.J.; Stark, N.M. Investigations of species effects in an injection molding grade wood filled polypropylene. In *Fourth International Conference on Woodfiber-Plastic Composites*; Forest Products Society: Madison, WI, USA, 1997; pp. 19–25. ISBN 0935018956.
9. Klyosov, A.A. *Wood-Plastic Composites*; John Wiley & Sons, Inc. Publication: Hoboken, NJ, USA, 2008; 702p, ISBN 978-0-470-14891-4.
10. Oksman Niska, K.; Sain, M. *Wood-Polymere Composites*; Woodhead Publishing, Ltd: Cambridge, UK, 2008; 384p, ISBN 978-1-84569-457-9.
11. Kim, J.W.; Harper, D.P.; Taylor, A.M. Effect of wood species on the mechanical properties of wood-plastic composites. *J. Appl. Polym. Sci.* **2009**, *112*, 1378–1385. [CrossRef]
12. Migneault, S.; Koubaa, A.; Erchiqui, F.; Chaala, A.; Englund, K.; Wolcott, M.P. Effect of processing method and fiber size on the structure and properties of wood-plastic composites. *Compos. Part A Appl. Sci.* **2009**, *40*, 80–85. [CrossRef]
13. Tisserat, B.; Reifschneider, L.; Gravett, A.; Peterson, S.C. Wood-plastic Composites Utilizing Wood Flours Derived from Fast-growing Trees Common to the Midwest. *BioResources* **2017**, *12*, 7898–7916. [CrossRef]

14. Shebani, A.N.; Van Reenen, A.J.; Meincken, M. The Effect of Wood Species on the Mechanical and Thermal Properties of Wood–LLDPE Composites. *J. Compos. Mater.* **2009**, *43*, 1305–1318. [CrossRef]
15. Gardner, D.J.; Han, Y.; Wang, L. Wood-Plastic Composite Technology. *Curr. For. Rep.* **2015**, *1*, 139–150. [CrossRef]
16. Merela, M.; Thaler, N.; Balzano, A.; Plavčak, D. Optimal Surface Preparation for Wood Anatomy Research of Invasive Species by Scanning Electron Microscopy. *Drv. Ind.* **2020**, *71*, 117–127. [CrossRef]
17. Merhar, M.; Gornik Bučar, D.; Merela, M. Machinability Research of the Most Common Invasive Tree Species in Slovenia. *Forests* **2020**, *11*, 752. [CrossRef]
18. Findlay, G.W.D.; Levy, J.F. Scanning electron microscopy as an aid to the study of wood anatomy and decay. *J. Inst. Wood Sci.* **1969**, *5*, 57–63.
19. Jansen, S.; Piesschaert, F.; Smets, E. Wood anatomy of Elaeagnaceae, with comments on vestured pits, helical thickenings, and systematic relationships. *Am. J. Bot.* **2000**, *87*, 20–28. [CrossRef] [PubMed]
20. Jansen, S.; Pletsers, A.; Sano, Y. The effect of preparation techniques on SEM-imaging of pit membranes. *IAWA J.* **2008**, *29*, 161–178. [CrossRef]
21. Collett, B.M. Scanning electron microscopy: A review and report of research in wood science. *Wood Fiber Sci.* **2007**, *2*, 113–133.
22. Vek, V.; Balzano, A.; Poljanšek, I.; Humar, M.; Oven, P. Improving Fungal Decay Resistance of Less Durable Sapwood by Impregnation with Scots Pine Knotwood and Black Locust Heartwood Hydrophilic Extractives with Antifungal or Antioxidant Properties. *Forests* **2020**, *11*, 1024. [CrossRef]
23. Chun, K.S.; Fahamy, N.M.Y.; Yeng, C.Y.; Choo, H.L.; Ming, P.M.; Tshai, K.Y. Wood plastic composites made from corn husk fiber and recycled polystyrene foam. *J. Eng. Sci. Technol.* **2018**, *13*, 3445–3456.
24. Kallakas, H.; Shamim, M.A.; Olutubo, T.; Poltimäe, T.; Süld, T.M.; Krumme, A.; Kers, J. Effect of chemical modification of wood flour on the mechanical properties of wood-plastic composites. *Agron. Res.* **2015**, *13*, 639–653.
25. Pratheep, V.G.; Priyanka, E.B.; Thangavel, S.; Gousanal, J.J.; Antony, P.B.; Kavin, E.D. Investigation and analysis of corn cob, coir pith with wood plastic composites. *Mater. Today Proc.* **2020**, *7*. [CrossRef]
26. Segerholm, B.K.; Ibach, R.E.; Wålinder, M.E. Moisture sorption in artificially aged wood-plastic composites. *BioResources* **2012**, *7*, 1283–1293.
27. Leu, S.-Y.; Yang, T.-H.; Lo, S.-F.; Yang, T.-H. Optimized material composition to improve physical and mechanical properties of extruded wood-plastic composites (WPCs). *Constr. Build. Mater.* **2012**, *29*, 120–127. [CrossRef]
28. EN 323. *Wood-Based Panels—Determination of Density*; CEN: Brussels, Belgium, 1993; p. 7.
29. EN ISO 178. *Plastics—Determination of Flexural Properties*; CEN: Brussels, Belgium, 2019; p. 25.
30. EN ISO 527-1. *Plastics—Determination of Tensile Properties—Part 1: General Principles*; CEN: Brussels, Belgium, 2019; p. 26.
31. EN 317. *Particleboards and Fibreboards—Determination of Swelling in Thickness after Immersion in Water*; CEN: Brussels, Belgium, 1993; p. 12.
32. Wheeler, E.A. Inside Wood–A web resource for hardwood anatomy. *IAWA J.* **2011**, *32*, 199–211. [CrossRef]
33. Medved, S.; Vilamn, G.; Merela, M. Alien wood species for particleboards. In Proceedings of the International Conference "Wood Science and Engineering in the Third Millennium", Braşov, Romania, 7–9 November 2019; pp. 321–328.
34. Callister, D.W., Jr. *Materials Science and Engineering: An Introduction*; John Wiley & Sons, Inc. Publication: Hoboken, NJ, USA, 2007; 201p, ISBN-13: 978-0-471-73696-7.
35. Schwarzkopf, M.; Muszynski, L. Strain distribution and load transfer in the polymer-wood particle bond in wood plastic composites. *Holzforschung* **2015**, *69*, 53–60. [CrossRef]
36. Sretenovic, A.; Müller, U.; Gindl, W. Mechanism of stress transfer in a single wood fibre-LDPE composites by means of electronic laser speckle interferometry. *Compos. Part A-Appl. Sci. Manuf.* **2006**, *37*, 1406–1412. [CrossRef]
37. IAPWS home page. Why does Water Expand When It Freezes? Why does Liquid Water have a Density Maximum? Available online: http://www.iapws.org/faq1/freeze.html (accessed on 10 October 2020).
38. Wimmer, R.; Steyrer, B.; Woess, J.; Koddenberg, T.; Mundigler, N. 3D printing and wood. *Pro Ligno* **2015**, *11*, 144–149.

Article

Comparative Studies on Two Types of OSB Boards Obtained from Mixed Resinous and Fast-growing Hard Wood

Aurel Lunguleasa [1,*], Adela-Eliza Dumitrascu [2] and Valentina-Doina Ciobanu [3]

1 Department of Wood Processing and Design of Wood products, Faculty of Wood Engineering, Transilvania University of Brasov, 1 Universitatii, 500068 Brasov, Romania
2 Department of Manufacturing Engineering, Transilvania University of Brasov, Mihai Viteazul, 500174 Brasov, Romania; dumitrascu_a@unitbv.ro
3 Department of Silviculture and Forest Engineering, Transilvania University of Brasov, Sirul Beethoven, 500123 Brasov, Romania; ciobanudv@unitbv.ro
* Correspondence: lunga@unitbv.ro

Received: 2 September 2020; Accepted: 17 September 2020; Published: 23 September 2020

Featured Application: The research carried out in this paper has an applicative role in the field of oriented strand board (OSB) manufacturing in the fields of both raw material recipes and their properties. In the field of raw material recipes, the paper recommends putting an increased emphasis on fast-growing deciduous species, which can replace a good part of increasingly expensive and hard-to-find resinous wood. In the field of the mechanical strength properties of OSBs, an optimum density of 714–740 kg/m^3 has been found to condition high strength without unjustifiably increasing their density.

Abstract: The paper aims to compare the oriented strand boards (OSBs) made in the laboratory from a mixture of softwood species to those made from hardwood species, followed by their comparison to European and industry standards. In this regard, the main properties of the panels made in the laboratory were determined, including density, absorption, and swelling in thickness, modulus of elasticity, modulus of rupture, and internal bond. The analysis of the properties of swelling (24 h) and absorption (24 h) revealed that the mixture of softwood species was slightly better thanthe hardwood one. It was also shown that the panels manufactured from the mixture of hardwood species had better mechanical properties than those made of the softwood mixture (modulus of rupture (MOR) – 43.48 N/mm^2, modulus of elasticity (MOE) = 7253 N/mm, and internal bond (IB) = 1.57 N/mm^2). Additionally, the comparative analysis of properties indicates that the density is highly significant in determining the MOE values of the OSBs. This will allow softwood speciestobe replaced with other species of soft and fast-growing deciduous trees such as willow, birch, and poplar in the manufacture of oriented strand boards.

Keywords: oriented strand boards (OSBs); fast-growing species, modulus of rupture (MOR); modulus of elasticity (MOE); internal bond (IB); swelling (S); water absorption (A)

---

1. Introduction

Oriented strands boards (OSBs) are increasingly sought after on the construction materials market due to its excellent properties, especially due to the increasingly competitive price. As a composite material, these boards have wood chips as their matrix and adhesives as reinforcement. European standards in the field of OSBs [1,2] classify them according to the field of use as indoor boards (OSB/1 and OSB/2) and outdoor boards (OSB/3 and OSB/4) and from the point of tenacity in light-load-bearing boards and high-load-bearing boards. OSBs are usually made of three layers: two surface layers with large chips and a core layer with smaller chips. The three layers are symmetrical, although the core layer is thicker, thus ensuring high dimensional stability and balanced strength in the direction parallel to the fibers of the surface chips (in the direction of board length or major axis), and perpendicular to them (in the direction of board width or minor axis). From this last point of view, OSBs are almost similar to plywood; one difference could be thatthe price of OSBs is much lower than plywood, which it successfully replaces today in the construction field.

Some authors [3–5] evaluated the possibility of making OSBs without using the classic strands after obtaining OSBs with an outer layer of chips that is four times smaller than those in current factory technology; they were compared with commercial OSB/3 boards. The results led to the idea that a new standard and category of OSBs can be considered. On the other hand, the air permeability rate, having as variable parameters the board thickness of 12 and 18 mm and the board type of OSB/3 and OSB/4, were measured. Test specimens made of OSB/3 had a low permeability resistance of 61% for both thicknesses. The possibility of making low-density OSB/3 boards from industrial pine chips was also evaluated. Boards with a minimum density of 425 kg/m^3 were obtained (with a compression coefficient of 0.85 for the analyzed pine species), which had the physical and mechanical properties above those required by EN 300:2006. In 2020, experimental studies on OSB-LVL joints [6], using a 6 mm thick OSB, were conducted. A simplified 3D nonlinear finite element model was also used to find the behavior of the structure subjected to vertical and lateral loads. Good results using OSBs were obtained. OSBs from balsa wood scraps [7] and the use of an adhesive based on castor oilwas also obtained. The boards obtained with different densities and percentages of adhesive were analyzedby comparison to the standard EN 300:2006. A better compression was confirmed by a density of 650 kg/m^3 and an internal bond of 0.46 MPa. An analysis between OSBs and other composite boards such as Medium Density Fibreboard and classic chipboard [8] was made in order to study its capabilityof use in the furniture and interior decoration industry. Other research [9,10] studied the influence of isocyanine adhesive content on the properties of OSBs treated at high temperatures of 160 and 175 °C. In 2012, other groups of researchers [10] studied the use of OSBs as a substrate in wood flooring, noting that there are minimal distortions in the use of OSBs. Other authors made an ecological OSB [11] from chips of cypress and pine wood and replaced the synthetic adhesive with a mixture of lignin and tannin. The OSB made of cypress chips was more resistant to termite insects, and putrefaction resistance increased with anincrease in the ratio of cypress chips to pine chips. A group managed by Arnould [12] tested an ecological OSB obtained in the laboratory against fungal attack. From the research, an optimum resin–wood combination, from the point of view of biological and mechanical properties, was obtained. Other authors [13] wanted to obtain rigid OSBs to be used in floors. During the research, two types of chips were used: one with 90% aspen and 10% birch, and another with 100% pine. The best boards were those with the aspen–birch mixture, with values of mechanical properties (Modulus of Rupture—MOR, Modulus of Elasticity—MOE, and Internal bond—IB) close to birch veneer plywood. This research has shown that a mixture of two species provides better quality OSBs than single species usage [14]. Two strand boards with low density (aspen) and one with medium density (birch) were compared using strands with three lengths of 78, 105 and 142 mm and two chip thicknesses of 0.55 and 0.75 mm [15]. The best strengths (MOR, MOE, compressive strength, and IB) were obtained for birch and long and thin strands, demonstrating that high strengths are obtained from softwood with a low or medium density. The mixture of three chip types (red cedar, eucalyptus, and pine) for OSBswas investigated. Research by this author [16] showed that eucalyptus can replace the more expensive pine wood, but red cedar reduces its mechanical properties. Another group managed by Chiromito [17] evaluated the influence of chip length on OSBs made from *Pinus taeda* species. The results showed that the obtained board's properties exceeded the requirements of the EN 310:1999 [18] standard, and the increase in the length of the strands positively influenced the properties of the obtained boards. Other authors [19] have investigated the influence of the juvenile wood of *Pinus radiata* D., the chips being obtained from several 26-year-old trees in Chile. There was a strong influence of strands of juvenile wood after exceeding the proportion of 70%, which means it is a challenge to use this type of wood in the composition of OSBs. Del Menezzi and Tomaselli [20] proposed a heat treatment of an OSB of pine chips produced in the laboratory as a method of dimensional stabilization in order to reduce hygroscopicity and eliminate pressing stresses. A heat treatment, similar to the fibreboard defibration procedure, was performed at 250 °C for three periods of 4, 7, and 10 min, and the results obtained from the research showed that the swelling in thickness, water absorption, and equilibrium humidity was significantly reduced. Dixon and his group [21] found a process for modeling a bamboo OSB from the point of view of the modulus of elasticity on the major and minor axes. Ferro and group [22], as a solution, replaced mature softwood species with juvenile woods and used castor oil-based polyurethaneas an adhesive. All determinations were in accordance with EN 300:2006 [2] and showed higher strengths than the reference ones; the boards made of high-density species showed better physical performance, and the boards made of wood species with higher density, added little by little, had high mechanical properties. The Fabrianto group [23] investigated the influence of three fast-growing tropical species on OSB/1 boards, with a predilection for mechanical properties and dimensional stability. The result was that the boards derived from the combination of high-density species and other species with low density had better dimensional stability than those produced from a single species. De Freitas and his group [24] demonstrated the potential of two low-density tropical species when used in making OSBs, using castor oil polyurethane resin as the adhesive. All the boards obtained were classified in the OSB/4 category (EN 300), namely, boards with high mechanical strength and low hygroscopicity. Han's group [25] investigated the mechanical properties of OSBs made of small-diameter logs of pine and willow containing substantial portions of juvenile wood. The researchers showed that with the increase in chip thinness, the internal bond increased by up to 20%. They investigated the effect of wood species and the orientation of chip layers by creating three types of orientations [26]—perpendicular, parallel and random—and the use of three wood species with anhydrous wood densities of 0.36, 0.41 and 0.46 g/cm^3. OSBs made from lower density wood species had better mechanical properties and high dimensional stability. The boards with a perpendicular middle layer had better properties than the random or parallel ones. Mantanisand his group [27] analyzed the typology of adhesives used in the manufacture of OSBs, their main characteristics being evaluated asthe low emission of formaldehyde and the lower price. There was also an evaluation of the production of OSBs per year at the European level (Figure 1), the top producers being Germany, followed by Poland and Romania [27].

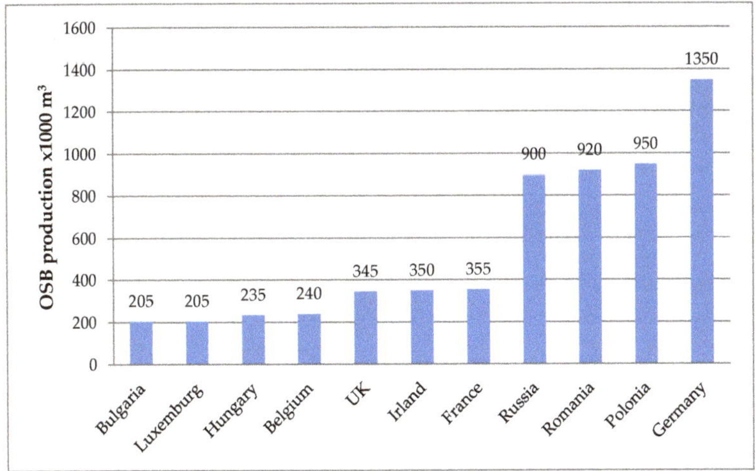

Figure 1. Statistical data compared to the production of oriented strand boards (OSBs).

Other researchers [28,29] studied the influence of the type of adhesive (urea–formaldehyde (UF) and phenol–formaldehyde (PF)) on the properties of OSBs made of pine. OSBs with PF adhesive on the surfaces and UF adhesive in the core were just as good as those that used only PF adhesive for both surface and core. A similar group evaluated the effect of pretreatment of strand particles and OSBs (post-treatment) on physical and mechanical properties. Heat treatment of large chips led to better dimensional stability and, compared to post-thermal treatment, had a more pronounced effect. The Okino group [30], who evaluated OSB properties of 750 kg/m^3 density made of Brazilian pine wood, with a chip length of 80 mm and the use of 5% and 8% urea–formaldehyde or phenol–formaldehyde resin, found that a higher adhesive content does not produce a significant increase in properties. They also evaluated the effect of high chip-drying temperature on the mechanical properties of OSBs [31]. The results showed that the strength to internal cohesion ratio was higher for dry chips at 150 °C than at other higher temperatures. Wang and Lam [32] evaluated a regression model of the slenderness ratio of strands and flake orientation to obtain superior characteristics of OSBsin MOE, MOR, IB, and thickness swelling. Other authors [33] investigated the behavior of pine chips on the properties of OSBs. Some impact of hot pressing on the surface layer (higher compression) was observed compared to the core layer.

As a conclusion of the analysis of the specialized articles in the field of the themes, an initiation of solutions to find new sources of raw materials for the production of OSBs, so as to decrease the pressure on softwood species that currently seem to be the basis of good quality raw material, is observed. Additionally, both OSB manufacturers and researchers have reported their work results to the requirements of the European standard EN 300:2006 [2] and chip geometry to factory conditions. Another pertinent conclusion of the bibliographic study is that the mixture of wood species is more advantageous than the use of individual species, and soft species with medium and low densities are most often used in OSB technology. In this sense, the main objective of the paper is to comparatively analyze OSBs obtained from a hardwood mixture andasoftwood mixture obtained under laboratory conditions (but respecting factory technology, including the chips used, the density of the board, and the adhesive type). In this sense, the physical properties (swelling and water absorption, density), as well as the mechanical ones (strength and modulus of elasticity to bending, internal cohesion) of the boards obtained in the laboratory, will be taken into account and compared to the OSB/3 board specifications required by factories.

2. Materials and Methods

For experiments, wet hardwood and softwood chips supplied by the Kronospan Trading Romania company, Brasov, Romania [33], were used in the manufacture of OSBs. The place of collection in the manufacturing flow was immediately after cutting them into chips, after a few weeks, thus obtaining a clear differentiation of wood species. Separate softwood chips (fir, spruce, and pine) and separate fast-growing hardwood chips (poplar, willow, and birch) were obtained, with an average geometry of 80 × 12 × 0.9 mm and minor differences from one species to another. The mixture of species was dictated by the properties of future boards, as specified by other authors [14]. The chips were dried under laboratory conditions up to the humidity of 6–8%, after which the hardwood and softwood species were mixed separately, thus obtaining the hardwood and softwood strands. The dried chips were subjected to sorting, after which, according to the factory technology, surface chips (F), core chips (core C), and chips with dimensions smaller than 5 × 5 mm^2 were eliminated. The chips were mixed with aLupranet-type adhesive (currently used by the Kronospan company), respecting the percentage and factory recipe. To obtain repeatability of the mixing conditions of the chips with the adhesive, a mechanical mixer was used, and the materials were weighed in grams on an electronic scale with an accuracy of 0.1 g. The operation was performed very carefully. The required amount of dry chips was calculated, taking into account the thickness of the 12 mm plate and the format of the press of the final plate. The formation of the chip carpet was performed on a metal support sheet and a forming frame, and, between the sheet and the mat, a heat-resistant foil that did not allow the OSB to stick to the support sheet was placed. The mat was three-layered, using the chips corresponding to the layers, such as the percentage of surface and core chips and the perpendicular arrangement of the core layer to the surface layers. The pressing took place at a temperature of 170 °C and a pressure of 4 MPa for a period of 15 min. The board obtained was an OSB/3 outdoor type, also made by Mirski and Dziurka [3], and it was conditioned for 48 h, after which it was edged, and the test pieces necessary for density, moisture content, static bending on major and minor axes, internal bond, swelling in thickness, and water absorption were created. The characteristics of OSB/3 boards made by the Kronospan Company, Brasov, Romania (OSB/3–Krono) are in accordance with the EN 300 standard (Table 1).

Table 1. Properties of boards with high-load-bearing capacity for the outside (OSB/3–Krono).

Items	Properties	Thickness 8–25 mm	Standard
1	Volume mass (kg/m^3)	≥620	EN 323
2	Bending strength—major axis (longitudinal) (N/mm^2)	≥20	EN 310
3	Bending strength—minor axis (transversal)(N/mm^2)	≥11	EN 310
4	Modulus of elasticity—longitudinal(N/mm^2)	≥4000	EN 310
5	Modulus of elasticity—transversal (N/mm^2)	≥1600	EN 310
6	Internal bond (N/mm^2)	≥0.35	EN 319
7	Swelling thickness (24 h) (%)	≤15	EN 317
8	Moisture content (%)	2–12	EN 322
9	Formic aldehyde emission Super Class E0 (ppm)	<0.03	EN 717-1
10	Internal cohesion after hot water test (N/mm^2)	≥0.15	EN 1087-1

2.1. Determination of Chip Moisture Content

The moisture of the OSBs was determined by the gravimetric method—the method ofweighing–drying–weighing of the prismatic pieces (10 specimens) with the dimensions of 50 × 50 mm^2. Prior to determination, the boards were conditioned for 48 h at a relative air humidity of 65% and a temperature of 20 °C. To determine the moisture, a Memmert drying oven (Germany) and a Kern electronic balance (Germany) were used for weighing the test pieces. Ten parts were weighed for each type of board, with an accuracy of 0.1 g. After that, the test parts were placed in the laboratory oven at 105 °C for a period of 10 h in order to dry them until constant mass. During drying, multiple weighings were made, and it was considered that an absolutely dry mass, m_0, was obtained when the difference between two successive weightings was insignificant (less than 0.2 g). Absolute moisture content (relative to the dry mass of the test piece) was determined as a percentage ratio between the mass of water in the wood and the mass of absolutely dry wood with the following relationship (Equation (1)):

$$M_C \;=\; (m_w - m_0) \div m_0 \times 100 \ (\%) \tag{1}$$

where M_c—moisture content (%); m_w—wet mass of test parts (g); m_0—absolutely dry mass of test parts (g).

2.2. Determination of Board Density

The determination of board density, conforming to EN 323:1993 [34], was made as a ratio between the mass and the volume of the test parts with dimensions of 50 × 50 mm^2, after the parts were conditioned. Considering that the test parts had a regular parallelepiped shape, the calculation relationship used to determine the density (ρ) was as follows (Equation (2)):

$$\rho \;=\; \left(m \times 10^6\right) \div (l_1 \times l_2 \times t) \left(\frac{\mathrm{kg}}{\mathrm{m}^3}\right) \tag{2}$$

where m—test piece mass (g); l_1—the longitudinal size of the test piece (mm); l_2—perpendicular dimension of the test piece (mm); t—the test piece thickness (mm).

2.3. Determination of Water Absorption and Swelling of the Boards Conforming to EN 317

Both determinations were performed at the same time on parallelepiped test parts with a section of 50×50 mm^2 [35]. The conditioned pieces were initially weighed, and their thickness measured, after which they were immersed for 24 h, and then the dimensions were weighed and measured again. The calculation relationships used for absorption (A) and swelling (S) were the following (Equation (3)):

$$A = (m_f - m_i) \div m_i \times 100 \ [\%]; \ S = (t_f - t_i) \div t_i \times 100 \ (\%) \tag{3}$$

where m_i—mass of test pieces before immersion (g); m_f—mass of test pieces after immersion in water (g); t_i—initial thickness of test pieces before immersion (mm); t_f—final thickness of test pieces after immersion in water (mm).

2.4. Determination of Modulus of Rupture (MOR) and Modulus of Elasticity (MOE) According to EN 310

Both tests (MOR and MOE) were determined simultaneously on the same universal test machine, an IMAL IB 600 (Italy). For this test, the corresponding devices were attached, respectively, to the two lower supports and the upper die for force application with a diameter of 30 mm [36]. The MOR and MOE determination relationships used by the test machine software were as follows (Equation (4)):

$$\begin{aligned} MOR &= (3 \times P \times l) \div \left(2 \times w \times h^2\right)\left(\tfrac{N}{mm^2}\right); \\ MOE &= \left(l^3 \times (P_2 - P_1)\right) \div \left(4 \times w \times h^3 \times (f_2 - f_1)\right)\left(\tfrac{N}{mm^2}\right) \end{aligned} \tag{4}$$

where P—the maximum stress (N); l—the distance between the supports (mm); w—the width of the cross-section of the test pieces (mm); h—the height of the test piece in the cross-section (mm); P_2—the force corresponding to a percentage of 40% of the breaking force (N); P_1—the force corresponding to a percentage of 10% of the total breaking force (N); f_2—the deformation corresponding to the force of P_2 (mm); f_1—the deformation corresponding to the force of P_1 (mm).

2.5. Determination of Internal Bond of Boards According to EN 319

The internal bond is a complex property of boards that depends on the type of adhesive, the size and particle size of the chips, the specific pressure applied to the press, and many other technical and technological elements. Internal bond was performed on 50×50 mm^2 test pieces [37], glued on special jaws, on the same universal test machine, an IMAL IB 600. The calculation ratio of this strength (IB) was as follows (Equation (5)):

$$IB = P_{max} \div l_1 \div l_2 \left(\frac{N}{mm^2}\right) \tag{5}$$

where P_{max}—maximum breaking force (N); l_1—the size of one side of the test piece (mm); l_2—the size of the other side of the test piece (mm).

In the study, different types of OSB boardswere analyzed: OSB boards obtained from mixed softwood (pine, fir, and spruce), mixed hardwood (poplar, willow, and birch), and OSB/3 manufactured by Kronospan. The mixture of species is approximately in equal proportions. The analyzed properties of OSB boards were density, swelling (24 h), absorption (24 h), internal bond (IB), modulus of rupture (MOR), and modulus of elasticity (MOE) on bending strength.

In order to compare the properties of OSB boards, the mean values were considered (displayed by interval plots diagrams). The analyses of the physical and mechanical properties of OSBs were statistically analyzed using Minitab 17 software (Minitab LLC, State College, PA, USA). The analysis consists of comparing the OSBs made of softwood and hardwood mixtures by applying interval plots with a confidence interval (CI) of 95% in order to underline the differences between the main properties of the OSBs under study.

3. Results

3.1. Analysis of Physical Properties of OSBs

The moisture content of the boards, determined in accordance with European standards (EN 322), was around 10%. The density of OSBs made of mixed softwood species was higher than the density of OSBs made of mixed hardwood, although the estimated mean values of the densities were similar for a confidence interval of 95%: the estimated median values for softwood OSBs were 753.8 kg/m^3 and for mixed hardwood OSBs 719.1 kg/m^3. Similar densities were found by the Barbirato group [7] and Okino [30], with a 0.8–1.0 compression coefficient related to wood species density. Additionally, it can be noticed that the differences between board properties are not statistically significant, and a high variation of density for OSBs made from mixed hardwood is caused by the difference in the analyzed sample sizes. The interval plot of density for each type of analyzed OSBs is plotted in Figure 2.

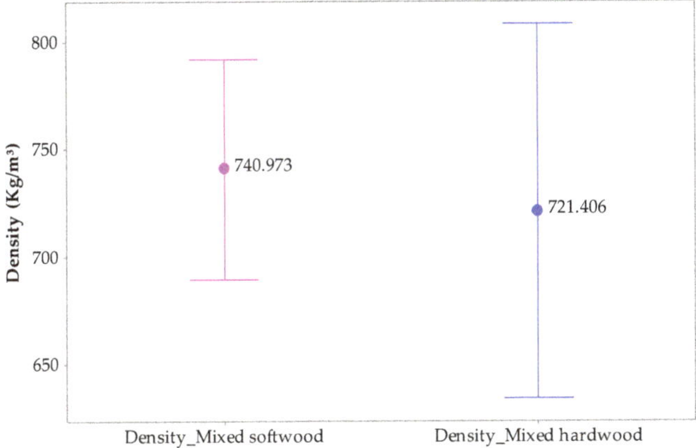

Figure 2. Interval plot of density for the analyzed types of OSBs.

Figure 3 shows that the physical properties of swelling and absorption for softwood OSBs are superior to hardwood boards. In the case of swelling, the values of the estimated means are comparable, but the values of the medians differ. Additionally, extreme values were recorded for OSBs from softwoods (minimum = 2.87%, maximum = 8.62%). Although the swelling for OSBs from softwood had a high estimated value of median compared to the OSBs from hardwoods, a large variation of data was observed. In terms of absorption, the estimated values of the means are higher for OSBs made of softwood. For analyzed types of OSBs, the estimated statistical parameters reveal the same trend of data distribution. The analysis of both types of OSBs indicates that the difference is not significant.

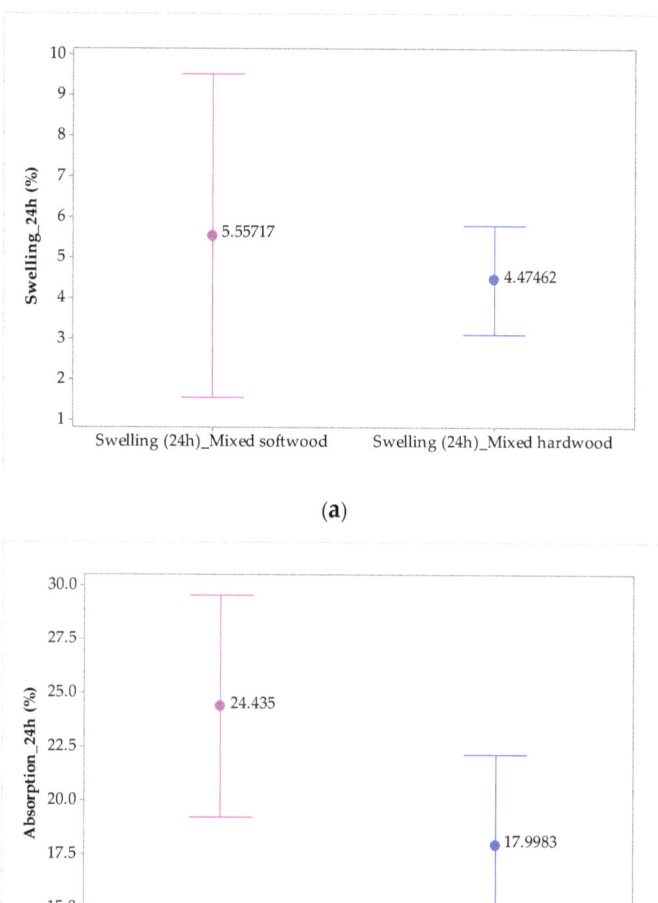

(a)

(b)

Figure 3. Interval plot of analyzed properties for OSB types: (**a**) swelling; (**b**) absorption.

3.2. Analysis of Mechanical Properties of OSBs

A comparative assessment of the MOR and MOE properties of OSBs by the implementation of interval plots with 95% CI indicates that the differencesare not significant.

Analyzing the MOR properties, the estimated average value for OSBs from softwoods is 54.50 N/mm^2, with 43.48 N/mm^2 for hardwood OSBs. From the point of view of the estimated median, the difference between them shows that the MOR properties of the softwood OSBs are superior to the MOR properties of the hardwood boards, with a difference of 25.3%. Similar results were obtained by other authors [15]. Additionally, the confidence interval of MOR values for softwood OSBs shows a high variance of data comparing to hardwood OSBs (Figure 4a).

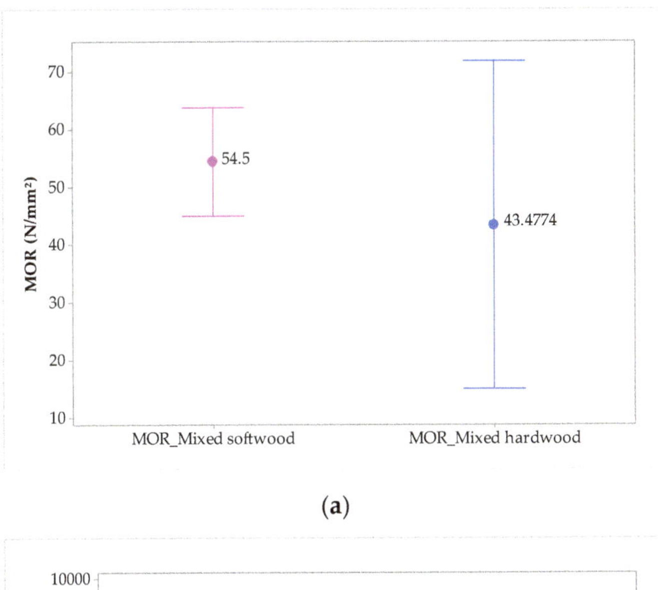

(**a**)

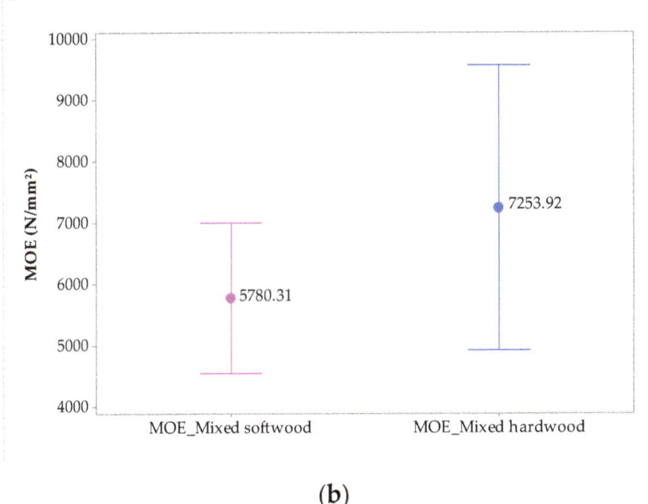

(**b**)

Figure 4. Interval plot of analyzed properties for OSB types: (**a**) modulus of rupture (MOR); (**b**) modulus of elasticity (MOE).

Regarding the MOE properties, the hardwood OSBs present superior MOE values, with a mean of 7253 N/mm^2, while softwood OSBs reveal lower properties (MOE = 5780 N/mm^2). This fact is shown by the estimated mean values for the analyzed types of OSBs (Figure 4b).

The comparative analysis of the MOR and MOE properties that are specific to the major and minor axes is presented graphically in Figures 5 and 6. Analyzing the modulus of rupture, it can be observed that the obtained values exceed the specified value of OSB/3–Krono. Additionally, higher values were obtained for OSBs made of mixed softwood species.

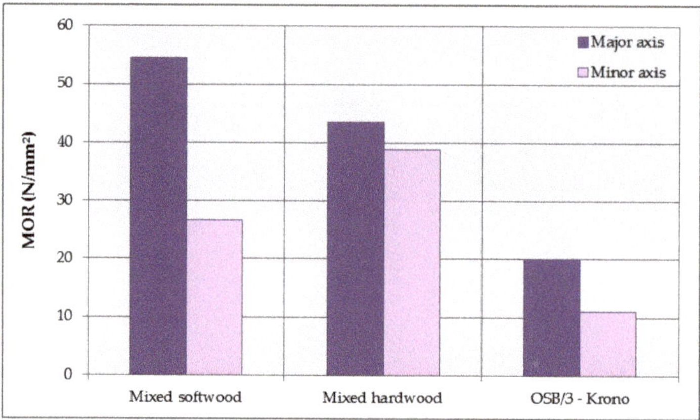

Figure 5. MOR of OSB panels from mixed species compared to OSB/3–Krono.

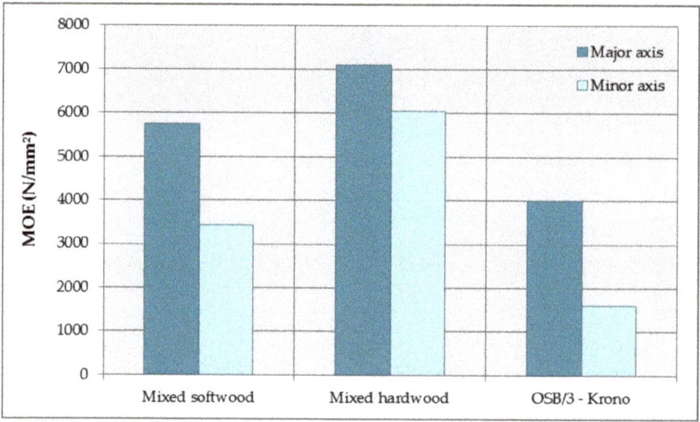

Figure 6. MOE of OSB panels from mixed species compared to OSB/3–Krono.

In the case of the modulus of elasticity of different types of manufactured OSBs, it can be seen that the modulus of elasticity at static bending on the major axis exceeded a value of 4000 N/mm^2, the manufactured board from the mixture of hardwood having the maximum value of 7110 N/mm^2 (Figure 6).

Outdoor OSBs are usually tested by internal bond, both for dry boards and for boards boiled in water for 2 h. The interval plot of internal bond properties for analyzed OSBs indicates that hardwood OSBs display higher IB properties than softwood OSBs. High values of mean (1.57 N/mm^2) and median (1.629 N/mm^2) for IB properties were estimated in the case of mixed hardwood OSBs. Based on the estimated statistical parameter and interval plots with 95% CI, the analysis indicates that the difference is not significant (Figure 7).

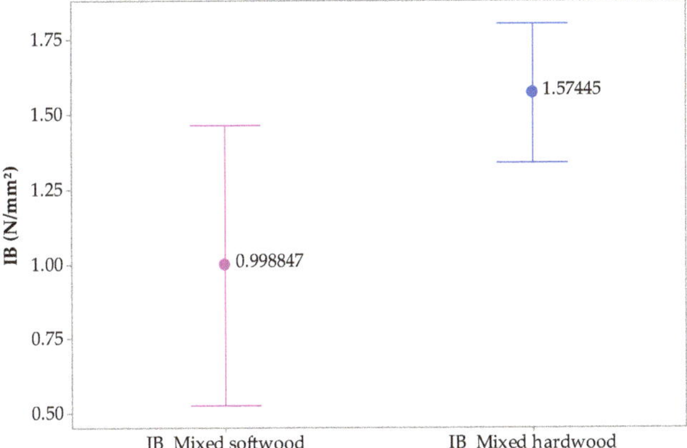

Figure 7. Interval plot of internal bond for the analyzed types of OSBs.

The diminishing strength of the internal bond after boiling in water was different from one group of species to another, as seen in Figure 8. It is observed that the decrease in IB was 38% for the softwood species mixture and 28% for the hardwood species mixture. These small differences show that the loss of strength by boiling does not depend on the group of species, but on the adhesive and other hydrophobic substances used. In this paper, the same adhesive recipe was used for all analyzed OSBs, which is why the adhesive factor was removed from the analysis. Additionally, in the case of these determinations, the obtained experimental values exceeded the minimum values of the Kronospan factory (OSB/3), and they are similar to the values obtained by other authors [26].

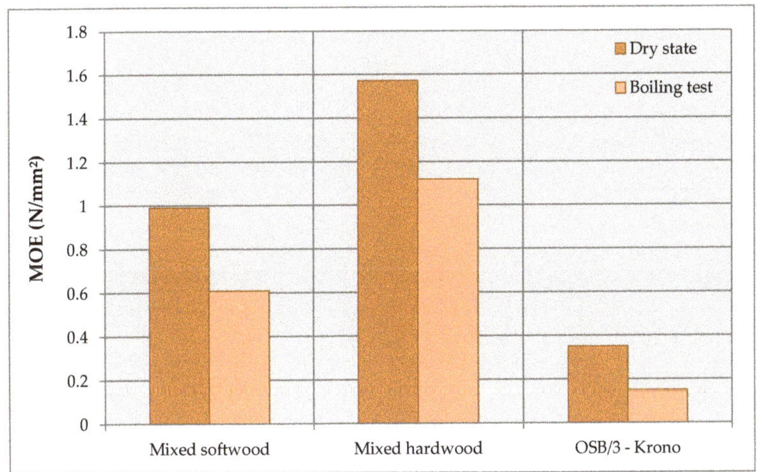

Figure 8. Internal bond of OSB panels from mixed species compared to OSB/3–Krono.

4. Discussion

Although some authors [15,26,27] have established that pine wood (or softwoods, in general) is the most used species in the OSB industry, through the experiments performed in this paper, it was found that the mixture of some fast-growing species has properties as good as those of softwoods, sometimes even better than them. This is explained by the fact that fast-growing species are soft species with a medium density, and their mixture leads to superior chip compaction within the OSB. These species approach a structure with the juvenile wood of some woody species, such as eucalyptus [16]. In the same sense, some authors [10] have shown that the mixture of aspen and birch creates much stronger boards than pine, and other authors [23] have observed that boards obtained from a mixture of species are more dimensionally stable than those obtained from a single wood species.

The density of OSBs in this work was 740 kg/m^3 for the softwood mix and 714 kg/m^3 for the hardwood mix, noting that at the same pressure, the hardwoods are less compressible by about 3.5%. These values are comparable to those found by other authors [30], with 750 kg/m^3 for pine OSBs or 650 kg/m^3 for balsa wood [7]. Absorption and swelling are up to 36% and 25% better in the hardwood mix than in the softwood mix, respectively. Regarding the resistance module, the OSBs resulting from softwoods are up to 25.5% better than those obtained from the hardwood mix, but the homogeneity of the resistance (obtained by the difference between the resistance on the major and minor axes) is much better in the case of the hardwood mix (4 N/mm^2) than the softwood mix (27 N/mm^2).

One of the major problems of the chipboard industry is that an increase in the density of the boards will cause the mechanical properties of the boards to increase accordingly. Therefore, an analysis of the correlation of mechanical properties with board density was considered appropriate. The influence of density on the physical and mechanical properties of OSB panels is presented in Figure 9. Multiple comparisons of measured properties were made, and the applied method is Dunnett's, which allows us to determine the confidence levels for each individual comparison.

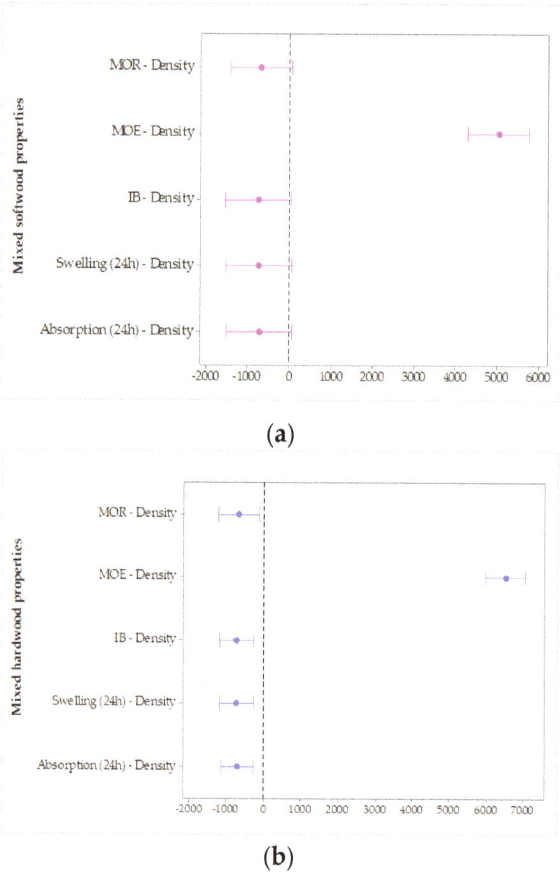

(a)

(b)

Figure 9. Comparative analysis of OSB panel properties from mixed species of softwood (**a**) and hardwood (**b**).

The analysis indicates that the density is highly significant in determining MOE values of OSBs manufactured from both mixed species. This can be explained by the fact that the elasticity of the boards is also dependent on other factors (wood elasticity, adhesive content, chip length) than the strength and thickness properties [38]. When the strength is higher, the modulus of elasticity is lower. For the other analyzed properties, there is no significant difference. Similar results indicate that MOR, MOE, IB, and WA were well correlated with board density, whereas the relationship between TS and density was less certain.

As a general conclusion of the whole work, it can be mentioned that the mixture of hardwood species has similar properties to those of the softwood mixture and is, with regards to the mechanical properties, a little better than them. This makes it possible to use these hardwood species in large quantities in the case of the raw material recipe that is used in the current conditions at Kronospan Brasov (over the percentage of 5–8%). Additionally, less expensive alternative solutions can be found [15], adopting the gradual or total replacement of softwood species with a mixture of fast-growing hardwoods, thus reducing the pressure currently exerted on softwood species.

5. Conclusions

1. The characteristics of OSBs obtained in laboratory conditions from mixed species of softwood or hardwood showed very good values, above those required by the standards in the field (EN 300) and higher compared to those required in the manufacturing flow.

2. Analyzing the estimated statistical parameters in the case of swelling and absorption properties after immersion in water for 24 h for OSBs obtained in the laboratory, it can be concluded that softwood boards are slightly superior to hardwood OSBs.

3. The statistical comparative analysis indicated that fast-growing species exhibited superior mechanical properties (MOR, MOE, and IB) compared to mixed softwood species.

4. The general conclusion of the paper is that the hardwood mixture can be used in large proportions in the OSB recipe without significantly diminishing the properties of the boards, but with favorable effects on raw material costs, also reducing the pressure on the acquisition and consumption of softwood species.

Author Contributions: Conceptualization, A.L. and V.-D.C.; methodology, A.L.; software, A.-E.D.; validation, A.L., A.-E.D., and V.-D.C.; formal analysis, A.-E.D.; investigation, A.L.; resources, V.-D.C.; data curation, A.-E.D.; writing—original draft preparation, A.L.; writing—review and editing, A.-E.D.; visualization, A.-E.D.; supervision, A.L.; project administration, V.-D.C.; funding acquisition, A.-E.D. All authors have read and agreed to the published version of the manuscript.

Funding: This research received no external funding.

Acknowledgments: The authors are grateful to Kronospan Trading Brasov Company for its support in the development of laboratory tests and for providing technical assistance, especially to Iancu Bogdan and Mastan Aurelian. We would also like to thank Transilvania University of Brasov for financial support.

Conflicts of Interest. The authors declare no conflict of interest.

References

1. ISO. *Wood-Based Panel—Oriented Strand Board (OSB)—Definitions, Classification, Specifications*; ISO 16864; International Organisation for Standardization: Geneva, Switzerland, 2016.

2. European Committee for Standardisation. *Oriented Strand Boards (OSB)—Definitions, Classification and Specifications*; EN 300; European Committee for Standardisation: Brussels, Belgium, 2006.

3. Mirski, R.; Dziurka, D.; Derkowski, A. Properties of oriented strandboards with externalla yers made of non-strand chips. *BioRes* **2016**, *11*, 8344–8354. [CrossRef]

4. Hodoušek, M.; Böhm, M.; Lemaster, R.L.; Bureš, M.; Beránková, J.; Cvach, J. Air permeation rate of oriented strand boards (OSB/3andOSB/4). *BioRes* **2015**, *10*, 1137–1148.

5. Mirski, R.; Dziurka, D. Low-density oriented strandboard. *BioRes* **2015**, *10*, 6388–6394. [CrossRef]

6.   Hassanieh, A.; Valipour, H. Experimental and numerical study of OSB sheathed-LVL studwall with stapled connections. *Constr. Build. Mater.* **2020**, *233*, 117373. [CrossRef]

7.   Barbirato, G.; Fiorelli, J.; Mejia, J.; Sarasini, F.; Ferrante, L. Quasi-static and dynamic response of oriented strand boards based on balsa wood waste. *Compos. Struct.* **2019**, *219*, 83–89. [CrossRef]

8.   Rebollar, M.; Pérez, R.; Vidal, R. Comparison between oriented strandboards and otherwood-based panels for the manufacture of furniture. *Mater. Des.* **2007**, *28*, 882–888. [CrossRef]

9.   Direske, M.; Bonigut, J.; Wenderdel, C.; Scheiding, W.; Krug, D. Effects of MDI content on properties of thermally treated oriented strandboard (OSB). *Eur. J. Wood Prod.* **2018**, *76*, 823–831. [CrossRef]

10.   Barbuta, C.; Blanchet, P.; Cloutier, A.; Yadama, V.; Lowell, E. OSB as substrate for engineered wood flooring. *Eur. J. Wood Prod.* **2012**, *70*, 37–43. [CrossRef]

11.   Amusant, N.; Arnould, O.; Pizzi, A.; Depres, A.; Mansouris, R.H.; Bardet, S.; Baudassé, C. Biological properties of an OSB eco-product manufactured from a mixture of durable and non durable species and natural resins. *Eur. J. Wood Prod.* **2009**, *67*, 439–447. [CrossRef]

12.   Arnould, O.; Sturzenbecher, R.; Bardet, S.; Hofstetter, K.; Guibal, D.; De Borst, K.; Nadine, A.; Pizzi, A.P. Mechanical potential of eco-OSB produced from durable and non durable species and natural resins. *Holzforschung* **2010**, *64*, 791–798. [CrossRef]

13.   Barbuta, C.; Cloutier, A.; Blanchet, P.; Yadama, V.; Lowell, E. Tailor made OSB for special application. *Eur. J. Wood Prod.* **2011**, *69*, 511–519. [CrossRef]

14.   Akrami, A.; Barbu, M.C.; Fruhwald, A. European hardwoods for reducing dependence on pine for oriented strandboard. *Int. Wood Prod. J.* **2014**, *5*, 133–135. [CrossRef]

15.   Beck, K.; Cloutier, A.; Salenikovich, A.; Beauregard, R. Effect of strand geometry and wood species onstrand board mechanical properties. *Wood Fiber Sci.* **2009**, *41*, 267–278.

16.   Bufalino, L. Alternative compositions of Oriented Strand Boards (OSB) made with commercial woods produced in Brazil. *Maderas Cienc. Technol.* **2015**, *17*, 105–116. [CrossRef]

17.   Chiromito, E.M.S.; Campos, C.I.; Ferreira, B.S.; Christoforo, A.L.; Lahr, F.A.R. Mechanical properties of wood panels produced with wood strands with three different lengths. *Sci. For.* **2016**, *44*, 175–180.

18.   European Committee for Standardization. *Wood-Based Panels: Determination of Modulus of Elasticity in Bending and of Bending Strength*; EN 310; European Committee for Standardization: Brussels, Belgium, 1999.

19.   Cloutier, A.; Ananias, R.A.; Ballerini, A.; Pecho, R. Effect of radiate pine juvenile wood on the physical and mechanical properties of oriented strandboard. *Holz Roh Werks.* **2007**, *65*, 157–162. [CrossRef]

20.   DelMenezzi, C.H.S.; Tomaselli, I. Contact thermal post-treatment of oriented strandboard to improve dimensional stability: A preliminary study. *Holz Roh Werks.* **2006**, *64*, 212–217. [CrossRef]

21.   Dixon, P.G.; Malek, S.; Semple, K.E.; Zhang, P.K.; Smith, G.D.; Gibson, L.J. Multi scale modeling of moso bamboo oriented strand board. *BioRes* **2017**, *12*, 3166–3181. [CrossRef]

22.   Ferro, F.S.; Souza, A.M.; de Araujo, I.I.; Van Der Neutde Almeida, M.M.; Christoforo, A.L.; Rocco Lahr, F.A. Effect of alternative wood species and first thinning wood on oriented strandboard performance. *Adv. Mater. Sci. Eng.* **2018**. [CrossRef]

23.   Febrianto, F.; Hidayat, W.; Samosir, T.P.; Lin, H.C.; Soong, H.D. Effect of strand combination on dimensional stability and mechanical properties of oriented strandboard made from tropical fastgrowing treespecies. *J. Biol. Sci.* **2010**, *10*, 267–272.

24.   DeFreitas, J.F.; de Souza, A.M.; Granco, L.A.M.N.; Chahud, E.; Christoforo, A.L.; Lahr, F.A.R. Production of structural OSB with cajueiro (*Anacardium* sp.) and a mescal (*Trattinikia* sp.)—APreliminaryStudy. *Int. J. Mater. Eng.* **2017**, *7*, 17–20.

25.   Han, G.; Wu, Q.; Lu, J.Z. Selected properties of wood strand and oriented strand board from small-diameter southern pine trees. *Wood Fiber Sci.* **2006**, *38*, 621–632.

26.   Hidayat, W.; Sya'bani, M.; Purwawangsa, H.; Hiswanto, A.; Febrianto, F. Effect of wood species and layer structure on physical and mechanical properties of strand board. *J. Trop. Wood Sci. Technol.* **2011**, *9*, 134–140.

27.   Mendes, R.F.; Mendes, L.M.; Carvalho, A.G.; Silva, A.F.A.; Guimarães, J.B. Effect of laminate inclusion and the type of adhesive in the properties of OSB panels of the wood from Pinus oocarpa. *Brazil. J. Wood Sci.* **2012**, *3*, 116–127.

28.   Mendes, R.F.; Bortoletto, J.R.G.; Almeida, N.F.; Surdi, P.G.; Barbeiro, I.N. Effect of thermal treatment on properties o f OSB Panels. *Wood Sci.Technol.* **2013**, *47*, 243–256. [CrossRef]

29. Mantanis, G.I.; Athanassiadou, E.; Barbu, M.C.; Wijnendaele, K. Adhesive systems used in the European particleboard, MDF and OSB industries. *Wood Mater. Sci. Eng.* **2018**, *13*, 104–116. [CrossRef]

30. Okino, E.Y.A.; Teixeira, D.E.; Souza, M.R.; de Santana, M.A.E.; Sousa, M.E. Properties of oriented strandboard made of wood species from Brazilian planted forests: Part 1: 80 mm-long strands of *Pinus taeda*, L. *Holz Roh Werks.* **2004**, *62*, 221–224. [CrossRef]

31. Plagemann, W.; Price, E.W.; Johns, W.E. The response of hardwood flakes and flakeboard to high temperature drying. *J Adhes.* **2006**, *16*, 311–338. [CrossRef]

32. Wang, K.Y.; Lam, F. Quadratic RSM models of processing parameters fort three-layer oriented flakeboards. *Wood Fiber Sci.* **1999**, *31*, 173–186.

33. Kronospan Trading. Available online: https://ro.kronospan-express.com/ro (accessed on 20 August 2016).

34. European Committee for Standardisation. *Wood-Based Panels—Determination of Density*; EN 323; European Committee for Standardization: Brussels, Belgium, 1993.

35. European Committee for Standardisation. *Particleboards and Fibreboards. Determination of Swelling in Thickness after Immersion in Water*; EN 317; European Committee for Standardization: Brussels, Belgium, 1993.

36. European Committee for Standardisation. *Particleboards and Fibreboards—Determination of Tensile Strength Perpendicular to the Plane of the Board*; EN 319; European Committee for Standardization: Brussels, Belgium, 1993.

37. Wu, Q.Z.; Lee, J.N. Tensile and dimensional properties of wood strands made from plantation southern pine lumber. *For. Prod. J.* **2005**, *52*, 1–6.

38. Chen, S.; Du, C.; Wellwood, R. Effect of panel density on major properties of oriented strandboard. *Wood Fiber Sci.* **2010**, *42*, 177–184.

applied sciences

Article

Walnut and Hazelnut Shells: Untapped Industrial Resources and Their Suitability in Lignocellulosic Composites

Marius Cătălin Barbu [1,2] , **Thomas Sepperer** [1], **Eugenia Mariana Tudor** [1,2,*] and **Alexander Petutschnigg** [1]

[1] Forest Products Technology and Timber Construction Department, Salzburg University of Applied Sciences, Markt 136a, 5431 Kuchl, Austria; cmbarbu@unitbv.ro (M.C.B.); thomas.sepperer@fh-salzburg.ac.at (T.S); alexander.petutschnigg@fh-salzburg.ac.at (A.P.)

[2] Faculty of Wood Engineering, University of Transilvania in Brasov, Romania, Bld. Eroilor nr.29, 500036 Brasov, Romania

* Correspondence: eugenia.tudor@fh-salzburg.ac.at

Received: 24 August 2020; Accepted: 10 September 2020; Published: 11 September 2020

Abstract: Walnut and hazelnut shells are agricultural by-products, available in high quantities during the harvest season. The potential of using these two agricultural residues as raw materials in particleboard production has been evaluated in this study. Different panels with either walnut or hazelnut shells in combination with melamine-urea formaldehyde or polyurethane at the same level of 1000 kg/m^3 density were produced in a laboratory hot press and mechanical properties (modulus of elasticity, bending strength, and Brinell hardness) and physical properties (thickness swelling and water absorption) were determined, together with formaldehyde content. Although Brinell hardness was 35% to 65% higher for the nutshell-based panels, bending strength and modulus of elasticity were 40% to 50% lower for the melamine-urea formaldehyde bonded nutshells compared to spruce particleboards, but was 65% higher in the case of using polyurethane. Water absorption and thickness swelling could be reduced significantly for the nutshell-based boards compared to the spruce boards (the values recorded ranged between 58% to 87% lower as for the particleboards). Using polyurethane as an adhesive has benefits for water uptake and thickness swelling and also for bending strength and modulus of elasticity. The free formaldehyde content of the lignocellulosic-based panels was included in the E0 category (≤2.5 mg/100 g) for both walnut and hazelnut shell raw materials and the use of polyurethane improved these values to super E0 category (≤1.5 mg/100 g).

Keywords: hazelnut; walnut; shells; lignocellulosic composites; UF; PUR; formaldehyde content

---

1. Introduction

The continuous interest in the efficient use and reuse of resources in the wood and agricultural sector for upcycled applications [1] is of great interest nowadays [2,3] in the context of the circular economy [4].

Steered by the paucity of non-renewable resources, the interest for wood could exceed its sustainable supply within the next few decades [5]. The wood demand has increased steadily, not only in the industry or for energy production (more than 50%), while the supply of wood is limited in specific regions of the world [6]. This leads to the need for substitutes for wood in engineered wood products (e.g., particleboards, PB). A promising alternative for wood in composites is provided by agricultural residues. A lot of research on agricultural waste has been carried out for PB based on wheat straw [7,8], rice straw [9,10], rapeseed [11], hemp shives [12,13], cotton dust [14], or sunflower stalks [15] and topinambour [16]. Brewer's spent grain is also a raw material for PB [17] together with tree bark [18–20]. The advantages of these agricultural and forestry by-products include reduced costs, ample availability, biodegradability, and renewability, followed by an enlarged flexibility and sound insulation [21,22]. Some drawbacks in using agricultural residues in lignocellulosic composites include unequal availability over the year, the manufacture of products with these raw materials cannot run year-round, no industrialized processing yet, big storage facilities, and different necessary pre-treatments [23–25]. Walnut and hazelnut shells are agricultural residues available in high quantities, but despite their thermal utilization [26], with no industrial use yet. Nut shells can exhibit high hardness and toughness [27]. Worldwide walnut production in 2019/2020 was roughly 965,400 tons and 528,070 tons for hazelnut [28]. Taking in account that roughly 67 % of the total fruit weight is comprised of the shell leads to 646,818 tons of walnut shells and roughly 353,807 tons of hazelnut shells each year [29]. Especially in Iran and Turkey research on the solely use of nutshells and nutshells in combination with wood has been done. The authors in [30] studied the properties of particleboard from hazelnut husks and combined with European black pine (*Pinus nigra* Arnold) [31,32]. Walnut shells were studied by [33,34]. Other research refers to particleboards manufactured with peanut hulls mixed with European black pine [35], peanut shell flour [36], and almond shells [37–39].

The aim of this study was to compare and evaluate the influence of the nutshell type (hazelnut and walnut) and resin (bonded with melamine urea formaldehyde (MUF) and polyurethane (PUR) 10% each) on the mechanical and physical properties and on the free formaldehyde content of particleboards produced solely from the above-mentioned nutshells. Other studies have dealt with PB bonded only with UF, so this study brings a novel process of gluing the nut shell panels with MUF and PUR as the properties of these boards including the Brinell hardness and formaldehyde content have not been reported yet.

2. Materials and Methods

The raw materials used for the particleboard production consisted of walnut and hazelnut shells. The hazelnut (*Corylus avellana* L.) shells were provided by the Faculty of Forestry at the University of Zagreb (Croatia). The walnut (*Juglans regia* L.) shells were provided by a family-owned walnut cracking company in Carinthia (Maria Rojach, Austria). Melamine urea formaldehyde (MUF) resin (Prefere 10G268) was provided by metaDynea (Krems, Austria) and polyurethane (PUR 501.0) was provided by Kleiberit Klebchemie (Kleiberit Klebchemie M. G. Becker GmbH & Co. KG, Weingarten, Germany).

A total of 40 kg walnut shells and 30 kg hazelnut shells were available for the project. The shells were at first manually cleaned from impurities and afterward shredded in an R40 industrial four shaft shredder at Untha Company (Kuchl, Austria) using an 8-mm screen. The shredded nutshells were dried in a Brunner-Hildebrand High VAC-S, HV-S1 (Hannover, Germany) kiln dryer for three days to reach a moisture content of about 5%. After drying, the particles were sorted into three main size classes (fine-grained, middle-grained and coarse-grained) using a sieve shaker Retsch AS 200 (Haan, Deutschland). Particles in the middle-sized (3–6 mm) and fine (<3 mm) fraction were used for the board production. The percentage of particles in each group and nutshell type is listed in Table 1.

Table 1. Weight distribution according to the fraction size of the raw material.

Raw Material	% Fine-Grained	% Middle-Sized Grained	% Coarse-Grained
Hazelnut	13.4	67.1	19.5
Walnut	12.4	73.4	14.2

A total of twelve boards (two for each group) was produced, of which the compositions are listed in Table 2.

Table 2. Manufacturing parameters of particleboards made from hazelnut, walnut, and spruce.

Board	Adhesive	Raw Material	Density (kg/m^3)	Press Time (min)	Press Temp. (°C)
A	MUF	Hazelnut	1000	6	160
B	MUF	Walnut	1000	6	160
C	MUF	Spruce	1000	6	160
D	PUR	Hazelnut	1000	20	60
E	PUR	Walnut	1000	20	60
F	UF	Spruce	700	6	180

The 320 mm × 320 mm boards were produced with a thickness of 10 mm and a density of 1000 kg/m^3 for nutshells and 700 kg/m^3 for the spruce control boards. The density of 1000 kg/m^3 could not be reached with wood particles due to the technical limitations of the hydraulic press Höfer HLOP 280. The walnut and hazelnut boards were composed of 70% middle-sized and 30% coarse-grained particles.

For all panels, the particles were mixed with resin manually and then formed into a mat and pre-compressed. This mat was pressed afterward in a hydraulic laboratory press (Höfer HLOP 280, Taiskirchen, Austria) with a pressure of 3 N/mm^2. For the UF and MUF bonded boards, 10% resin based on weight was used including 1% ammonium sulfate as a hardener. These boards (Figure 1) were pressed at 160 °C for 6 min. The PUR bonded boards were also produced with 10% adhesive, but these were pressed at 60 °C for 20 min. After pressing, the panels were conditioned at 20 °C and 65% relative air humidity for one week before cut to test specimen size, which was done according to [40].

Figure 1. Hazelnut and walnut 10 mm thick samples bonded with PUR; 1000 kg/m^3 density; composition: 70% fine-grained particles (<3 mm) and 30% coarse-grained particles (3–6 mm).

In terms of mechanical properties, bending strength (MOR) and modulus of elasticity (MOE) according to [41] and Brinell Hardness [42] were tested.

For the bending properties [41], the three-point bending test was employed.

The physical properties such as thickness swelling and water absorption after 24 h water immersion [43] and density [44] were also measured.

To determine the formaldehyde content of the panels, 250 mm × 250 mm boards with fine-grained particles (<3 mm) and 10 mm thickness and a density of 1000 kg/m^3 were manufactured with walnut and hazelnut shells. Moisture content (m.c.) was measured for each type of board (Table 3).

Table 3. Moisture content of the nutshell boards prior to determining the formaldehyde content with the perforator method.

Board	Glue	m.c. (%)	Particle Size (mm)
Walnut	UF	1.79	<3 mm
Walnut	PUR	5.25	<3 mm
Hazelnut	UF	2.28	<3 mm
Hazelnut	PUR	6.28	<3 mm

Each board was cut into 2.5 × 2.5 mm samples after cooling. The test specimens were placed in airtight bags and delivered to the Kaindl Company, Wals, Salzburg, Austria, where formaldehyde content was measured according to [45]. This method is recommended for nonlaminated and uncoated wood-based panels.

3. Results

For all boards described earlier, the mechanical and physical properties were evaluated according to the corresponding standard and statistically analyzed using mean separation tests and ANOVA.

3.1. Mechanical Properties

In terms of mechanical properties, MOE, MOR and Brinell hardness have been evaluated. The mean value, standard deviation, and minimal and maximal values are listed in Table 4. Results of ANOVA are indicated by letters a–e and u–z. The first letter refers to the different raw material (for values with the same letter, the raw material has no significant influence), the second letter refers to the used adhesive (again, and for values with the same letter, the adhesive has no influence).

Table 4. Mechanical properties of the hazelnut, walnut, and spruce particle boards.

Properties	Board	Mean	Std. Dev.	Min	Max
MOE [GPa]	A	1.13 [a,u]	0.42	0.73	1.79
	B	1.15 [b,v]	0.29	1.20	1.96
	C	2.57 [c,w]	0.36	2.12	3.00
	D	1.32 [d,u]	0.30	0.92	1.60
	E	1.30 [d,x]	0.44	0.68	1.70
	F	0.86 [e,y]	0.13	0.65	1.00
MOR [N/mm^2]	A	4.25 [a,u]	1.21	2.55	5.47
	B	5.20 [a,w]	1.02	3.79	6.53
	C	13.26 [b,y]	2.04	11.09	16.26
	D	7.43 [c,v]	1.34	5.36	8.65
	E	8.86 [c,x]	2.75	5.35	12.18
	F	5.47 [d,z]	0.75	4.33	6.32
Brinell hardness [N/mm^2]	A	62.50 [a,u]	12.55	39.95	82.65
	B	43.70 [b,v]	7.07	36.57	57.72
	C	28.00 [c,x]	5.48	22.26	35.39
	D	53.90 [d,u]	12.70	27.82	70.59
	E	55.50 [d,w]	8.51	37.30	63.89
	F	18.00 [e,y]	3.08	15.26	22.82

[a,b,c,d,e] values with the same letter were not significantly different (raw material), [u,v,w,x,y,z] values with the same letter were not significantly different (adhesive).

3.1.1. MOE and MOR

For the MUF bonded boards, the walnut shell ones (B) performed better in terms of MOE (1.51 GPa) and a MOR of 5.20 N/mm^2 compared to boards made from hazelnut shells (A) with an MOE of 1.30 GPa and a MOR of 4.25 N/mm^2 (Table 4). Although ANOVA showed that the material had a statistically significant influence in terms of MOE, it did not have an influence on the MOR. When PUR was used as an adhesive, it did not make a difference whether the boards were made of hazelnut (D) or walnut (E) shells. MOE was roughly 1.3 GPa for both, MOR 7.43 and 8.86 N/mm^2 for hazelnut and walnut shell boards, respectively. Furthermore, ANOVA showed that for walnut shell boards, it did not make a difference if MUF or PUR (both at a concentration of 10 w.t. %) was used as an adhesive when it came to MOE, while it did influence the MOR. When hazelnut shells were used as a raw material, the adhesive had a statistically significant influence on both MOE and MOR. The reference boards made from spruce and MUF (C) achieved the highest values for MOE and MOR (2.57 GPa and 13.26 N/mm^2, respectively). Surprisingly, the PUR bonded spruce boards (F) performed weaker than the nutshell panels (0.87 GPa for MOE and 5.47 N/mm^2 for MOR). The results for both hazelnut and walnut panels bonded with PUR were similar to the values obtained by [34], namely a MOR between 6 and 9 N/mm^2 and an MOE between 0.8 and 1.3 GPa. When the same amount of hazelnut shells was mixed with wood particles (50% + 50%), the values of the MOE and MOR increased significantly [32].

In terms of MOE (Table 4), the walnut shell boards showed higher values than the ones produced by [46], which reached 1.15 GPa on average compared to 1.51 GPa and might have been caused by the adhesive used. The study by [46] applied UF, while for these boards was utilized a less brittle MUF [47]. Another reason might be that [46] produced boards with a density of 700 kg/m^3 while those boards had a target density of 1000 kg/m^3 [48].

The results for MOE and MOR of the nutshell boards were expected to range below the results of spruce particleboards. This can be traced back to the fact that the mechanical properties are strongly influenced by different properties of the raw material like density, chemical composition, and particle size. A low-density raw material allows a higher compression in the panel, leading to improved performance in bending tests [49]. Given that the density of the used nutshells was very high (between 700 and 1030 kg/m^3 depending on the fraction) and spruce particles have a density of 450 kg/m^3, the lower performance of the MUF bonded nutshell boards can be traced back to a lower compression rate [50]. The weak behavior of the PUR bonded spruce boards can be traced back to unsatisfying bonding, caused by an insufficient moisture content of the raw material (3%) for the adhesive to react [51].

3.1.2. Brinell Hardness

Boards made from hazelnut shells and MUF (A) showed the highest average Brinell hardness with 62.5 N/mm^2 while the lowest hardness value was achieved by PUR bonded spruce particles with only 18 N/mm^2 (Table 4). The highest value was achieved by hazelnut with 82.6 N/mm^2 while the lowest for those panels was 39.9 N/mm^2. The big difference between the highest and lowest value and the high standard deviation can be explained as a result of the different measuring spots and the manual scattering of raw material in the press mold. This showed that the hardness was very high when measured on a big piece of the nutshells, while it was much lower when determined on the fine-grained particles that are filling the gaps between the coarse ones. This was not only valid for MUF, but also for PUR. The highest Brinell hardness value for a hazelnut board glued with PUR (D) was 70.59 N/mm^2 while the lowest was 27.82 N/mm^2. The mean was determined at 54 N/mm^2. Walnut panels showed similar results this time, PUR bonded boards (E) performed a little better with an average Brinell hardness of 55.5 N/mm^2, while MUF boards (B) only reached 43.7 N/mm^2. The highest value for the walnut boards was 63.9 N/mm^2 for those produced with PUR. ANOVA showed that when it came to PUR bonded boards, the nutshell type had no influence. There was also no statistically significant difference for hazelnut boards concerning the adhesive. Compared to spruce particleboards that showed an average result of 18 N/mm^2 for PUR and 28 N/mm^2 for MUF, the nutshells were much harder. However, the results for the wood-based boards were closely distributed to the mean value, with a standard deviation of only 3 N/mm^2. This means that the spot where the force is applied is less relevant compared to the nutshell boards.

3.2. Physical Properties

In terms of physical properties, thickness swelling (TS) and water absorption (WA) both after 24 h water immersion were evaluated.

Thickness swelling after 24 h water immersion for MUF bonded hazelnut husk panels (A) was determined with 17.5% while it was a little lower for walnut shells (B) with 13.2% (Table 5). Compared to the spruce reference board (C) with an average of 52%, the increase in thickness was much lower. The results for hazelnut (A) glued with MUF were a little better compared to those obtained by [31] and [35] (29.3% TS with a standard deviation of 3.5%), while the results for walnut (B) were higher compared to [34], who evaluated the thickness swelling with 10.2%.

Table 5. Physical properties of the hazelnut, walnut, and spruce particle boards.

Properties	Board	Mean	Std. dev.	Min	Max
TS 24 h [%]	A	17.47 [a,u]	2.88	13.56	21.91
	B	13.23 [b,w]	1.35	10.98	15.21
	C	52.09 [c,y]	7.84	42.69	64.84
	D	9.66 [d,v]	1.48	8.33	12.66
	E	8.37 [d,x]	1.26	6.48	10.29
	F	44.48 [e,z]	11.01	30.92	64.39
WA 24 h [%]	A	26.96 [a,u]	2.96	23.32	32.54
	B	24.54 [a,w]	3.55	19.29	28.78
	C	64.00 [b,y]	9.64	45.24	75.02
	D	12.44 [c,v]	2.05	9.60	15.28
	E	11.49 [c,x]	2.56	7.40	14.40
	F	89.99 [d,z]	7.04	81.49	104.09

[a,b,c,d,e] values with the same letter were not significantly different (raw material), [u,v,w,x,y,z] values with the same letter were not significantly different (adhesive).

Thickness swelling after 24 h was significantly reduced when PUR was used to produce the boards. This means that it was only 9.7% for hazelnut (D) and 8.4% for walnut (E). The spruce boards showed an increase in thickness of nearly 52% for MUF and still 44.5% for PUR. The nutshells in combination with PUR performed better compared to the spruce particles. It was found that when MUF was used as the adhesive, there was a significant difference between the walnut and hazelnut shells, while the material did not have a big influence when bonded with PUR.

The results for water absorption after 24 h water submersion were similar to those for thickness swelling after 24 h (Table 5). This means that the nutshell boards glued with MUF had a higher water uptake, namely 27% for hazelnut (A) and 25% for walnut (B), compared to the PUR bonded ones with 12.5% (D) and 11.5% (E), respectively. The values were much lower compared to the results of spruce particles with 64% for MUF (C) and almost 90% for PUR (F). Regarding water uptake over 24 h, there was no influence of the used nutshell type in combination with either MUF or PUR.

The same trend of blocking and reduction of TS and WA was observed in the case of using hazelnut shells combined with wood particles [32], with a mean of 20% for TS and 70% for WA and from [34], in the case of walnut shells with the same value for TS as reported by [32] and lower WA of 37%.

It was expected that thickness swelling and water absorption will be lower, compared to the spruce particleboards. The water uptake of lingo-cellulosic materials is strongly influenced by the number of free hydroxyl groups where water is bonded to. These hydroxyl groups are mainly present in the natural polymer cellulose. Hemicellulose is amorphous and has a hydrophilic character that is additionally increasing the water uptake. Lignin, however, is totally hydrophobic, which means that water cannot be absorbed within [52]. The high amount of lignin in the walnut shells (49.1%) compared to roughly 35% in softwood indicated a decreased amount of water absorption. Furthermore, hazelnut husks contain almost 42% of lignin and only 55% of holocellulose compared to 65% in softwood.

PUR is more hydrolytically stable than MUF, which explains the big differences for thickness swelling and water absorption when these two adhesives are compared. The high value for water absorption for the PUR glued spruce particleboards was again caused by the low (3%) moisture content of the raw material and a non-complete reaction of the adhesive.

3.3. Formaldehyde Content

The corrected values of free formaldehyde content varied depending on the type of adhesive formulation for the board (Figures 2 and 3).

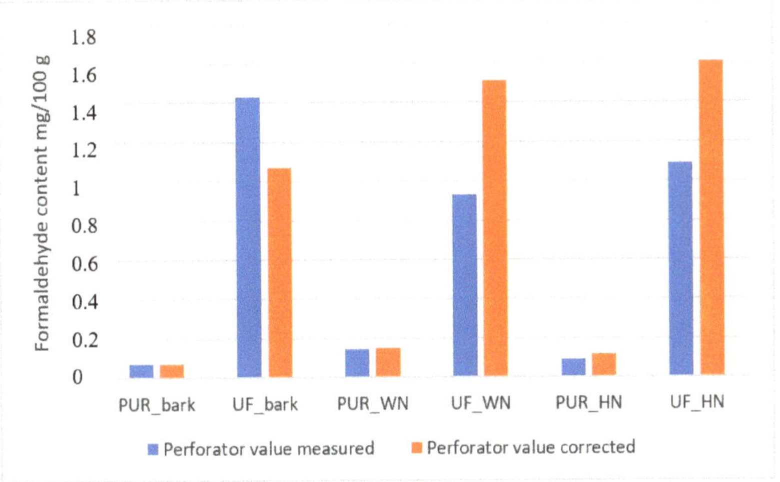

Figure 2. Free formaldehyde content for both the measured and corrected perforator values (EN 120:2011) of samples of walnut and hazelnut shell boards compared with larch bark panels from [53].

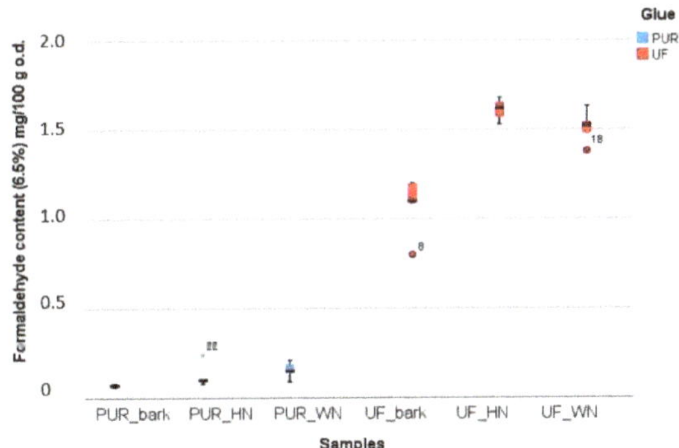

Figure 3. Free formaldehyde content measured according to EN 120:2011 of 10 mm walnut and hazelnut panels bonded with UF and MUF and compared with the values of larch bark boards glued with the same adhesives from [53].

The values for walnut and hazelnut shell panels were compared with similar 10 mm boards manufactured with the same adhesives reported by [53]. The lowest formaldehyde content of 0.07 mg/100 g oven dry was measured for the board with larch bark bonded with PUR (PUR_bark). This is included in the super E0 classification (≤1.5 mg/100 g). In the same category were comprised the values of the walnut (PUR_WN) and hazelnut (PUR_HN) shell boards glued with PUR and the larch bark board bonded with UF. To the E0 class (≤ 2.5 mg/100 g) belonged the values for the nutshell boards bonded with UF (UF_HN and UF_WN). These results are consistent with those obtained by [54] regarding formaldehyde release from low-emission wood-based panels using the perforator method, which ranged from 0.71 to 2.99 mg/100 g and was slightly lower than that of [34].

The role of bark to decrease the formaldehyde content of wood-based composites was also studied by [55,56]. It was expected that the level of free formaldehyde content of the panels glued with PUR would be significantly lower than that of the boards bonded with UF, but the combination of both adhesives and raw materials resulted in panels that reached at least the E0 classification.

4. Conclusions

This study has shown that the nutshells can be used for the manufacture of PBs with improved performances in terms of physical properties (dimensional stability), but also with higher Brinell hardness compared to the spruce particleboards.

MOR and MOE are lower by roughly 50% and cannot meet the requirements for P1 or P2 PBs, according to EN 310:2005. Future research should consider the improvement potentials by choosing the proper percentage of nutshells chips, combining the coarse-grained particles with the fine-grained ones, in order to fill voids in-between, combined with a proper bonding system.

This study has also shown that, when MUF is used to bond the boards, the raw material has no influence on MOR and water absorption, when PUR is used, the raw material has no influence on MOE, but with an improvement in terms of MOR (35% higher compared to spruce particles), Brinell hardness (up to 67%), thickness swelling (65–75% lower), and water absorption (58–87% lower). There is no statistically significant influence of the adhesive for hazelnut shells in terms of MOE and Brinell hardness.

The values of the formaldehyde content are all included in the E0 emission class. In the case of the panels bonded with PUR, the measured values were less than 1.5 mg/100 g, which means that the super E0 category was reached. All these perforator results for the panels manufactured with hazelnut and walnut shells recommend them as low-emissions lignocellulosic composites.

The nutshells, utilized only for energy purposes [57], could be considered as a raw material in particleboard production. Further research can focus on the properties of improved formulations for adhesives and lignocellulosic particles, eventually with reinforcements [58] to achieve similar properties to P1 and P2 [59].

Author Contributions: Conceptualization, T.S.; Methodology, T.S. and E.M.T. Validation, E.M.T.; Formal analysis, A.P. and M.C.B.; Investigation, T.S.; Resources, T.S.; Writing—original draft preparation, T.S. and E.M.T.; Writing—review and editing, T.S. and E.M.T.; Visualization, T.S., and M.C.B.; Supervision, A.P. and M.C.B. All authors have read and agreed to the published version of the manuscript.

Funding: This research received no external funding.

Acknowledgments: The authors would like to thank Frederick Kamke from Oregon State University (Oregon, USA) for making the project possible and his support throughout the project; Danijela Domljan from the University of Zagreb (Croatia) for providing the hazelnut shells used for the board production in this project and to Kaindl Co. in Salzburg for supporting this research with the determination of the formaldehyde content of the walnut and hazelnut samples.

References

1. Tudor, E.M.; Zwicl, C.; Eichinger, C.; Petutschnigg, A.; Barbu, M.C. Performance of softwood bark comminution technologies for determination of targeted particle size in further upcycling applications. *J. Clean. Prod.* **2018**, *269*, 122412. [CrossRef]

2. Queirós, C.; Cardoso, S.; Lourenço, A.; Ferreira, J.M.I.; Lourenço, M.; Pereira, H. Characterization of walnut, almond, and pine nut shells regarding chemical composition and extract composition. *Biomass Convers. Biorefinery* **2019**, *10*. [CrossRef]

3. de Carvalho Araújo, C.; Salvador, R.; Piekarski, C.; Sokulski, C.; de Francisco, A.; Camargo, S. Circular Economy Practices on Wood Panels: A Bibliographic Analysis. *Sustainability* **2019**, *11*, 1057. [CrossRef]

4. Fan, Y.; Lee, C.; Lim, J.; Klemeš, J. Cross-disciplinary approaches towards smart, resilient and sustainable circular economy. *J. Clean. Prod.* **2019**, *232*, 1482–1491. [CrossRef]

5. Risse, M.; Weber-Blaschke, G.; Richter, K. Resource efficiency of multifunctional wood cascade chains using LCA and exergy analysis, exemplified by a case study for Germany. *Resour. Conserv. Recycl.* **2017**, *126*, 141–152. [CrossRef]

6. Eurostat. Wood Products-Production and Trade. Available online: https://ec.europa.eu/eurostat/statistics-explained/index.php/Wood_products_-_production_and_trade (accessed on 19 August 2020).

7. Boquillon, N.; Elbez, G.; Schönfeld, U. Properties of wheat straw particleboards bonded with different types of resin. *J. Wood Sci.* **2004**, *50*, 230–235. [CrossRef]

8. Wang, D.; Sun, S. Low density particleboard from wheat straw and corn pith. *Ind. Crop. Prod.* **2002**, *50*, 43–50. [CrossRef]

9. Luo, P.; Yang, C.; Li, M.; Wang, Y. Manufacture of thin rice straw particleboards bonded with various polymeric methane diphenyl diisocyanate/ urea formaldehyde resin mixtures. *BioResources* **2020**, *15*, 935–944.

10. Yasina, M.; Waheed Bhutto, A.; Bazmi, A.; Karim, S. Efficient Utilization of Rice-wheat Straw to Produce Value-added Composite Products. *J. Environ. Chem. Eng.* **2010**, *1*, 1–8.

11. Dukarska, D.; Czarnecki, R.; Dziurka, D.; Mirski, R. Construction particleboards made from rapeseed straw glued with hybrid pMDI/PF resin. *Eur. J. Wood Prod.* **2017**, *75*, 175–184. [CrossRef]

12. Balducci, F.; Harper, C.; Meinlschmidt, P.; Dix, B.; Sanasi, A. Development of Innovative Particleboard Panels. *Drv. Ind.* **2008**, *59*, 131–136.

13. Réh, R.; Vrtielka, J. Modification of the core layer of particleboard with hemp shives and its influence on the particleboard properties. *Acta Fac. Xylologiae* **2013**, *55*, 51–59.

14. Nemli, G.; Odabas Serin, Z.; Özdemir, F.; Ayrılmış, N. Potential use of textile dust in the middle layer of three-layered particleboards as an eco-friendly solution. *BioResources* **2019**, *14*, 120–127.

15. Bektas, I.; Guler, C.; Kalaycioğlu, H.; Mengeloglu, F.; Nacar, M. The Manufacture of Particleboards using Sunflower Stalks (*Helianthus annuus* L.) And Poplar Wood (*Populus alba* L.). *J. Compos. Mater.* **2005**, *39*, 467–473. [CrossRef]

16. Klímek, P.; Meinlschmidt, P.; Wimmer, R.; Plinke, B.; Schirp, A. Using sunflower (*Helianthus annuus* L.), topinambour (*Helianthus tuberosus* L.) and cup-plant (*Silphium perfoliatum* L.) stalks as alternative raw materials for particleboards. *Ind. Crop. Prod.* **2016**, *92*, 157–164. [CrossRef]

17. Klímek, P.; Wimmer, R.; Mishra, P.; Kúdela, J. Utilizing brewer's-spent-grain in wood-based particleboard manufacturing. *J. Clean. Prod.* **2017**, *141*, 812–817. [CrossRef]

18. Wang, X.; Fang, H.; Feng, M.; Zhang, Y.; Yan, N. Manufacturing Medium-Density Particleboards from Wood–Bark Mixture and Different Adhesive Systems. *For. Prod. J.* **2015**, *65*, 20–25. [CrossRef]

19. Lakreb, N.; As, N.; Gorgun, V.; Sen, U.; Gomes, G.; Pereira, H. Production and characterization of particleboards from cork-rich *Quercus cerris* bark. *Eur. J. Wood Prod.* **2018**, *76*, 989–997. [CrossRef]

20. Medved, S.; Tudor, E.M.; Barbu, M.C.; Jambreković, V.; Španić, N. Effect of Pine (*Pinus Sylvestris*) Bark Dust on Particleboard Thickness Swelling and Internal Bond. *Drv. Ind.* **2019**, *70*, 141–147. [CrossRef]

21. Merkel, K.; Rydarowski, H.; Kazimierczak, J.; Bloda, A. Processing and characterization of reinforced polyethylene composites made with lignocellulosic fibres isolated from waste plant biomass such as hemp. *Compos. B Eng.* **2014**, *67*, 138–144. [CrossRef]

22. Yu, X.; Xu, H. Lightweight Composites Reinforced by Agricultural Byproducts. In *Lightweight Materials from Biopolymers and Biofibers*; American Chemical Society: Omaha, NE, USA, 2014; pp. 209–238.

23. Bowyer, J.; Stockmann, V. Agricultural residues: An exciting bio-based raw material for the global panels industry. *Forest Prod. J.* **2001**, *51*, 10–21.

24. Siti Suhaily, S.; Khalil, A.; Asniza, M.; Nurul Fazita, M.; Mohamed, A.; Dungani, R.; Zulqarnain, W.; Syakir, M. Design of green laminated composites from agricultural biomass. In *Lignocellulosic Fibre and Biomass-Based Composite Materials*; Elsevier: Sawston, UK, 2017; pp. 291–311.

25. Papadopoulou, E.; Chrissafis, C. Particleboards from agricultural lignocellulosics and biodegradable polymers prepared with raw materials from natural resources. In *Natural Fiber-Reinforced Biodegradable and Bioresorbable Polymer Composites*; Woodhead Publishing: Sawston, UK, 2017; pp. 19–30.

26. Figen-Antmen, Z. Exploitation of Peanut and Hazelnut Shells as Agricultural Industrial Wastes for Solid Biofuel Production. *Fresenius Environ. Bull.* **2019**, *28*, 2340–2347.

27. Flores-Johnson, E.; Carrillo, J.; Zhai, C.; Gamboa, R.; Gan, Y.S.L. Microstructure and mechanical properties of hard Acrocomia mexicana fruit shell. *Sci. Rep.* **2018**, *8*, 9668. [CrossRef] [PubMed]

28. Statista. Production of Tree Nuts Worldwide in 2019/2020, by Type (in 1,000 Metric Tons). Available online: https://www.statista.com/statistics/1030790/tree-nut-global-production-by-type/ (accessed on 19 August 2020).

29. Martínez, M.; Moiraghi, L.; Agnese, M.; Guzman, C. Making and some properties of activated carbon produced from agricultural industrial residues from Argentina. *J. Argent. Chem. Soc.* **2003**, *91*, 104–108. [CrossRef]

30. Gürü, M.; Aruntaş, Y.; Tüzün, F.; Bilici, I. Processing of urea-formaldehyde-based particleboard from hazelnut shell and improvement of its fire and water resistance. *Fire Mater.* **2009**, *33*, 413–419. [CrossRef]

31. Güler, C.; Çöpür, Y.; Büyüksari, U. Producing particleboards from hazelnut (*Corylus avellana* L.) husk and European Black Pine (*Pinus nigra Arnold*). *Wood Res.* **2009**, *54*, 125–132.

32. Kowaluk, G.; Kądziela, J. Properties of particleboard produced with use of hazelnut shells. *Ann. WULS SGGW Wood Technol.* **2014**, *85*, 131–134.

33. Gürü, M.; Atar, M.; Yıldırım, R. Production of polymer matrix composite particleboard from walnut shell and improvement of its requirements. *Mater. Des.* **2008**, *29*, 284–287. [CrossRef]

34. Pirayesh, H.; Khanjanzadeh, H.; Salarib, H. Effect of using walnut/almond shells on the physical, mechanical properties and formaldehyde emission of particleboard. *Compos. Part. B Eng.* **2013**, *45*, 858–863. [CrossRef]

35. Güler, C.; Cöpur, Y.; Tascioglu, C. The manufacture of particleboards using mixture of peanut hull (*Arachis hypoqaea* L.) and European Black pine (*Pinus nigra Arnold*) wood chips. *Bioresour. Technol.* **2008**, *99*, 2893–2897. [CrossRef]

36. Batalla, L.; Nuñez, A.; Marcovich, N. Particleboards from peanut-shell flour. *J. Appl. Polym. Sci.* **2005**, *97*, 916–923. [CrossRef]

37. Gürü, M.; Tekeli, S.; Bilici, I. Manufacturing of urea–formaldehyde-based composite particleboard from almond shell. *Mater. Des.* **2006**, *27*, 1148–1151. [CrossRef]

38. Pirayesh, H.; Khazaeian, A. Using almond (*Prunus amygdalus* L.) shell as a bio-waste resource in wood-based composite. *Compos. Part. B Eng.* **2012**, *43*, 1475–1479. [CrossRef]

39. Chaudhary, A.; Gope, P.; Singh, V. Effect of Almond Shell Particles on Tensile Property of Particleboard. *J. Mater. Environ. Sci.* **2013**, *4*, 109–112.

40. EN 326:2012. *Wood-Based Panels—Sampling, Cutting and Inspection—Part 1: Sampling and Cutting of Test Pieces and Expression of Test Results*; European Committee for Standardization: Brussels, Belgium, 2012.

41. EN 310:2005. *Wood-Based Panels: Determination of Modulus of Elasticity in Bending and of Bending Strength*; European Committee for Standardization: Brussels, Belgium, 2005.

42. EN 1534:2010. *Wood flooring—Determination of Resistance to Indentation—Brussels, Belgium*; European Committee for Standardization: Brussels, Belgium, 2010.

43. EN 317:2005. *Particleboards and fibreboards—Determination of Swelling in Thickness after Immersion in Water*; European Committee for Standardization: Brussels, Belgium, 2005.

44. EN 323:2015. *Wood-based Panels—Determination of Density*; European Committee for Standardization: Brussels, Belgium, 2015.

45. EN 120:2011. *Wood-Based Panels—Determination of Formaldehyde Release—Extraction Method (Called Perforator Method)*; European Committe for Standardization: Brussels, Belgium, 2011.

46. Pirayesh, H.; Khazaeian, A.; Tabarsa, T. The potential for using walnut (*Juglans regia* L.) shell as a raw material for wood-based particleboard manufacturing. *Compos. Part. B Eng.* **2012**, *43*, 3276–3280. [CrossRef]

47. Pizzi, A.; Mittal, K. *Handbook of Adhesive Technology*; Dekker: New York, MY, USA, 1994.

48. Dunky, M.; Niemz, P. *Holzwerkstoffe und Leime: Technologie und Einflussfaktoren, Mit 150 Tabellen*; Springer: Berlin, Germany, 2002.

49. Tabarsa, T.; Ashori, A.; Gholamzadeh, M. Evaluation of surface roughness and mechanical properties of particleboard panels made from bagasse. *Comp. Part. B Eng.* **2011**, *42*, 1330–1335. [CrossRef]

50. Kollmann, F. *Technologie des Holzes und der Holzwerkstoffe*, 2nd ed.; Springer: Berlin, Germany, 1982.

51. Zeppenfeld, G.; Grunwald, D. *Klebstoffe in der Holz- und Möbelindustrie*, 2nd ed.; Weinbrenner, DRW-Verlag: Leinfelden-Echterdingen, Germany, 2005.

52. Gwon, J.; Lee, S.; Chun, S.; Doh, G.; Kim, J. Effects of chemical treatments of hybrid fillers on the physical and thermal properties of wood plastic composites. *Compos. Part. A Appl. Sci. Manuf.* **2010**, *41*, 1491–1497. [CrossRef]
53. Tudor, E.M.; Barbu, M.C.; Petutschnigg, A.; Réh, R.; Krišťák, Ľ. Analysis of Larch-Bark Capacity for Formaldehyde Removal in Wood Adhesives. *Int. J. Environ. Res. Public Health* **2020**, *17*, 764. [CrossRef]
54. Roffael, E.; Johnsson, B.; Engström, B. On the measurement of formaldehyde release from low-emission wood-based panels using the perforator method. *Wood Sci. Technol.* **2010**, *44*, 369–377. [CrossRef]
55. Medved, S.; Gajšek, U.; Tudor, E.M.; Barbu, M.C.; Antonović, A. Efficiency of bark for reduction of formaldehyde emission from particleboards. *Wood Res.* **2019**, *64*, 307–316.
56. Réh, R.; Igaz, R.; Krišťák, Ľ.; Ružiak, I.; Gajtanska, M.; Božíková, M.; Kučerka, M. Functionality of Beech Bark in Adhesive Mixtures Used in Plywood and Its Effect on the Stability Associated with Material Systems. *Materials* **2019**, *12*, 1298. [CrossRef] [PubMed]
57. Şenol, H. Biogas potential of hazelnut shells and hazelnut wastes in Giresun City. *Biotechnol. Rep.* **2019**, *24*, e00361. [CrossRef] [PubMed]
58. Nicolao, E.S.; Leiva, P.; Chalapud, M.C.; Ruseckaite, R.A.; Ciannamea, E.M.; Stefani, P.M. Flexural and tensile properties of biobased rice husk-jute-soybean protein particleboards. *J. Build. Eng.* **2020**, *30*, 101261. [CrossRef]
59. EN 312:2005. *Particleboards—Specifications—Part 3: Requirements for Boards for Use in Interior Fitments (Including Furniture) in Dry Conditions*; European Committee for Standardization: Brussels, Belgium, 2005.

 *applied sciences*

Article

Impact of Structural Defects on the Surface Quality of Hardwood Species Sliced Veneers

Vasiliki Kamperidou [1],* , Efstratios Aidinidis [2] and Ioannis Barboutis [1]

[1] Department of Harvesting and Technology of Forest Products, Aristotle University of Thessaloniki, 541 24 Thessaloniki, Greece; jbarb@for.auth.gr

[2] Department of Forestry and Natural Environment Management, Agricultural University of Athens, 118 55 Athens, Greece; eaidinidis@aua.gr

* Correspondence: vkamperi@for.auth.gr; Tel.: +30-2310998895

Received: 20 August 2020; Accepted: 7 September 2020; Published: 9 September 2020

Abstract: The surface roughness constitutes one of the most critical properties of wood and wood veneers for their extended utilization, affecting the bonding ability of the veneers with one another in the manufacturing of wood composites, the finishing, coating and preservation processes, and the appearance and texture of the material surface. In this research work, logs of five significant European hardwood species (oak, chestnut, ash, poplar, cherry) of Balkan origin were sliced into decorative veneers. Their surface roughness was examined by applying a stylus tracing method, on typical wood structure areas of each wood species, as well as around the areas of wood defects (knots, decay, annual rings irregularities, etc.), to compare them and assess the impact of the defects on the surface quality of veneers. The chestnut veneers presented the smoothest surfaces, while ash veneers, despite the higher density, recorded the highest roughness. In most of the cases, the roughness was found to be significantly lower around the defects, compared to the typical structure surfaces, probably due to lower porosity, higher density and the presence of tensile wood. The results reveal that the presence of defects does not affect the roughness of the veneers and increases neither the processing requirements of the veneer sheets before finishing, nor the respective production cost of veneers and the veneer-based wood panels. The high utilization prospects of the examined wood species in veneer production, even those bearing various defects, is highlighted.

Keywords: chestnut; decay; defect; density; knot; roughness; surface; texture; quality; veneer

1. Introduction

The surface of each material or final product consists of a miniature of peaks and valleys, the size and distribution of which determine the surface properties of the material, such as roughness, texture, etc. The surface roughness is estimated in order to predict the surface behavior of the material during its application in various uses. As regards wood, roughness greatly affects its aesthetics and the structures in which it participates, and should be in line with the criteria and requirements of consumers in terms of quality. As regards wood veneer sheets, rough surfaces of veneers not only negatively affect the appearance of the finished products, but also affect manufacturing processes such as coatings and adhesion appliance, and adhesion strength, since they reduce the contact between them, resulting, according to the literature, in weak interactions between glue and wood and, therefore, low-strength properties of laminated veneer lumber, plywood and several other wood-based composites [1–3].

Due to its structure and anatomical features, the wood surface is a multidimensional and complex substrate, and its roughness is influenced by various factors such as the wood species (hardwood versus softwood), wood density and porosity (denser wood corresponds to lower porosity and smoother surfaces), annual rings' width, ratio of early wood to late wood, the log temperature during slicing/peeling and wood storage conditions (temperature, relative humidity), the moisture content, wood anisotropy, structure and types of cells and the kinetics of liquids–gases into its mass, as well as several mechanical and machine processing operations (sawing, sanding, planning, etc.) parameters, such as the cutting means type, knife angle and marks per centimeter, cutterhead speed, tool wear, cutting direction (longitudinal, radial and tangential), etc. [1,3–8]. Tanritanir et al. [9] revealed that steaming for 20 h is an ideal pre-treatment of veneers to provide smooth surfaces of both heartwood and sapwood. In general, the use of coarse-grained veneers can reduce the bonding quality by 1/3, compared to smooth surface veneers [10–12]. Less rough wood surfaces exhibit better performance in the application of finishing agents, more uniform distribution of adhesive, require much lower amounts of paint/dye to cover the whole surface, while the phenomenon of resin bleeding through the face veneer is avoided [11,12]. Furthermore, according to the literature, the surface roughness of wood material decreases as the grit number of sandpaper increases from 60 to 240 [13]. Usually, the veneer production industries apply a sanding of 80–100 grit number, to keep the production cost at low levels, while the woodworkers and manufacturers further apply additional sanding processes to the veneer-based panels, once or twice, using 180 or 220 sandpapers.

The defects generally affect the appearance of the veneer sheets, making them usually less preferable for face-side application in furniture and structures [3]. A high number of defects makes the veneer be categorized as low-value and it is usually applied in back-side applications. However, the wood defects correspond to the natural appearance of wood and, especially in recent years, there has been a phenomenon of asking for artificially aged or intensely rough furniture, precisely because they refer to and remind the customer of something special and unique. Furthermore, in the recent years in which the wood of high quality has been in short supply, the rational utilization of woody biomass, even the low-value wood species bearing a high number of defects, as well as the high-quality raw-lumber saving strategy, seems to be of crucial importance.

Currently, there is no comprehensive information available concerning the way that several different wood defects affect the smoothness and surface quality of sliced veneer sheets manufactured from different hardwood species. Therefore, the purpose of this study is to examine the surface roughness parameters values of sliced veneers made of five different species significant for veneer production European hardwood species (ring-porous, semi-ring porous and diffuse porous), are investigated, for the first time according to the literature, in terms of how their roughness level is influenced by the presence of various structural defects in the mass and surface of veneers, such as knots, irregularities of annual rings structure (spiral grain), decay, discoloration etc., compared to typical structure surfaces of each wood species' veneers. Additionally, it is investigated how the surface roughness of sliced veneer sheets, continuous in row and successively cut, differentiates in different wood depths, observing the evolution of the whole defect as it is encountered in the trunk, in areas of typical and non-typical wood structure.

2. Materials and Methods

The raw material of this experimental work consisted of logs of five European hardwood species of Greek and Balkan origin of large diameter (350 mm mean diameter). Specifically, one log was examined per examined forest species, which were oak (*Quercus robur* L.), chestnut (*Castanea sativa*), ash (*Fraxinus excelsior* L.), hybrid poplar (*Populous* spp.) and wild cherry (*Prunus avium*). Only the poplar wood was of Greek origin, while the rest of the species used were obtained from 3 different Balkan countries (Romania, Croatia and Serbia), and they were all commercially converted into decorative sliced veneer sheets, using the veneer slicing method, applied in the infrastructures of a Greek industry of sliced veneer production, located in central Greece (Chalkida, Evia), so that the processing conditions, the cutting means and the slicing method applied would be common for the five wood species. The veneer production fulfilled the requirements of the industrial sliced veneer production standards of this certified company, as regards the absence of thickness inequalities, defects attributed to mechanical processing failures, like burning, etc. The trunks were initially peeled to remove the bark, cut into logs, and then steamed under the same conditions (duration, temperature, pressure) followed by the industry, prior to the slicing process. The cutting machine used for the slicing of veneers was of horizontal operation, since this is suitable for the veneer slicing of hardwood species and they were all cut into plain cut (flat cut) veneers. After a visual assessment of the defects on the produced veneers, for each case of wood species and defect species, a package of 10 sliced veneer sheets, continuous in a row and successively cut, was obtained, aiming to observe the evolution of the whole defect as it is encountered in the trunk mass. The veneer sheets produced were of 0.55 ± 0.1 mm thickness, mainly cut from the heartwood part of trunks and for the purposes of this experiment, they were cut in our laboratory in smaller dimensions (350 mm length × 250 mm width) to be easily handled, and were left to be conditioned at $20 \pm 2\,°C$ and $65 \pm 5\%$ relative humidity, until constant weight. All veneer samples were conditioned to equilibrium moisture content (EMC), which ranged at low levels (5.5–10%) [14]. Twenty days before the roughness measurements' implementation, the veneers were anchored tightly on flat surfaces and, subsequently, the veneer sheets' surfaces were slightly sanded with 80-grit sand paper for 15 s. under the same laboratory conditions, since the sanding process creates a new and fresh surface by removing the material and, therefore, can improve the surface quality of veneers before finishing [15]. Afterwards, the veneers were removed from the abovementioned flat surfaces and left for approximately three weeks to be conditioned at $20 \pm 2\,°C$ and $65 \pm 5\%$ relative humidity, until constant weight. At the end of the conditioning duration, the EMC was measured again by applying the drying method of the veneer samples [14] and recording similar EMC values with those prior to the sanding and conditioning processes (<10%). The mean density of the veneers was also measured after their conditioning process (calculated as dry mass/ wet volume, with volume measured in the state of the EMC), following the respective international standard process [16], with the only difference that specimens of different dimensions were measured (20 mm × 20 mm × 0.5 mm). For the dry mass measurement, a weight of high accuracy (of 4 decimals) was used, and for the volume determination, a digital caliper was used. The density of oak wood was found to be 0.742 g/cm^3, of chestnut wood was 0.554 g/cm^3, of ash wood was 0.705 g/cm^3, of poplar wood was 0.385 g/cm^3 and of cherry wood was 0.627 g/cm^3.

On the veneers, 10–12 measurements of roughness parameters were randomly implemented on the surface of typical wood structure areas and, respectively, another 10–12 measurements of roughness were conducted on non-typical wood structures of veneer surfaces in the peripheral area of each defect (10–30 mm radius around the defect). The defects were different for each wood species, including knots (Figure 1), tensile wood, irregular annual rings, deflection of wood fibers (spiral grain), discoloration (Figure 2), decay (Figure 3) etc., since, as with the material of wood, its defects, as well, are unique. The number of 3 to 10 different veneer sheets were measured from each veneer's package and for each wood species, in order to investigate the potential differentiation of roughness as a function of different wood depths on the defects' development areas.

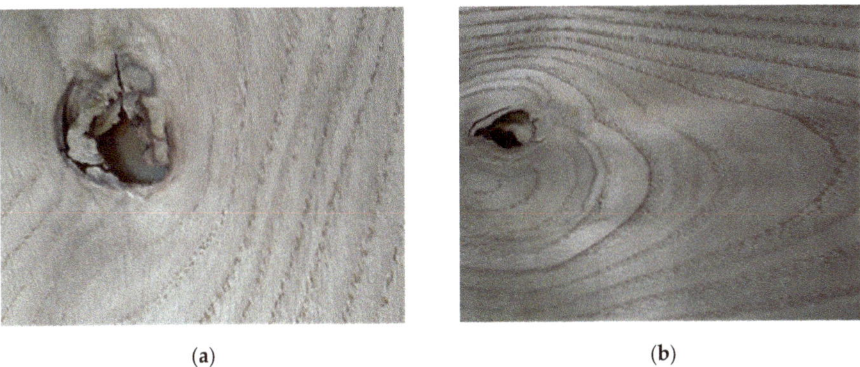

(**a**)   (**b**)

Figure 1. Dead knots in oak wood sliced plain-cut veneer (**a**), and in chestnut wood sliced veneer (**b**).

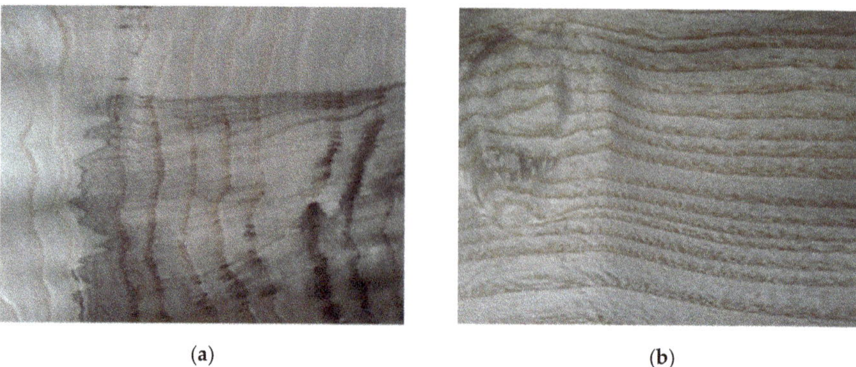

(**a**)   (**b**)

Figure 2. Ash wood sliced veneer with discoloration and irregularities of annual rings (**a**), and live internal knot with spiral grain around the knot (**b**).

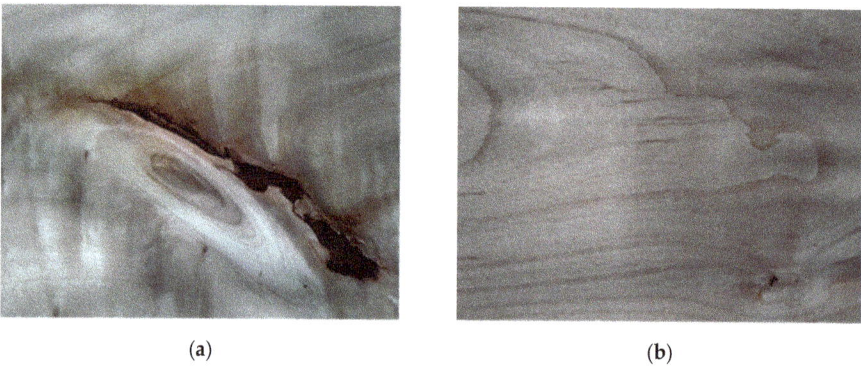

(**a**)   (**b**)

Figure 3. Poplar wood sliced plain-cut veneer sheet with live knot and decay (**a**), and cherry wood veneer with irregularities of annual rings (**b**).

The roughness parameters of the prepared veneer surfaces were evaluated using a fine stylus type profilometer, Mitutoyo Surftest SJ-301 (Figure 4), with the profile tracing method using the diamond stylus of the device, according to ISO 4287:1997 [17]. The stylus technique was determined to be used, since compared to the other methods, such as pneumatic, laser, and acoustic emission, it is accurate, practical, and repeatable [5]. The measuring speed, the diameter of the pin and the upper angle of the pin tool were 10 mm / min, 4 μm, and 90°, respectively. The sampling length was of 2.5 mm, and the evaluation length was of 12.5 mm (five times of the sampling length). The values of the surface roughness parameters were determined to be within ±0.01 μm. The measurements were implemented in a direction perpendicular to the direction of grain orientation.

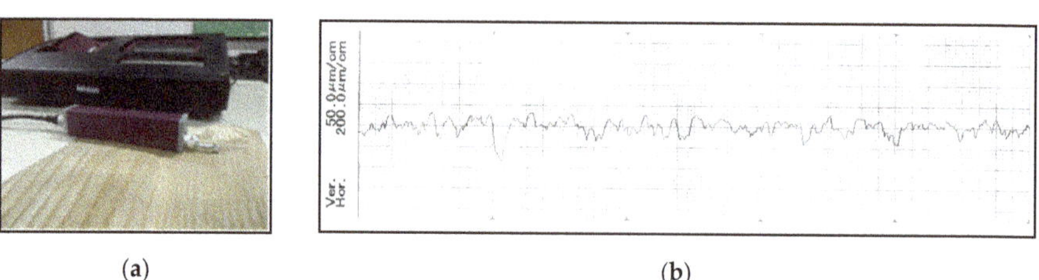

(**a**)   (**b**)

Figure 4. Roughness test using the Mitutoyo Surftest SJ-301 profilometer (**a**); an example of the roughness spectrum on a poplar sliced veneer sheet of typical structure (**b**).

The roughness measurement points were randomly selected by marking them on the surface of the samples, in order to cover the whole area. Three roughness parameters, the mean numerical deviation from the midline profile along the entire length of the stylus movement (mean arithmetic deviation of profile—Ra), the average height between the peak-valley derived from five identical lengths of the profile (mean peak-to-valley height—Rz), and the distance between peak and valley points of the profile, which can be used as an indicator of the maximum defect height within the assessed profile (maximum roughness—Ry), have been widely used in previous studies [18–20], where detailed information about these roughness parameters has been presented. These parameters, employed also in the current study, have also been used previously in the quantification of surface quality of veneers [4,21], and other wood composites, and they are defined by the respective roughness standards [17]. Prior to each measurement, the instrument was calibrated and the roughness measurements were performed at room temperature (20 ± 2 °C) [22,23].

For the processing and statistical analysis of the test results, the statistical package SPSS Statistics PASW 18 was used to determine the variability of the roughness parameters' mean values, and the effect of two different independent variables, "Veneers" (referring to the different veneers from V1 to V10, obtained continually and successively cut/produced as it is found in the trunk), and "Structure" (referring to the typical and non-typical structure of wood), and the potential interaction of these two factors upon the dependent variable of roughness parameter Ra (chosen as the most significant one and representative), using two way analysis of variance (ANOVA) with a significance level of 0.05 ($p < 0.05$).

3. Results and Discussion

3.1. Veneer Surfaces of Typical Wood Structure

According to the results (Figure 5), all the three surface roughness parameters (Ra, Rz, Ry), measured on typical wood structure surfaces, follow similar routes concerning the different wood species. The veneers of ash wood, despite their high wood density (0.705 g/cm^3), exhibited the highest surface roughness parameter values among the five species examined in this study, presenting statistically significant differences from the respective roughness values of all the other wood species veneers. The lowest roughness parameter values, and therefore, the smoother surfaces, were observed in veneers' surfaces of chestnut, whose roughness parameters' mean values were also found to differ significantly from the rest of wood species' respective values. Oak, poplar and cherry veneers recorded similar values of surface roughness parameters, even though they are characterized by much different wood density (oak 0.742 g/cm^3 and poplar 0.385 g/cm^3) and different structure (ring-porous/diffuse-porous/semi-ring-porous).

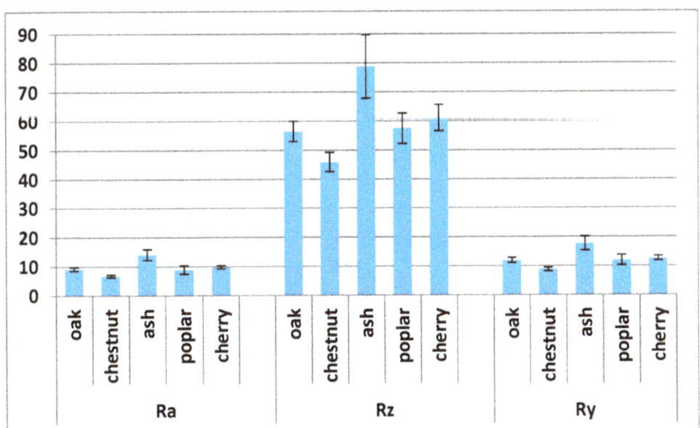

Figure 5. Mean values of surface roughness parameters Ra, Rz and Ry (μm), measured on typical wood structure areas of the sliced veneer sheets of the five different species studied.

3.2. Oak Veneers

In the case of oak wood, the roughness parameters of three veneers from the oak veneers package that corresponds to the evolution of the whole knot, were chosen to be measured (Figure 6). According to the results, all the three surface roughness parameters (Ra, Rz, Ry) were found to be, from the statistical analysis point of view, significantly lower in the case of the area around the knot, indicating a smoother area around the knot, compared to the typical structure of wood surface veneers. Generally, oak wood is a ring-porous hardwood species, whose structure, as expected, results in higher roughness levels. The lower roughness of the wood areas around the dead knot that had been felled, could be attributed to their higher density, lower porosity and the presence of tensile wood that was also visually detected. As it is widely known, tensile wood is formed on hardwoods on the upper side of logs and branches in places that are under tension. Tensile wood is characterized by lighter color and fibers that have thick walls and very small cavities. The cell walls in tensile wood areas are characterized by the presence of a gelatinous layer, which consists of concentrated microfiber substrates arranged almost parallel to the fiber axis, and can be deposited on the layer S_3 or even on layer S_2, causing the cell walls to be of higher thickness and the surface of tensile wood to become glossy [24]. Therefore, the tensile wood areas were easily recognized on the sliced veneers of this study.

According to the statistical analysis of the results, in all cases examined, the Levene tests revealed that the null hypothesis that the error variance of the dependent variable (Ra) is equal across the groups, was accepted (6th requirement of a successful ANOVA), recording a significance level > 0.05 (0.12–0.278). Investigating the roughness in different depths and areas around the knot (radius of 30 mm from the knot), going from veneer V1 to veneer V10, it is apparent (Table 1), that the roughness level records a slight decrease, marginally not statistically significant. The tests of Between-Subjects effects revealed that the factor of "veneers", referring to this progress of the different cutting depths from V1 to V10, affected the Ra variance by 16.4%. In addition, 83.4% of Ra variance is attributed to the factor of "Structure", referring to the typical and non-typical structure of wood (the latter around the knot), while the interaction between the factors "Veneers" and "Structure" affects the Ra variance by 21.4%.

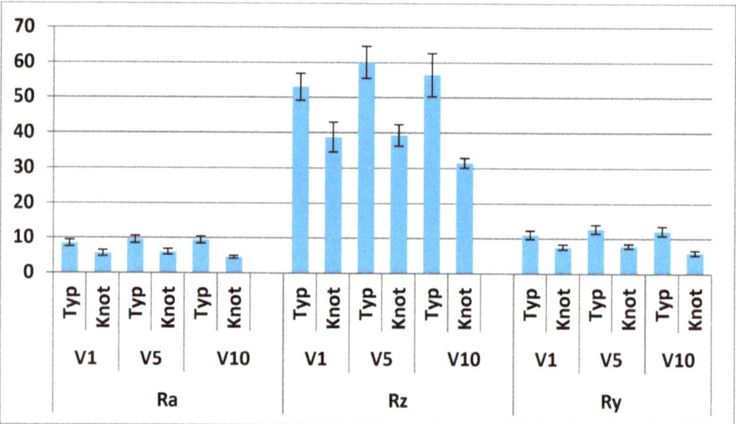

Figure 6. Surface roughness parameters Ra, Rz and Ry (µm) of oak wood veneers (V1, V5 and V10) on typical wood structure areas (Typ) and areas around dead knot (Knot).

Table 1. Descriptive statistics of dependent variable Ra measured on oak veneers of typical and non-typical structure around the knot.

Dependent Variable: Ra				
Veneers	**Structure**	**Mean**	**Std. Deviation**	**N**
V1	Typical	8.4445	0.97158	11
	Knot	5.6367	0.84131	12
	Total	6.9796	1.68515	23
V5	Typical	9.5570	1.02813	11
	Knot	6.0482	0.84163	11
	Total	7.8026	2.01623	22
V10	Typical	9.4528	0.97849	11
	Knot	4.5440	0.38021	10
	Total	7.1153	2.61815	21
Total	Typical	9.1514	1.08814	33
	Knot	5.4427	0.94755	33
	Total	7.2971	2.12520	66

3.3. Chestnut Veneers

The results of roughness measurements carried out on veneers of the ring-porous hardwood species of chestnut wood (V1–V8) (Figure 7, Table 2), reveal that in six of the eight veneers studied, the roughness around the defect area was found to be lower than the respective mean roughness parameter values of typical wood structure surfaces, with only three of them corresponding to statistically significant differences. Only in the case of V6, the area around the dead knot was found to be of higher roughness than typical wood structure surfaces, but without marking a statistically significant difference. Investigating the roughness in different depths and areas around the dead knot, going from veneer V1 to veneer V10, it is evident that the roughness level records a gradual, though statistically significant, increase.

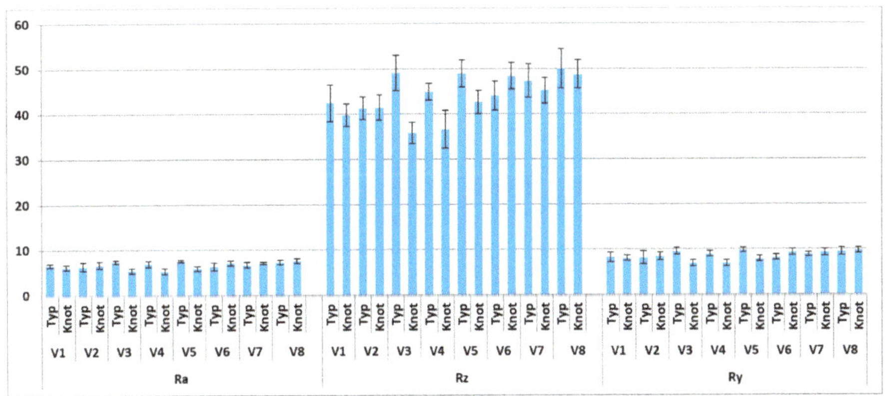

Figure 7. Surface roughness parameters Ra, Rz and Ry (μm) of chestnut wood veneers (V1–V8) on typical wood structure areas and areas around dead knot.

The tests of Between-Subjects effects demonstrated that the factor of "veneers", referring to this progress of the different depths from V1 to V8, affected the Ra variance by 29.9%. In addition, 15.2% of Ra variance is attributed to the factor of "Structure", referring to the typical and non-typical structure of wood (around the knot), while the interaction between the factors "Veneers" and "Structure" affects the Ra variance by 47.5%.

Table 2. Descriptive statistics of dependent variable Ra measured on chestnut veneers of typical and non-typical structure around the knot.

Dependent Variable: Ra				
Veneers	**Structure**	**Mean**	**Std. Deviation**	**N**
	Typical	6.5417	0.40739	10
V1	Knot area	6.0357	0.55518	10
	Total	6.2887	0.54037	20
	Typical	6.2597	0.91630	10
V2	Knot area	6.5904	0.74865	10
	Total	6.4251	0.83185	20
	Typical	7.2360	0.37384	10
V3	Knot area	5.2980	0.49109	10
	Total	6.2670	1.08112	20
	Typical	6.8160	0.67526	10
V4	Knot area	5.2220	0.65752	10
	Total	6.0190	1.04375	20
	Typical	7.4400	0.27227	10
V5	Knot area	5.7360	0.56386	10
	Total	6.5880	0.97459	20
	Typical	6.2220	0.77334	10
V6	Knot area	7.0332	0.52258	10
	Total	6.6276	0.76539	20
	Typical	6.5920	0.63405	10
V7	Knot area	7.0416	0.24691	10
	Total	6.8168	0.52202	20
	Typical	7.0700	0.52286	10
V8	Knot area	7.4718	0.53123	10
	Total	7.2709	0.55286	20
	Typical	6.7722	0.71659	80
Total	Knot area	6.3036	0.96310	80
	Total	6.5379	0.87820	160

3.4. Ash Veneers

Ash is a wood species with a ring-porous wood structure, with the apertures of vessels to potentially increase the roughness of the wood surfaces [3]. The results from the roughness parameter measurements on ash veneers' surfaces (Figure 8, Table 3) demonstrate that the surface areas with irregularities of annual rings, in four of the total five cases examined (veneers V2–V5), exhibited significantly lower roughness parameters and smoother surfaces than typical ash wood surfaces, while in only one case (veneer V1), the typical and non-typical wood structure surfaces displayed similar roughness values. The factor of "veneers", referring to this progress of the different depths from V1 to V8, statistically significantly affected the Ra variance by 23.7%. In addition, 80.8% of Ra variance is attributed to the factor of "Structure", referring to the typical and non-typical structure of wood around and on the irregularities of annual rings, while the interaction between the factors "Veneers" and "Structure" affects the Ra variance by 56.6%.

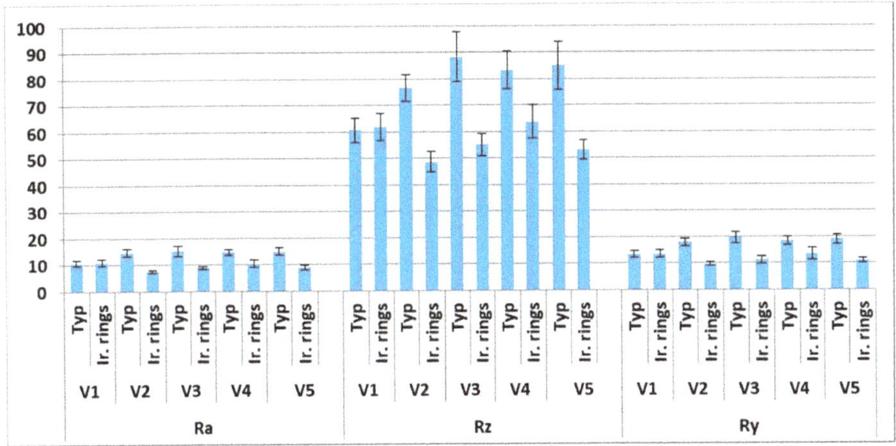

Figure 8. Surface roughness parameters Ra, Rz and Ry (μm) of ash wood veneers (V1–V5) on typical wood structure areas and areas of irregular annual rings and discoloration.

Table 3. Descriptive statistics of dependent variable Ra measured on ash veneers of typical and non-typical structure on areas of irregular annual rings and discoloration.

Dependent Variable: Ra				
Veneers	**Structure**	**Mean**	**Std. Deviation**	**N**
V1	Typical	10.8020	1.02040	10
	Knot area	11.0020	1.17776	10
	Total	10.9020	1.07740	20
V2	Typical	14.6460	1.45356	10
	Knot area	7.5680	0.60913	10
	Total	11.1070	3.78950	20
V3	Typical	15.1980	1.84820	10
	Knot area	8.9640	0.61100	10
	Total	12.0810	3.46726	20
V4	Typical	14.7300	1.19957	10
	Knot area	10.4760	1.47529	10
	Total	12.6030	2.54457	20
V5	Typical	15.0300	1.26777	10
	Knot area	8.9480	0.77944	10
	Total	11.9890	3.28382	20
Total	Typical	14.0812	2.13308	50
	Knot area	9.3916	1.55828	50
	Total	11.7364	3.00125	100

In the case of ash veneers obtained from the area near the trunk base, bearing discoloration and eccentric annual rings (spiral grain), statistically significant differences were not recorded between the roughness of typical and non-typical structure areas on the surface of veneers (Figure 9, Table 4). The factor of "veneers", referring to this progress of the different depths from V1 to V10, statistically significantly affected the Ra variance by 41.4%. In addition, 15.9% of Ra variance is attributed to the factor of "Structure", referring to the typical and non-typical structure of wood around and on the irregularities of annual rings, while the interaction between the factors "Veneers" and "Structure" affects the Ra variance by 28.3%.

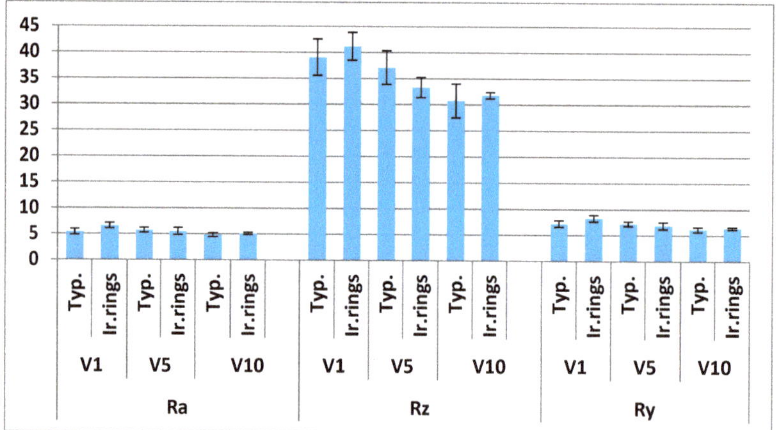

Figure 9. Surface roughness parameters Ra, Rz and Ry (µm) of ash heartwood veneers (V1, V5, V10) on typical wood structure areas and in areas of eccentric annual rings (spiral grain) of the wood near the trunk base, also bearing discoloration.

Table 4. Descriptive statistics of dependent variable Ra measured on ash stump veneers of typical wood structure areas and areas of eccentric annual rings and discoloration.

Dependent Variable: Ra				
Veneers	**Structure**	**Mean**	**Std. Deviation**	**N**
V1	Typical	5.3760	0.54929	10
	Irr.rings	6.5480	0.51903	10
	Total	5.9620	0.79498	20
V5	Typical	5.6860	0.45634	10
	Irr.rings	5.4680	0.67166	10
	Total	5.5770	0.56995	20
V10	Typical	4.8020	0.36039	10
	Irr.rings	5.1000	0.20219	6
	Total	4.9137	0.33728	16
Total	Typical	5.2880	0.58079	30
	Irr.rings	5.7985	0.80812	26
	Total	5.5250	0.73531	56

3.5. Poplar Veneers

Poplar is a fast-growing diffuse-porous hardwood species of high availability, whose utilization is restricted by its low density [3]. As regards the poplar wood veneers (Figure 10), the surface roughness parameters in the areas around the decay were found to be, in most of the cases (veneers V1, V3, V4 and V5), of lower surface roughness, compared to the wood surface of typical structures. Specifically, in the case of veneers V1 and V5, the differences were found to be statistically significant, with the roughness values of areas around decay to be the lowest ones (Table 5). The tests of Between-Subjects effects revealed that the factor of "veneers", referring to this progress of the different depths from V1 to V5, significantly affected the Ra variance, by 39%. In addition, 52.4% of Ra variance is attributed to the factor of "Structure", referring to the typical and non-typical structure of wood (around the decay area), while the interaction between the factors "Veneers" and "Structure" significantly affects the Ra variance, by 59.1%.

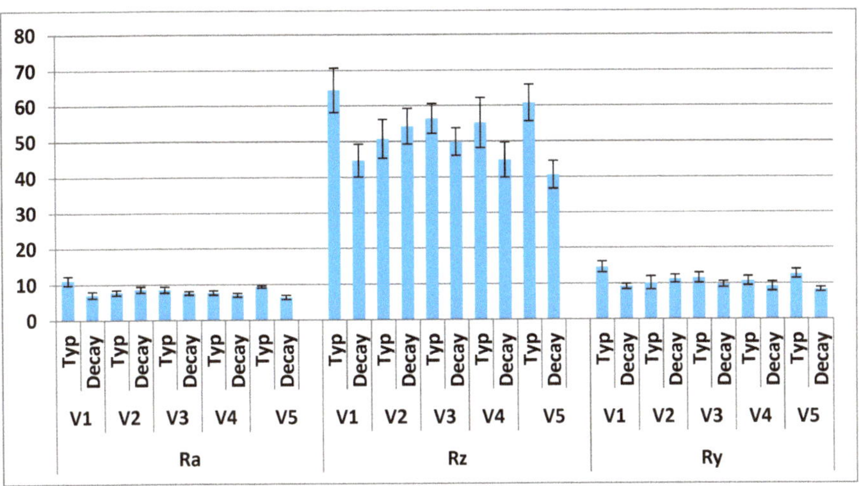

Figure 10. Surface roughness parameters Ra, Rz and Ry (µm) of poplar wood veneers (V1–V5) on typical wood structure areas and around areas of decay.

Table 5. Descriptive statistics of dependent variable Ra measured on poplar veneers of typical wood structure areas and areas of decay.

Dependent Variable: Ra				
Veneers	**Structure**	**Mean**	**Std. Deviation**	**N**
	Typical	11.0760	1.25466	10
V1	Decay	7.0540	0.89370	10
	Total	9.0650	2.31969	20
	Typical	7.7180	0.70577	10
V2	Decay	8.5400	0.83865	10
	Total	8.1290	0.86424	20
	Typical	8.4760	0.80899	10
V3	Decay	7.5660	0.52812	10
	Total	8.0210	0.81244	20
	Typical	7.6540	0.61401	10
V4	Decay	7.0120	0.53705	10
	Total	7.3330	0.65089	20
	Typical	9.2900	0.31035	6
V5	Decay	6.3250	0.59409	8
	Total	7.5957	1.59551	14
	Typical	8.8039	1.54594	46
Total	Decay	7.3400	0.99082	48
	Total	8.0564	1.48116	94

Concerning the roughness parameter values measured on the surface of poplar wood veneers (Figure 11, Table 6), around typical and non-typical wood structure areas, it was also revealed that the areas around the live knot detected were found to be, in each case, of significantly lower surface roughness compared to those of typical structure. The factor of "veneers", referring to this progress of the different depths from V1 to V5, significantly affected the Ra variance, by 70.4%. In addition, 60.9% of Ra variance is attributed to the factor of "Structure", referring to the typical and non-typical structure of wood around and on the irregularities of annual rings, while the interaction between the factors "Veneers" and "Structure" was not found to be statistically significant and it affects the Ra variance only by 3.2%.

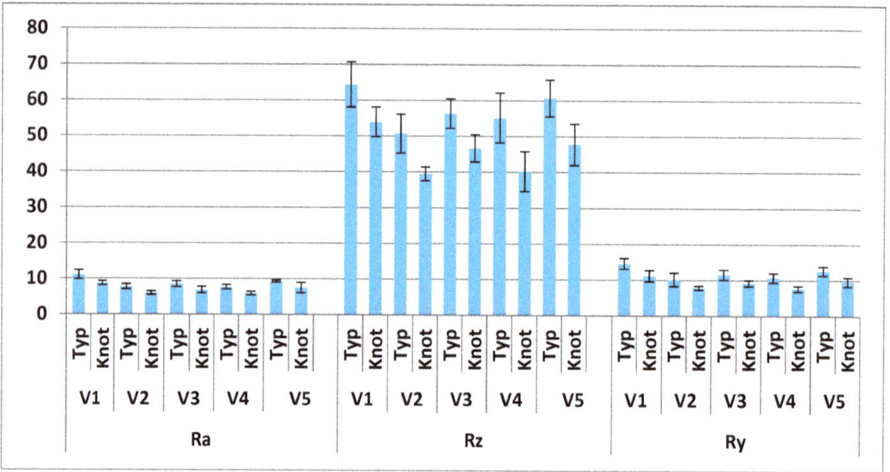

Figure 11. Surface roughness parameters Ra, Rz and Ry (μm) of poplar wood veneers (V1–V5) on typical wood structure areas and areas around live knots.

Table 6. Descriptive statistics of dependent variable Ra measured on poplar veneers of typical wood structure areas and areas around a live knot.

Dependent Variable: Ra				
Veneers	**Structure**	**Mean**	**Std. Deviation**	**N**
	Typical	11.0760	1.25466	10
V1	Knot area	8.7100	0.55913	10
	Total	9.8930	1.53847	20
	Typical	7.7180	0.70577	10
V2	Knot area	5.9720	0.50037	10
	Total	6.8450	1.07554	20
	Typical	8.4760	0.80899	10
V3	Knot area	6.8320	0.91609	10
	Total	7.6540	1.19113	20
	Typical	7.6540	0.61401	10
V4	Knot area	5.9200	0.48885	10
	Total	6.7870	1.04069	20
	Typical	9.2900	0.31035	6
V5	Knot area	7.1913	1.25355	8
	Total	8.0907	1.42999	14
	Typical	8.8039	1.54594	46
Total	Knot area	6.9140	1.28380	48
	Total	7.8388	1.70045	94

3.6. Cherry Wood Veneers

Furthermore, the roughness parameters of three veneers (V1, V5, V10) made of the semi-ring porous cherry wood were investigated (Figure 12, Table 7), revealing in each case lower surface roughness in the areas of annual rings with irregularities, compared to surface areas of typical wood structures. Nevertheless, only in the case of veneer V1, the difference between the typical structure's wood surface and irregular annual rings' surface was found to be statistically significant. Statistically significant differences between the roughness parameter values of the defect areas of different veneers (going from veneer V1 to V10) were not found. More specifically, the tests of Between-Subjects effects revealed that the factor of "veneers", referring to this progress of the different depths from V1 to V5, insignificantly affected the Ra variance, by 5%. The factor of "Structure" (typical and non-typical structure) around and on the irregular annual rings, affects the Ra variance significantly by 62.4%, while the interaction between the factors "Veneers" and "Structure" does not significantly affect the Ra variance (14.1%).

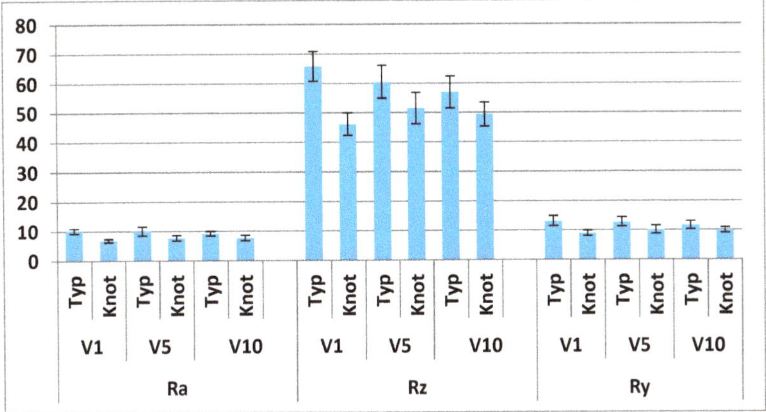

Figure 12. Surface roughness parameters Ra, Rz and Ry (μm) of cherry wood veneers (V1, V5, V10) on typical wood structure areas and around areas of irregular rings.

Table 7. Descriptive statistics of dependent variable Ra measured on cherry veneers of typical wood structure areas and areas around areas of irregular rings.

Dependent Variable: Ra				
Veneers	**Structure**	**Mean**	**Std. Deviation**	**N**
	Typical	10.1200	0.82847	10
V1	Knot area	6.7960	0.54060	10
	Total	8.4580	1.83608	20
	Typical	10.0760	1.56220	10
V5	Knot area	7.6860	0.89091	10
	Total	8.8810	1.74217	20
	Typical	9.1450	0.78999	8
V10	Knot area	7.6800	0.90488	10
	Total	8.3311	1.11873	18
	Typical	9.8257	1.18180	28
Total	Knot area	7.3873	0.87864	30
	Total	8.5645	1.60158	58

3.7. Differences between Roughness of Typical Wood Structure and Non-Typical Structure of Defects Veneer Surfaces

Concerning the veneers of the five hardwood species examined, their roughness was found to be lower in each case in the areas around the defects, compared to typical structure wood areas (Figure 13). Even though oak veneers did not present the highest level of surface roughness among the species examined, they demonstrated the highest difference of roughness parameters, recorded between areas of typical structure and non-typical wood structure (mean decrease of 38.84% compared to the reference material). Ash wood veneers, which exhibited the higher roughness values among the five species, recorded a quite high difference between the roughness levels of typical and non-typical wood structure areas (mean decrease of 31.59%). Chestnut veneers, which generally recorded the lowest surface roughness among the five species studied, also presented the lowest difference between the roughness parameters of the typical and non-typical structure areas (mean decrease in roughness of 7.11% compared to control). Poplar and cherry wood veneers recorded a medium level decrease in roughness in areas around the defects in relation to the typical structure areas that ranged between 18.77% and 21.88%.

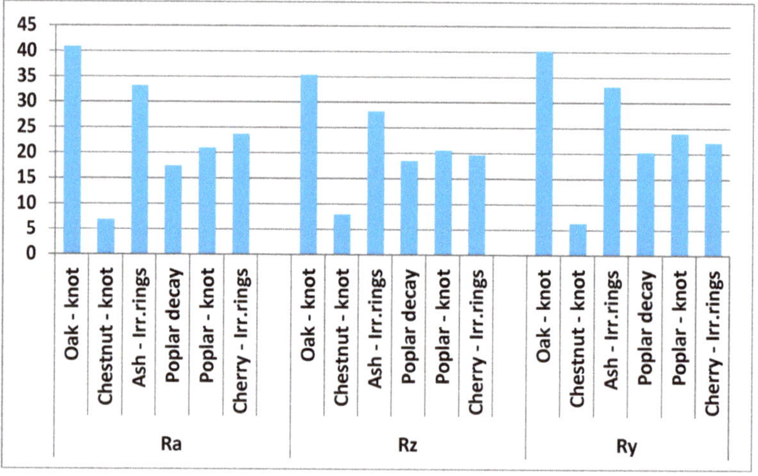

Figure 13. Percentage values depicting the differences between roughness parameters Ra, Rz and Ry of typical wood structure veneer surfaces and non-typical structure areas around the defects (decreased around the defects).

4. Conclusions

In this study, sliced veneers of five different European hardwood species of high significance were commercially produced and conditioned under the same conditions, to investigate the surface roughness of them in areas of typical wood structure and non-typical wood structure, in areas around defects. The chestnut species presented the lowest surface roughness among the five species studied, demonstrating the smoother surfaces, while ash wood veneers recorded the highest roughness, despite the high wood density. Although the chestnut wood studied in this experimental work was of lower density, its veneers presented smoother surfaces, compared to the other species, and this fact reveals the potential of utilizing this valuable species even more intensively in veneer production. Since the veneers were processed and conditioned under the same conditions, it is indicated by the results that density is a significant factor, but not the only one, affecting the smoothness of the veneer surfaces. Other morphological characteristics of wood, as well as the slightly different EMC of the different wood species veneers, probably have more influence on the smoothness and surface quality.

According to the results of the surface roughness parameter measurements, almost all the areas around the different defects recorded lower surface roughness values compared to typical structure areas, which could be possibly attributed to the different structure, lower porosity, higher density, presence of tensile wood, etc., in the areas around the defects. The smoothness of these surface areas around the defects indicates that the defects increase neither the roughness of surfaces, nor the processing requirements of the veneers, and therefore, do not increase the cost of veneer production. The veneers bearing several defects should not be considered as low-value and useless, but equally valuable, since they can be utilized in a wide range of applications, applying them on the backside of furniture and structures, or after cutting and removing the area of the defect and substituting it with another one of typical structure, or maintaining the unique appearance of the defect in the structure if possible. These data and findings were obtained through our first experimental attempt in this wide scientific field and the respective preliminary tests conducted in the frame of a project, while further studies will certainly follow from our research team, as well as the research community, in the near future, clarifying the impact of defects on the surface quality of sliced veneers and contributing to the comprehensive understanding of such veneers' final application and the respective manufactured veneer-based structures' and panels' performances. In the future, it is proposed that the effect of a single type of defect on the roughness and surface quality of a single wood species will be thoroughly investigated, examining several different logs.

Author Contributions: Conceptualization, E.A.; methodology, E.A. and I.B.; experimental work/measurements, V.K.; resources, E.A.; data curation, V.K.; writing—original draft preparation, V.K.; writing—review and editing, E.A. and I.B.; supervision, E.A. and I.B. All authors have read and agreed to the published version of the manuscript.

Funding: This research received no external funding.

Conflicts of Interest: The authors declare no conflict of interest.

References

1. Aydin, I.; Colakoglu, G.; Hiziroglu, S. Surface characteristics of spruce veneers and shear strength of plywood as a function of log temperature in peeling process. *Int. J. Solids Struct.* **2006**, *43*, 6140–6147. [CrossRef]

2. Coelho, C.L.; Carvalho, L.M.H.; Martins, J.M.; Costa, C.A.V.; Masson, D.; Meausoone, P.J. Method for evaluating the influence of wood machining conditions on the objective characterization and subjective perception of a finished surface. *Wood Sci. Technol.* **2008**, *42*, 181–195. [CrossRef]

3. Bao, M.; Huang, X.; Zhang, Y.; Yu, W.; Yu, Y. Effect of density on the hygroscopicity and surface characteristics of hybrid poplar compreg. *J. Wood Sci.* **2016**, *62*, 441–451. [CrossRef]

4. Ilter, E.; Camliyurt, C.; Balkiz, O. Research on the determination of the surface roughness values of bormulleriana fir (Abies bornmulleriana Mattf.). In *Technology Bulletin*; Central Anatolia Forestry Research Institute: Ankara, Turkey, 2002; p. 48.

5. Unsal, O.; Ayrilmis, Y.; Korkut, S. Effect of drying temperature on surface roughness of beech (*Fagus orientalis Lipsky* L.) Veneer. In Proceedings of the 9th International IUFRO Wood Drying Conference, Nanjing, China, 21–26 August 2005; pp. 316–319.

6. Magoss, E. General regularities of wood surface roughness. *Acta. Silv. Lign. Hung.* **2008**, *4*, 81–93. Available online: https://pdfs.semanticscholar.org/809d/96f71527243326b0e8d1102f3aaf5f8a2f5c.pdf (accessed on 10 July 2020).

7. Sofuoğlu, S.D.; Kurtoğlu, A. Some machining properties of 4 wood species grown in Turkey. *Turk. J. Agric. For.* **2014**, *38*, 420–427. [CrossRef]

8. Csanády, E.; Magoss, E.; Tolvaj, L. Surface Roughness of Wood. In *Quality of Machined Wood Surfaces*; Springer International Publishing: Cham, Switzerland, 2015; pp. 183–236. ISBN 978-319-22418-3.

9. Tanritanir, E.; Hiziroglu, S.; As, N. Effect of steaming time on surface roughness of beech veneer. *Build. Environ.* **2006**, *41*, 1494–1497. [CrossRef]

10. Faust, T.D. Real time measurement of veneer surface roughness by image analysis. *Forest Prod. J.* **1987**, *37*, 34–40.

11. Dundar, T.; Akbulut, T.; Korkut, S. The effects of some manufacturing factors on surface roughness of sliced Makoré (Tieghemella heckelii Pierre Ex A.Chev.) and rotary-cut beech (*Fagus orientalis* L.) veneers. *Build. Environ.* **2008**, *43*, 469–474. [CrossRef]

12. Bekhta, P.; Hiziroglu, S.; Shepelyuk, O. Properties of plywood manufactured from compressed veneer as building material. *Mater. Des.* **2009**, *30*, 947–953. [CrossRef]

13. Hendarto, B.; Shayan, E.; Ozarska, B.; Carr, R. Analysis of roughness of a sanded wood surface. *Int. J. Adv. Manuf. Tech.* **2006**, *28*, 775–780. [CrossRef]

14. ISO 13061-1:2014. *Physical and Mechanical Properties of Wood—Test Methods for Small ClearWood Specimens—Part 1: Determination of Moisture Content for Physical and Mechanical Tests*; ISO: Geneva, Switzerland, 2014.

15. Demirkir, C.; Aydin, I.; Colak, S.; Colakoglu, C. Effects of plasma treatment and sanding process on surface roughness of wood veneers. *Turk. J. Agric. For.* **2014**, *38*, 663–667. [CrossRef]

16. ISO 13061-2:2014. *Physical and Mechanical Properties of Wood—Test Methods for Small ClearWood Specimens—Part 2: Determination of Density for Physical and Mechanical Tests*; ISO: Geneva, Switzerland, 2014.

17. ISO 4287:1997. *Geometrical Product Specifications (GPS)—Surface Texture: Profile Method—Terms, Definitions and Surface Texture Parameters*; ISO: Geneva, Switzerland, 1997.

18. Budakci, M.; Cemil Ilce, A.; Gurleyen, T.; Uter, M. Determination of the surface roughness of Heat-treated wood materials planed by the cutters of the horizontal milling machine. *Bioresources* **2013**, *8*, 3189–3199. [CrossRef]

19. Kamperidou, V.; Barboutis, I. Mechanical strength and surface roughness of thermally modified poplar wood. *PRO Ligno* **2017**, *13*, 107–114. Available online: http://www.proligno.ro/en/articles/2017/201704.htm (accessed on 30 December 2017).

20. Li, G.; Wu, Q.; He, Y.; Liu, Z. Surface roughness of thin wood veneers sliced from laminated green wood lumber. *Maderas. Cienc. Tecnol.* **2018**, *20*, 3–10. [CrossRef]

21. Mummery, L. *Surface Texture Analysis*; Hommelwerke: Muhlhausen, Germany, 1993; p. 106.

22. Korkut, D.S.; Korkut, S.; Bekar, I.; Budakçı, M.; Dilik, T.; Çakıcıer, N. The effects of heat treatment on the physical properties and surface roughness of turkish hazel (*Corylus colurna* L.) wood. *Int. J. Mol. Sci.* **2008**, *9*, 1772–1783. [CrossRef] [PubMed]

23. Korkut, S.; Budakci, M. The effects of high-temperature heat-treatment on physical properties and surface roughness of Rowan (*Sorbus aucuparia* L.) wood. *Wood Res.* **2010**, *55*, 67–78. Available online: http://www.woodresearch.sk/wr/201001/08.pdf (accessed on 30 October 2010).

24. Filippou, I. *Chemistry and Chemical Technology of Wood*; Giahoudi-Giapouli Publications: Thessaloniki, Greece, 2014; p. 357.

Article

Utilization of Partially Liquefied Bark for Production of Particleboards

Wen Jiang [1], Stergios Adamopoulos [1,*], Reza Hosseinpourpia [1], Jure Žigon [2], Marko Petrič [2], Milan Šernek [2] and Sergej Medved [2,*]

1   Department of Forestry and Wood Technology, Linnaeus University, Lückligs Plats 1, 35195 Växjö, Sweden; wen.jiang@lnu.se (W.J.); reza.hosseinpourpia@lnu.se (R.H.)

2   Department of Wood Science and Technology, Biotechnical Faculty, University of Ljubljana, Rožna Dolina C VIII/34, SI-1000 Ljubljana, Slovenia; jure.zigon@bf.uni-lj.si (J.Ž.); marko.petric@bf.uni-lj.si (M.P.); milan.sernek@bf.uni-lj.si (M.Š.)

*   Correspondence: stergios.adamopoulos@lnu.se (S.A.); sergej.medved@bf.uni-lj.si (S.M.)

Received: 13 July 2020; Accepted: 28 July 2020; Published: 30 July 2020

Abstract: Bark as a sawmilling residue can be used for producing value-added chemicals and materials. This study investigated the use of partially liquefied bark (PLB) for producing particleboard with or without synthetic adhesives. Maritime pine (*Pinus pinaster* Ait.) bark was partially liquefied in the presence of ethylene glycol and sulfuric acid. Four types of particleboard panels were prepared with a PLB content of 4.7%, 9.1%, 20%, and 33.3%, respectively. Another five types of particleboard panels were manufactured by using similar amounts of PLB and 10 wt.% of melamine–urea–formaldehyde (MUF) adhesives. Characterization of bark and solid residues of PLB was performed by Fourier transform infrared spectroscopy (FTIR), thermogravimetric analysis (TGA), and automated vapor sorption (AVS). Mechanical and physical properties of the particleboard were tested according to the European standards EN 310 for determining modulus of elasticity and bending strength, EN 317 for determining thickness swelling after immersion in water, and EN 319 for determining internal bond strength. The results showed that the increase in PLB content improved the mechanical strength for the non-MUF boards, and the MUF-bonded boards with up to 20% of PLB met the requirements for interior uses in dry conditions according to EN 312. The non-MUF boards containing 33.3% of PLB and the MUF-bonded boards showed comparable thickness swelling and water absorption levels compared to the reference board.

Keywords: bark; bonding; partial liquefaction; MUF adhesives; water vapor sorption; thickness swelling; wood-based panels

1. Introduction

Particleboard is a panel product made from wood particles, originating from low value wooden raw material (e.g., chips and shavings) or other lignocellulosic materials, bonded by synthetic adhesives and pressed at high pressures and temperatures [1,2]. Particleboard is a low-cost panel product with adequate strength for furniture and interior applications and have been widely applied in flooring, wall and ceilings, flat-pack furniture, cabinets, and work surfaces such as speaker boxes, sewing machine tops, etc. [3,4]. Adhesives are an important element in the wood-based panel industry. Particleboard is traditionally produced with wood adhesives such as urea–formaldehyde (UF), melamine–urea–formaldehyde (MUF), and isocyanate-based adhesives. It is estimated that the adhesives used for particleboard production in Europe are split among UF (92%), MUF (7%), and isocyanates (1%) [5]. All these existing commercial adhesives are petroleum-based, and thus not sustainable [6]. At the same time, the concern about formaldehyde emissions from wood-based panels, especially in indoor applications, is currently the most important driving factor for wood panel manufacturers to move away from using formaldehyde-based synthetic adhesives [7]. Therefore, the development of natural binders and bio-based adhesives for wood panel production is needed. A common problem for bio-based adhesives is their industrialization, which requires stable qualities and quantities of raw materials and final products. Adhesives based on a variety of natural materials such as starch, proteins, lignin and tannin have been proved to be less reactive than their formaldehyde-based counterparts, and this leads to much longer press times and considerably higher production costs. Thus, the particleboard industry has not yet been able to use natural binders in the production and the penetration of such adhesive systems in the market is rather small [8].

Recently, the utilization of bio-based adhesives from liquefied biomass has received considerable attention. Liquefaction is a method that converts wood and other lignocellulosic biomass into liquids for obtaining oils, chemicals, and other value-added materials [9]. Liquefaction of biomass includes two methods: hydrothermal liquefaction (HTL) and moderate acid-catalyzed liquefaction (MACL). HTL is usually carried out in the water or organic solvents at a temperature of 200–400 °C and pressure of 5–20 MPa and MACL takes place at a lower temperature of 120–250 °C under atmospheric pressure with the assistance of acid catalysts [9–13]. The primary products are chosen by the different liquefaction methods and the liquefying conditions. HTL produces bio-oil as main products while MACL mainly produces bio-polyol or phenolated compounds depending on the solvents that are used [9]. Liquefied wood (LW) from MACL with polyhydric alcohols and phenols has been used in different adhesive systems, such as polyurethane, UF, MUF, phenol-formaldehyde, and epoxy systems [14–21]. LW has high reactivity with other adhesives precursors and reactive sites due to a large amount of phenolic and alcoholic hydroxyl groups in their compositions [16].

Kunaver et al. [16] produced particleboard from melamine–formaldehyde (MF) or MUF adhesives with added LW, where spruce was liquefied in glycerol–diethylene glycol mixture as a solvent and *p*-toluenesulfonic acid as a catalyst for 3 h at 180 °C. The results showed that the addition of 50% LW to the MF and MUF adhesives did not influence the mechanical properties of particleboard but significantly reduced the formaldehyde emissions. Čuk et al. [22] bonded particleboard with MF that was partially substituted by LW. Adhesive mixtures containing 20% of LW had the largest improvements in the mechanical properties of the particleboard compared to the reference boards. Substitution of MF by LW of up to 30% resulted in a comparable mechanical strength to board bonded only with MF, while significantly reducing formaldehyde emissions. LW worked as a plasticizer and increased the mobility of the MF resin molecules. However, the thermal stability of MF substituted by LW was reduced because LW prolonged the curing time of the final adhesive, decreased the cross-linking degree, and accelerated the thermal degradation. Janiszewska et al. [23] produced particleboard with a mixture of UF and LW (ratio 4:1) and investigated the influence of different liquefying solvents, e.g., a mixture of solvents from the polyhydroxy alcohol group, including glycerine, ethylene glycol, propylene glycol, diethylene glycol, and dipropylene glycol, on the chemical structure, physical and mechanical properties of the particleboard. For all liquefying solvents, boards exhibited comparable properties to the ones produced without LW.

Bark, as an industrial residual material, with heterogeneous structure and diverse chemical composition, can be used for the production of a variety of and bio-composites and bio-compounds such as tannin and polyphenols, bio-oil, antioxidants, and bio-based adhesives [24–31]. Particleboard based on larch bark has been reported by Tudor with a much lower formaldehyde emission content than the wood-based panels, which means bark work as formaldehyde scavenger in the particleboard [31]. Limited publications can be found related to the utilization of liquefied bark in wood adhesives and wood-based composites, while, all of them followed a complete liquefaction process. Janiszewska [24] liquefied bark in polyhydric alcohols and *p*-toluenesulfonic acid at 120 °C for 2 h, and then prepared three-layer particleboard by using an adhesive mixture of 80% MUF and 20% liquefied bark (LB) with the addition of 1 M NaOH or 25% ammonium hydroxide for neutralizing the adhesive mixtures. The boards made with MUF-LB adhesives had 10–20% lower modulus of elasticity, 22–29% lower bending strength, and 25% lower tensile strength in comparison with the ones made with commercial MUF adhesives. It was also determined that replacing of MUF with LB led to a slight reduction in formaldehyde content. Lee and Liu [32] prepared particleboard bonded by LB-based resol adhesives, which was prepared from LB of Taiwan Acacia (*Acacia confusa*) and China fir (*Cunninghamia lanceolata*) with two types of catalysts, i.e., sulfuric acid and hydrochloric acid. The results showed that the resol type adhesives prepared from bark, liquefied with sulfuric acid, had a higher viscosity compared to those made from liquefied bark with hydrochloric acid. Adhesives from liquefied China fir had a higher viscosity than those from liquefied Taiwan acacia with the same acid catalyst. The thermal analysis showed that the adhesives based on hydrochloric acid-catalyzed liquefied wood had a higher maximum temperature, a greater height of exothermic peak, and a larger quantity of exothermic heat at thermosetting than the adhesives based on sulfuric acid-catalyzed liquefied wood. Particleboard made with resol adhesives based on liquefied Taiwan acacia bark catalyzed by sulfuric acid had the best mechanical properties and the lowest thickness swelling among all particleboard panels.

A complete moderate-acid catalyzed liquefaction process of biomass in alcohols or phenols takes approximately 90–120 min under constant stirring and heating [9,12]. The obtained compounds require further purification stage to retrieve liquids for applying in adhesive formulations, which is time- and energy-consuming. Therefore, present work explores a novel approach at the production of partially liquefied bark (PLB) and its further utilization in manufacturing particleboard panels. Changes in the chemical characterization structure, thermal stability, and water vapor sorption behavior of bark due to partial liquefaction process were determined by Fourier transform infrared (FTIR) spectroscopy, thermogravimetric analysis (TGA), and automated vapor sorption (AVS) apparatus, respectively on the solid residues of PLB. The effect of PLB on the physical, mechanical, and microscopic structure of particleboard made with or without MUF adhesives were then analyzed. It was hypothesized that bark was partially solvolyzed after liquefaction generating PLB, which is a chemical-activated particle that can be incorporated for better compatibility with wood particles, and can thus enhance the physical and mechanical properties of the particleboard.

2. Materials and Methods

2.1. Materials

Bark of maritime pine (*Pinus pinaster* Ait.) was used for liquefaction. Bark was purchased from BVB Substrates (De Lier, The Netherlands), and then milled by a Condux mill CSK 360/N1 (Hanau, Germany) to particles. Chemicals used for liquefaction were ethylene glycol (EG) (Honeywell, Charlotte, NC, USA) as a solvent and 96% sulfuric acid (SA) (KEMIKA d.d., Zagreb, Croatia) as a catalyst. 1,4-dioxane, purchased from Honeywell GmbH (Seelze, Germany) was used for purification after liquefaction. Fresh wood particles from spruce (*Picea* spp.) were collected from a local sawmill in Ljubljana, Slovenia. Melamine–urea–formaldehyde (MUF) adhesives H97 was provided by Melamine Kočevje d.d. (Kočevje, Slovenia). Ammonium sulfate with 20% solid content was used as an adhesive hardener at a content of 3% in the adhesive formulations.

2.2. Partial Liquefaction Process

Oven-dried bark particles (103 °C for 24 h) together with the solvent and catalyst were added into a three-neck glass reactor submerged in an oil bath equipped with a mechanical stirrer and a water condenser. A weight ratio of 3:1 was used for solvent and bark. The catalyst concentration was 3% (*w/w*) based on the solvent mass. The liquefaction process was initiated at 180 °C with constant stirring under ambient atmosphere. After 30 min, the liquefaction was stopped by removing the reactor from the oil bath and transferring the liquefied bark to a clean beaker for cooling down to room temperature.

Figure 1 shows the bark, bark mixed with solvent and catalyst before liquefaction, and PLB after liquefaction. PLB (Figure 1c) used for the production of particleboard is a wet material with a high solid content (un-liquefied bark), unreacted EG and SA, and liquefaction liquid intermediates. PLB had a solid content of 41% (oven-dry mass of solids divided by total PLB mass). For further characterization, PLB was oven-dried at 103 °C for 24 as the solvent-containing solid residue of PLB. Purified PLB as a solvent-free solid residue of PLB was prepared by first dissolving wet PLB in a mixture solvent of 1,4-dioxane and water at a mass ratio of 4:1. Then the wet PLB and solvent mixture was centrifuged at 1000 rpm for 10 min by removing the residual solvents and the intermediate chemicals. The obtained solids as purified PLB were dried in the oven at 103 °C for 24 h. Bark, oven-dried PLB, and purified PLB were milled to powders with a size of 2 mm for subsequent analysis.

(a)   (b)   (c)

Figure 1. Bark (**a**), mixture of bark with the solvent and catalyst (**b**), and partially liquefied bark (**c**).

2.3. Particleboard Production

Single-layer particleboard panels were manufactured with a target thickness of 8 mm by following standard procedures that simulate industrial production in the laboratory. Wet PLB was mixed with dry wood particles (less than 4% moisture content) at different loading levels of 4.7%, 9.1%, 20%, and 33.3%, and then the corresponding panels were labelled as I, II, III, and IV. It should be mentioned that the PLB content was measured based on the total mass of the mixture of PLB and wood particles. Five panels were manufactured by adding 10 wt.% of MUF adhesives and PLB at a loading level of 0, 4.7%, 9.1%, 20%, and 33.3% to dry wood particles, and these panels were labelled as V, VI, VII, VIII, and IX. For MUF, 3% (*w/w* of dry adhesives) of hardener was used. The mats were then manually formed into frame dimensions of 500 × 500 mm^2. The hot-pressing temperature was set as 190 °C. The pressing speed was set to 52 s·mm^{-1} (pressing time including closing and opening was 420 s). Such as long-pressing time was needed due to high mat moisture content due to the usage of wet PLB, hence there is a degassing stage in the middle of the pressing schedule (Figure 2). The final density and thickness of the particleboard are shown in Table 1.

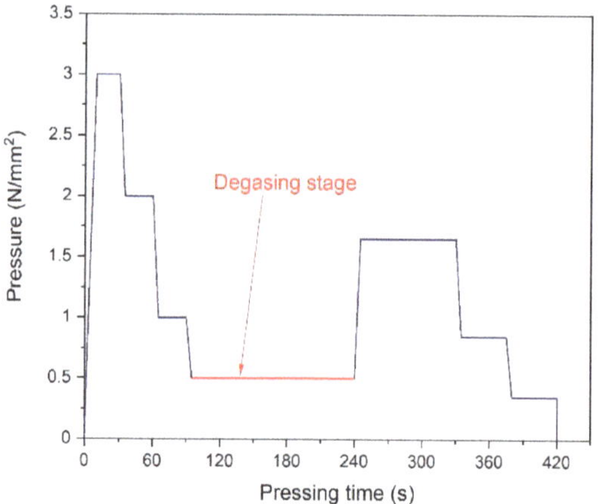

Figure 2. Pressing diagram.

Table 1. Density and thickness of the manufactured particleboard.

Panel	Description	Density (kg/m^3)	Thickness (mm)
I	4.7% PLB	399 ±19	7.98 ± 0.21
II	9.1% PLB	397 ± 65	8.05 ± 0.54
III	20.0% PLB	554 ± 18	8.32 ± 0.27
IV	33.3% PLB	540 ± 74	8.18 ± 0.21
V *	10.0% MUF	755 ± 17	7.63 ± 0.09
VI	10% MUF + 4.7% PLB	662 ± 23	7.70 ± 0.05
VII	10% MUF + 9.1% PLB	734 ± 22	7.60 ± 0.15
VIII	10% MUF + 20.0% PLB	686 ± 62	7.79 ± 0.11
IX	10% MUF + 33.3% PLB	632 ± 16	8.03 ± 0.12

* as a reference board made from melamine–urea–formaldehyde (MUF) adhesives and wood particles.

2.4. Characterizations

2.4.1. Fourier Transform Infrared (FTIR) Spectroscopy

The chemical structure of bark, oven-dried PLB, and purified PLB powders was analyzed with a Fourier Transform Infrared Spectrometer (Alpha FTIR spectrometer, Bruker, Karlsruhe, Germany) with a versatile high throughput ZnSe ATR crystal. The FTIR analysis was conducted in a wavelength region from 4000 to 800 cm^{-1} at room temperature, accumulating 64 scans with a resolution of 4 cm^{-1}.

2.4.2. Thermogravimetric Analysis (TGA)

The thermal stability of bark, oven-dried PLB, and purified PLB powders were analyzed using a NETZSCH STA 409PC instrument (Netzsch, Selb, Germany). Approximately 5 mg of dried samples (24 h at 105 °C) were heated from 30 to 800 °C at a rate of 10 °C/min under a flowing nitrogen atmosphere.

2.4.3. Automated Vapor Sorption (AVS)

The water vapor sorption behavior of bark, oven-dried PLB, and purified PLB powders was determined using an automated vapor sorption (AVS) apparatus (Q5000 SA, TA Instruments, New Castle, DE, USA) as reported previously [33,34]. Approximately 5 mg of grounded samples, passed through a 10-mesh sieve, were exposed to the relative humidity (RH) from 0 to 90% in step sequences of 15% and then continued to reach 95% at a constant temperature of 25 °C. The instrument maintained a constant target RH until the mass change in the sample (dm/dt) was less than 0.01% per min over a 10 min period. The equilibrium moisture content (EMC) for each sample was assessed based on their equilibrium weight at each given RH step throughout the adsorption run.

2.4.4. Mechanical Properties of Particleboard

The bending test for measuring the modulus of elasticity (MOE) and modulus of rupture (MOR) of the particleboard panels manufactured with PLB was performed according to EN 310:1993 [35] by using a universal testing machine (Zwick/Roell Z005, Zwick/Roell GmbH, Ulm, Germany). Six samples per board measuring 210 × 25 mm^2 were tested using a span of 160 mm and a cross-head speed of 7 mm min^{-1} (time to break was between 45 and 75 s). MOE was determined between 10 and 40% maximum load.

Six samples measuring 210 × 25 mm^2 (width × length) were cut from each panel and tested for their tensile strength parallel to the surface by using a universal testing machine (Zwick/Roell Z005, Ulm, Germany).

The internal bond (IB) strength test was conducted following EN 319:1993 [36]. Six samples per particleboard measuring 50 × 50 mm^2 were bonded with hot-melt glue and the test perpendicular to their surfaces were performed using a universal testing machine (Instron 4466, Darmstadt, Germany). A loading speed of 0.75 mm·min^{-1} was used for testing (time to break was between 50 and 70 s).

2.4.5. Thickness Swelling and Water Absorption of Particleboard

Samples measuring 50×50 mm^2 were cut from the particleboard and immersed in water at 20 ± 2 °C for 2 h and 24 h. Thickness swelling (TS) was evaluated by the difference between the final and initial thickness, and water absorption (WA) was evaluated by the weight difference. Four samples per panel were used for the determinations according to EN 317:1993 [37].

2.4.6. Scanning Electron Microscopy (SEM)

The formed bonds between wood particles, PLB and MUF adhesives were studied on the cross-section of the particleboard with a scanning electron microscope (SEM, FEI Quanta 250, FEI, Hillsboro, OR, USA), equipped with the energy dispersive X-ray spectrometer (EDX, AMETEK Inc., Berwyn, PA, USA). Before observations, the surfaces of the selected area of samples were evened by cutting on a Leica SM2010R microtome (Leica, Wetzlar, Germany). The SEM micrographs were taken with large field detector (LFD) at 100×, 500× and 1000× magnifications in a low vacuum (50 Pa), at a voltage of 5.0 kV, a spot size of 3.0, and a beam transition time of 45 µs.

2.5. Statistical Analysis

The SPSS version 25.0 statistical software package (IBM Corp., Armonk, NY, USA) was used for the statistical analysis. One-way ANOVA was performed on the mechanical and water-related results for the analysis of variance at a 95% confidence interval ($p < 0.05$). The statistical differences between mean values were assessed by using the Tukey's honestly significant difference (HSD) test.

3. Results and Discussion

3.1. Characterization of Bark and Solid Residues of PLB

FTIR spectroscopy analysis detected the chemical structure of bark and solid residues of PLB (oven-dried PLB and purified PLB) (Figure 3). Raw bark showed a strong absorption peak at 3350 cm^{-1}, which corresponds to -OH stretch vibration in cellulose, lignin, and hemicelluloses. Bark and oven-dried PLB illustrated two distinct peaks at 2918 and 2845 cm^{-1}, which are assigned to -CH stretch vibration in aromatic methoxyl groups and aliphatic methyl and methylene groups [38,39]. The spectra for oven-dried PLB appears similar to that of purified PLB. A slight decrease in the -OH and -CH bonds were observed in oven-dried PLB and purified PLB compared to bark. This might be related to the degradation and dehydration of bark, and the formation of alcohol-soluble intermediates [12,40–42]. A peak representing the C-O linkage of alcohol or ether that are typical polyol products from the complete liquefaction was found at around 1112 cm^{-1} in oven-dried PLB and purified PLB (Figure 3). This peak verified the production of polyols through the partial liquefaction [43]. The absorption band at 1030 cm^{-1} corresponding to -CO stretching for both PLB and purified PLB was weakened due to the cleavage of β-O-4 bonds of lignin [42]. The broad band between 1266 and 1030 cm^{-1} in the spectrum for oven-dried PLB and purified PLB indicates -CO stretching in primary alcohol, secondary alcohol, ethers, and esters. The changes in the absorbance of -CO groups in bark and PLB confirmed that the liquefaction has occurred with the formation of the above intermediates. The intensive vibration at 1727 and 1605 cm^{-1} in bark, oven-dried PLB and purified PLB correspond to -C=O in hemicelluloses and lignin, and to -C=C- stretching in lignin, which indicates the high content of lignin remained in the bark structure after partial liquefaction.

Thermal degradation behavior of bark, PLB, and purified PLB was examined by TGA (Figure 4a) and derivative thermogravimetric (DTG, Figure 4b) analyses of the samples after drying in the oven at 105 °C for 24 h. There were apparent differences in the thermal degradation pattern of bark after partial liquefaction and purification compared to raw bark. Figure 4a shows the weight loss of the three tested samples up to a maximum of 40–60% of their initial weight. It was previously reported by Yang et al. (2007) that cellulose pyrolysis occurs in a higher temperature range than lignin and hemicelluloses, and lignin is the most difficult polymer to decompose with a solid residue of 45.7 wt.% [44]. As shown in Figure 4b and Table 2, decomposition of bark (T_{onset}) initiated at 219 °C and reached a maximum mass loss (T_{max1}) at 361 °C, thus representing typical pyrolysis of lignocellulosic material [45,46]. The respective T_{onset} of PLB and purified PLB were, respectively, 152 and 148 °C and were both lower than that of bark. The decomposition of PLB and purified PLB started earlier than bark, which might be attributed to a large number of hemicelluloses and amorphous part of cellulose that decomposed in the liquefaction due to the effect of the acid catalyst [42]. The decomposition of the intermediate products such as alcohols and esters in the PLB and purified PLB caused a shifted second maximum degradation temperature T_{max2}. As shown in Table 2, the T_{onset}, T_{max1}, and T_{max2} of PLB and purified PLB were very close to each other but considerably lower than bark, which can be related to the lower content of volatiles in PLB and purified PLB [47].

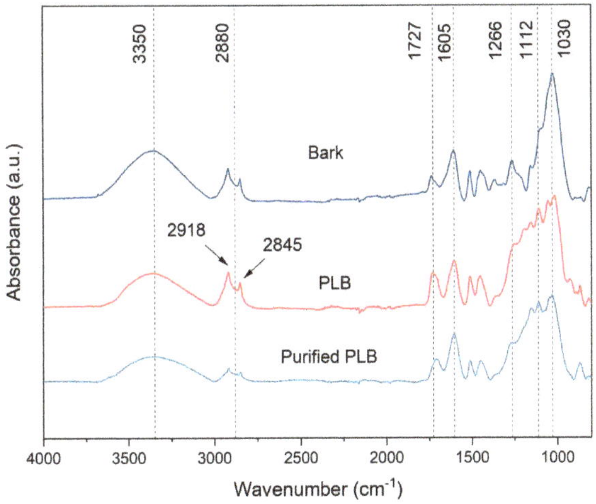

Figure 3. Fourier transform infrared (FTIR) spectrum of raw bark, oven-dried partially liquefied bark (PLB), and purified PLB.

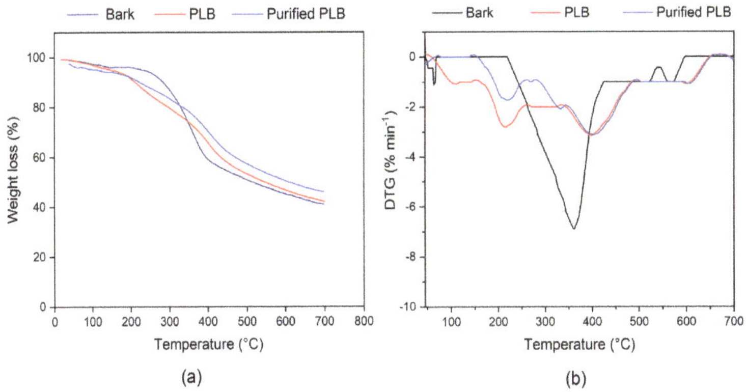

Figure 4. Mass loss (**a**) and first derivative (DTG) (**b**) of bark, PLB, and purified PLB.

Table 2. Thermal degradation properties of bark, PLB, and purified PLB.

Material	T_{onset} (°C)	T_{max1} (°C)	T_{max2} (°C)	T_{offset} (°C)
Bark	219	361	–	428
PLB	152	215	397	495
Purified PLB	148	218	403	495

The water vapor sorption isotherms of bark, PLB and purified PLB are presented in Figure 5a. PLB and purified PLB showed considerably lower EMC than bark in the RH range of 0% to 75%. A strong upward bend was observed in the EMC of PLB and purified PLB from 75% to 95% and surpassed bark at 95% RH. PLB showed an EMC of 33% at 95% RH, while the EMC values of purified PLB and bark at this RH level were 30% and 24%, respectively. Moisture increment (MI) of bark was found to vary little over the entire RH range (Figure 5b), as it was slightly decreased from 15% to 45% RH, then gradually increased from 45% to 90% RH, and then decreased from 90% to 95% RH. PLB and purified PLB, however, showed a different MI trend. The MI of PLB gradually increased with increasing the RH from 15% to 75% and then sharply increased from 75% to 95%. MI was decreased in purified PLB at the RH range of 15% to 60% but then increased greatly after 60% RH. PLB and purified PLB illustrated a more hydrophilic behavior than bark at the higher RH range over 75%. This can be related to more accessible hygroscopic hydroxyl sites after partial liquefaction from the degradation of cellulose due to alcohol hydrolysis catalyzed by strong acids [48,49]. It was reported previously that liquefaction of lignocellulosic materials in polyhydric alcohols initially hydrolyzes the glucoside linkage of the cellulose to produce glucoside monomers, which further decomposed to levulinic acid esters [41,50]. During the liquefaction process, the polymeric structure of lignin was degraded, and the obtained lignin monomers reacted with ethylene glycol to form a condensed lignin-based polymeric material with predominant aromatic hydroxyl groups, which can enhance the hydrophilicity of PLB [51]. The higher moisture sorption of PLB as compared to purified PLB can be attributed to the remaining ethylene glycol that provided extra hydroxyl groups in the PLB. Some part of these hydroxyl groups may have also been removed in purified PLB by the solvent during the purification step.

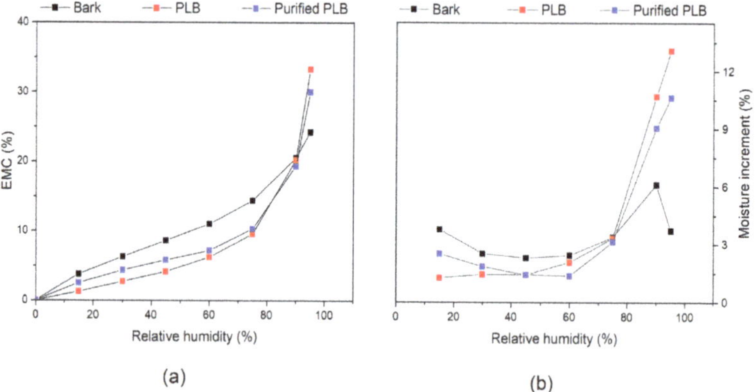

Figure 5. (a) Equilibrium moisture content (EMC) of bark, PLB and purified PLB samples exposed to increasing water vapor from relative humidity (RH) of 0 to 95%; (b) moisture increment during adsorption of bark, PLB, and purified PLB samples.

3.2. Performance of the Particleboard Containing PLB

The mechanical properties (i.e., MOE, MOR, IB, and tensile strength) of the particleboard panels containing PLB in the presence of MUF or not are presented in Figure 6. The non-MUF bonded boards containing 4.7% and 9.1% PLB (boards I and II) did not show any cohesion, and, thus, easily decomposed during cutting as illustrated in Figure 7. The MOR, MOE, and tensile strength values of the non-MUF boards were improved by increasing the PLB content. The particleboard containing wood particles and the highest amount of PLB (board IV) exhibited the highest values. A significant increment in the MOE, MOR and tensile strength of the non-MUF boards occurred when the PLB content increased from 9.1% to 20%. At least 20% of PLB was required for the non-MUF panels to be able to perform the IB test. It should be noted that the mechanical property values of all the non-MUF boards (I, II, III, and IV) were very low and did not meet the requirement of particleboard for interior uses in dry conditions according to EN 312:2010.

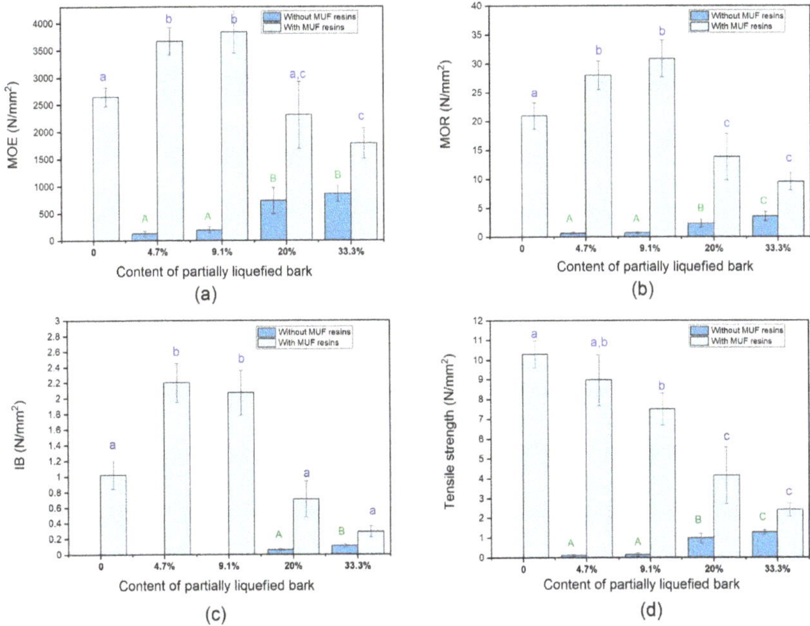

Figure 6. Average MOE (**a**), MOR (**b**), IB (**c**), and tensile strength (**d**) of particleboard containing partially liquefied bark, with or without MUF adhesives. Values labelled with the same letter (small for MUF boards, and capital for non-MUF boards) are not statistically different from each other (ANOVA, Tukey's HSD test, $p < 0.05$). Error bars represent standard deviations.

Figure 7. Particleboard samples containing 4.7%, 9.1%, 20%, and 33.3% (I–IV in order) partially liquefied bark without using MUF adhesives.

The mechanical properties of the particleboard were apparently changed by the synergistic effect of MUF adhesives and PLB. The MOE, MOR, and IB strength of the boards bonded with MUF adhesives increased by adding 4.7% and 9.1% PLB (boards VI and VII), and they were significantly higher than the reference one, board V. However, the differences between MOE, MOR, and IB values of the boards with 4.7% and 9.1% of PLB, i.e., boards VI and VII, were not statistically significant. Further, increasing the PLB content to 20% and 33.3% drastically decreases the MOE, MOR, and IB strength of the boards. This might be attributed to the decreasing proportion of the wood particles in the board content, which led to a decrease in the density (as shown in Table 1). The higher PLB amount may have also increased the inhomogeneity of the boards, and thus disturbed the equal distribution of the applied stresses during the mechanical tests. The results of mechanical properties showed that up to 20% of PLB can be used in the particleboard content to produce boards that meet the minimum requirements for interior use in dry conditions according to EN 312:2010, in terms of MOR, MOE, and IB values that are, respectively, 13, 1800, and 0.40 N/mm^2. The tensile strength parallel to surface was negatively affected by increasing the PLB content in the MUF-bonded boards. When over 20% PLB was applied, the tensile strength of the boards (VIII, IX) was lower than half of the strength of the reference (V). As a conclusion, 9.1% of PLB loading should be allowed for producing particleboard with good overall mechanical properties.

Thickness swelling (TS) and water absorption (WA) of the particleboard after 2 and 24 h immersion in water are shown in Figure 8. The boards I and II exhibited remarkably high TS and WA after 2 h immersion in water, and they were decomposed after 24 h. This can be related to the very low internal bond strength between wood particles and PLB. Low interaction of PLB and wood particles may also be an additional reason, which caused poor interfaces. The increase of the PLB level in the non-MUF panels from 4.7% to 33.3% decreased the TS and WA suggesting that the PLB protected the wood particles against water. Similar trends were observed in TS of the boards made with MUF adhesives. This might be attributed to the increasing amount of compact PLB particles in the particleboard mat, which resulted in a reduction of liquid water penetration into the board. The lowest TS value obtained in the MUF-bonded boards containing the highest amount of PLB (33.3%), which was 3.15% after 2 h and 4.04% after 24 h immersion in water. According to EN 312:2010, the maximum thickness swelling within 24 h of particleboard for non-load bearing applications in humid conditions is 17%. Therefore, the boards produced of the highest amount of PLB (33.3%) without MUF adhesives (panel IV) as well as the panels manufactured with MUF adhesives and 9.1, 20, and 33.3% PLB (panels VII, VIII, and IX) fulfilled the standard requirement. The WA values of the particleboard manufactured with MUF adhesives and 4.7–33.3% PLB (VI, VII, VIII, and IX) were statistically lower than the reference (board V). The current result from TS and WA test indicated that PLB acted as an excellent water-resistant reagent in the particleboard.

The overall results from the mechanical and water-related tests confirmed the hypothesis of this study that PLB particles provided an activated surface that can enhance its compatibility with wood particles. The SEM micrographs provided a visual explanation of the mechanical properties and TS changes due to the incorporation of different PLB levels in the particleboard. In detail, the SEM micrographs displayed that PLB caused a compact region with good interaction with wood particles (Figure 9). The compact PLB provided a less porous structure than wood particles and resulted in a reduction of water penetration in the particleboard. The PLB itself showed a homogeneous compact structure that can facilitate the transfer of stresses from the surface layers through the board [22]. An equal distribution of applied stresses at the PLB-wood particles interfaces could be achieved at a PLB level of 9.1% in the particleboard. Moreover, PLB has an advantage of modifying the surfaces of wood particles with the presence of unreacted solvent, acid and intermediates from partial liquefaction. Pressing of particleboard took place under the same temperature of 180 °C as partial liquefaction helps to transfer the liquid phases of PLB, containing remaining polyhydric alcohol and strong acids, to the wood surface for creating chemical bonding and self-adhesion ability. It can be seen from Figure 9b–d that cell walls of wood near the interface between PLB and wood particles are densified that can be caused by the chemical-active components of PLB.

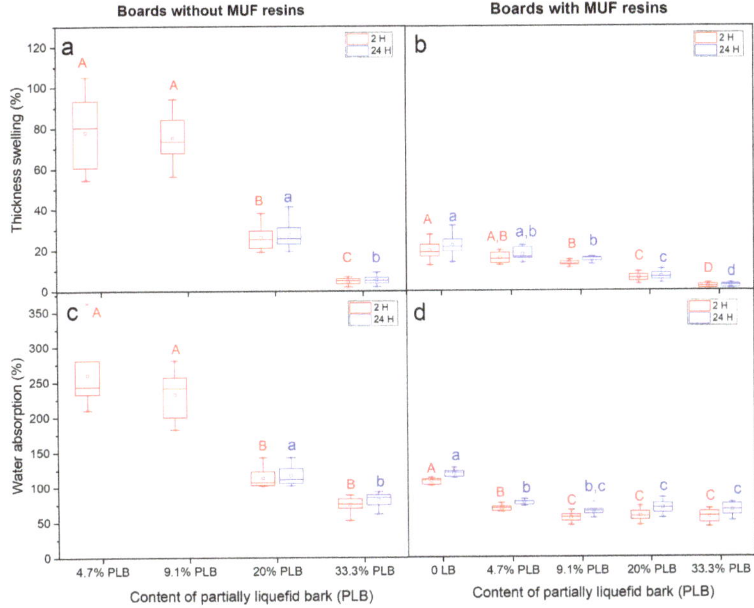

Figure 8. Results of thickness swelling and water absorption of particleboard without (**a,c**) and with MUF (**b,d**) after 2 and 24 h of testing. Values labelled in the same colored letter (small blue for 24 h, and capital red for 2 h) are not statistically different from each other (ANOVA, Tukey's HSD test, $p < 0.05$).

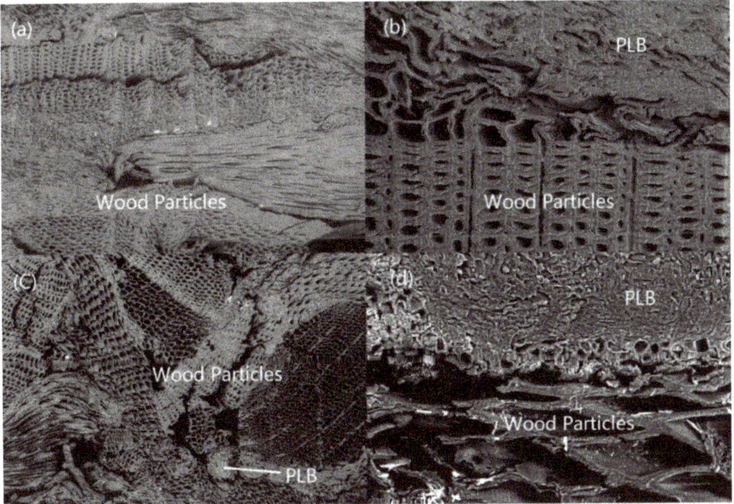

Figure 9. Scanning Electron Microscopy (SEM) micrographs showing the interaction of partially liquefied bark (PLB) and wood particles in the particleboard: (**a**) reference; (**b**) with 20% PLB and no MUF; (**c**) with 9.1% PLB and 10% MUF; and (**d**) with 20% PLB and 10% MUF.

4. Conclusions

Bark as an abundant and easily accessible industrial waste has not been economically and significantly used. The current paper provides a new method for developing bio-based panels from pine bark with the partial liquefaction technique. In this study, pine bark was partially liquefied and then used to produce single-layer particleboard. Four particleboard panels were produced (I–IV) with wood particles and different load levels of PLB (4.7%, 9.1%, 20%, and 33.3%), and five boards (V–IV) were manufactured with wood particles, 10% MUF adhesives and different load levels of PLB (0, 4.7%, 9.1%, 20%, and 33.3%).

Characterization of bark and solid residues of PLB (oven-dried PLB and purified PLB) revealed that the cleavage of the glucoside linkage of cellulose and the β-O-4 bonds of lignin occurred during the partial liquefaction for forming intermediates. PLB and purified PLB decomposed faster than bark during the thermogravimetric analysis because the bark was degraded in the partial liquefaction. PLB and purified PLB were more hydrophilic than bark at higher RHs, meaning that PLB has more hydroxyl groups accessible to water. Excessive hydroxyl groups can be attributed to the bonding between particles and PLB.

Production of particleboard by replacing wood particles with PLB was possible with or without adding MUF adhesives. However, the boards made without MUF adhesives exhibited inferior mechanical properties as compared to the ones with MUF. The mechanical properties and thickness swelling of the non-MUF boards were considerably improved by increasing the content of PLB from 9.1% to 20%. The best mechanical properties were obtained in the particleboard with MUF adhesives and a PLB content of 9.1%. However, up to 20% of PLB content can be used for manufacturing particleboard for meeting the standard requirements. The boards manufactured with 4.7% and 9.1% PLB (I and II) without MUF adhesives showed very poor water resistance, while further addition of PLB at 20% and 33.3% in boards III and IV significantly enhanced the hydrophobicity of the boards. The boards bonded with MUF and different load levels of PLB from 4.7% to 33.3% exhibited significantly lower TS and WA values than the reference board V. The above results indicated that PLB acted as a good water-resistant substance in the particleboard.

Author Contributions: Conceptualization, W.J., S.A., M.P., M.Š., and S.M.; methodology, S.M. and R.H.; software, W.J.; validation, W.J. and S.M.; formal analysis, W.J.; investigation, W.J., R.H., and J.Ž.; resources, M.P. and M.Š.; data curation, W.J.; writing—original draft preparation, W.J.; writing—review and editing, S.A., R.H., and J.Ž.; visualization, W.J.; supervision, S.A. and S.M.; project administration, S.A.; funding acquisition, S.A. All authors have read and agreed to the published version of the manuscript.

Funding: This research was funded by the Formas project 942-2016-2 (2017-21) titled "Utilization of renewable biomass and waste materials for production of environmental-friendly, bio-based composites".

Acknowledgments: The authors would like to thank Ove Eklund from Ikea Industry AB, Hultsfred, Sweden for his technical support in testing internal bond strength.

Conflicts of Interest: The authors declare no conflict of interest.

References

1. Nemli, G.; Demirel, S.; Gümüşkaya, E.; Aslan, M.; Acar, C. Feasibility of incorporating waste grass clippings (*lolium perenne* l.) in particleboard composites. *Waste Manag.* **2009**, *29*, 1129–1131. [CrossRef] [PubMed]

2. Rowell, R.M. *Handbook of Wood Chemistry and Wood Composites*; CRC Press: Boca Raton, FL, USA, 2013; Volume 2.

3. Jones, D.; Brischke, C. *Performance of Bio-Based Building Materials*; Woodhead Publishing: Cambridge, UK, 2017.

4. Wang, D.; Sun, X.S. Low density particleboard from wheat straw and corn pith. *Ind. Crop. Prod.* **2002**, *15*, 43–50. [CrossRef]

5. Sandberg, D. Additives in wood products—Today and future development. In *Environmental Impacts of Traditional and Innovative Forest-Based Bioproducts*; Kutnar, A., Muthu, S.S., Eds.; Springer: Singapore, 2016; pp. 105–172.

6. Ferdosian, F.; Pan, Z.; Gao, G.; Zhao, B. Bio-based adhesives and evaluation for wood composites application. *Polymers* **2017**, *9*, 70. [CrossRef] [PubMed]

7. Hemmilä, V.; Adamopoulos, S.; Karlsson, O.; Kumar, A. Development of sustainable bio-adhesives for engineered wood panels—A review. *RSC Adv.* **2017**, *7*, 38604–38630. [CrossRef]

8. Solt, P.; Konnerth, J.; Gindl-Altmutter, W.; Kantner, W.; Moser, J.; Mitter, R.; van Herwijnen, H.W.G. Technological performance of formaldehyde-free adhesive alternatives for particleboard industry. *Int. J. Adhes. Adhes.* **2019**, *94*, 99–131. [CrossRef]

9. Jiang, W.; Kumar, A.; Adamopoulos, S. Liquefaction of lignocellulosic materials and its applications in wood adhesives—A review. *Ind. Crop. Prod.* **2018**, *124*, 325–342. [CrossRef]

10. Huang, H.-J.; Yuan, X.-Z.; Wu, G.-Q. Liquefaction of biomass for bio-oil products. In *Waste Biomass Management—A Holistic Approach*; Singh, L., Kalia, V.C., Eds.; Springer International Publishing: Cham, Switzerland, 2017; pp. 231–250.

11. Dimitriadis, A.; Bezergianni, S. Hydrothermal liquefaction of various biomass and waste feedstocks for biocrude production: A state of the art review. *Renew. Sustain. Energy Rev.* **2017**, *68*, 113–125. [CrossRef]

12. Hu, S.; Luo, X.; Li, Y. Polyols and polyurethanes from the liquefaction of lignocellulosic biomass. *ChemSusChem* **2014**, *7*, 66–72. [CrossRef]

13. Chen, H. *Lignocellulose Biorefinery Engineering: PRINCIPLES and Applications*; Woodhead Publishing: Cambridge, UK, 2015.

14. Kishi, H.; Akamatsu, Y.; Noguchi, M.; Fujita, A.; Matsuda, S.; Nishida, H. Synthesis of epoxy resins from alcohol-liquefied wood and the mechanical properties of the cured resins. *J. Appl. Polym. Sci.* **2011**, *120*, 745–751. [CrossRef]

15. Kobayashi, M.; Hatano, Y.; Tomita, B. Viscoelastic properties of liquefied wood/epoxy resin and its bond strength. *Holzforschung* **2001**, *55*, 667. [CrossRef]

16. Kunaver, M.; Medved, S.; Čuk, N.; Jasiukaitytė, E.; Poljanšek, I.; Strnad, T. Application of liquefied wood as a new particle board adhesive system. *Bioresour. Technol.* **2010**, *101*, 1361–1368. [CrossRef] [PubMed]

17. Tohmura, S.-I.; Li, G.-Y.; Qin, T.-F. Preparation and characterization of wood polyalcohol-based isocyanate adhesives. *J. Appl. Polym. Sci.* **2005**, *98*, 791–795. [CrossRef]

18. Lee, W.-J.; Lin, M.-S. Preparation and application of polyurethane adhesives made from polyhydric alcohol liquefied taiwan acacia and china fir. *J. Appl. Polym. Sci.* **2008**, *109*, 23–31. [CrossRef]

19. Poljanšek, I.; Likozar, B.; Čuk, N.; Kunaver, M. Curing kinetics study of melamine–urea–formaldehyde resin/liquefied wood. *Wood Sci. Technol.* **2013**, *47*, 395–409. [CrossRef]

20. dos Santos, R.G.; Carvalho, R.; Silva, E.R.; Bordado, J.C.; Cardoso, A.C.; do Rosário Costa, M.; Mateus, M.M. Natural polymeric water-based adhesive from cork liquefaction. *Ind. Crop. Prod.* **2016**, *84*, 314–319. [CrossRef]

21. Pan, H. Synthesis of polymers from organic solvent liquefied biomass: A review. *Renew. Sustain. Energy Rev.* **2011**, *15*, 3454–3463. [CrossRef]

22. Čuk, N.; Kunaver, M.; Poljanšek, I.; Ugovšek, A.; Šernek, M.; Medved, S. Properties of liquefied wood modified melamine-formaldehyde (mf) resin adhesive and its application for bonding particleboards. *J. Adhes. Sci. Technol.* **2015**, *29*, 1553–1562. [CrossRef]

23. Janiszewska, D.; Frąckowiak, I.; Bielejewska, N. Application of selected agents for wood liquefaction and some properties of particleboards produced with the use of liquefied wood. *Drew. Pr. Naukowe Doniesienia Komun.* **2016**, *59*, 223–230.

24. Janiszewska, D. Bark liquefaction for use in three-layer particleboard bonding. *Drew. Pr. Naukowe Doniesienia Komun.* **2018**, *61*, 119–127.

25. González-García, S.; Lacoste, C.; Aicher, T.; Feijoo, G.; Lijó, L.; Moreira, M.T. Environmental sustainability of bark valorisation into biofoam and syngas. *J. Clean. Prod.* **2016**, *125*, 33–43. [CrossRef]

26. Lazar, L.; Talmaciu, A.I.; Volf, I.; Popa, V.I. Kinetic modeling of the ultrasound-assisted extraction of polyphenols from picea abies bark. *Ultrason. Sonochem.* **2016**, *32*, 191–197. [CrossRef]

27. Zhao, Y.; Yan, N.; Feng, M.W. Bark extractives-based phenol–formaldehyde resins from beetle-infested lodgepole pine. *J. Adhes. Sci. Technol.* **2013**, *27*, 2112–2126. [CrossRef]

28. D'Souza, J.; Yan, N. Producing bark-based polyols through liquefaction: Effect of liquefaction temperature. *ACS Sustain. Chem. Eng.* **2013**, *1*, 534–540. [CrossRef]

29. Feng, S.; Cheng, S.; Yuan, Z.; Leitch, M.; Xu, C. Valorization of bark for chemicals and materials: A review. *Renew. Sustain. Energy Rev.* **2013**, *26*, 560–578. [CrossRef]

30. Sillero, L.; Prado, R.; Labidi, J. Optimization of different extraction methods to obtaining bioactive compounds from larix decidua bark. *Chem. Eng. Trans.* **2018**, *70*, 1369–1374.

31. Tudor, E.M.; Barbu, M.C.; Petutschnigg, A.; Réh, R.; Krišťák, Ľ. Analysis of larch-bark capacity for formaldehyde removal in wood adhesives. *Int. J. Environ. Res. Public Health* **2020**, *17*, 764. [CrossRef] [PubMed]

32. Lee, W.-J.; Liu, C.-T. Preparation of liquefied bark-based resol resin and its application to particle board. *J. Appl. Polym. Sci.* **2003**, *87*, 1837–1841. [CrossRef]

33. Hosseinpourpia, R.; Echart, A.S.; Adamopoulos, S.; Gabilondo, N.; Eceiza, A. Modification of pea starch and dextrin polymers with isocyanate functional groups. *Polymers* **2018**, *10*, 939. [CrossRef]

34. Hosseinpourpia, R.; Adamopoulos, S.; Parsland, C. Utilization of different tall oils for improving the water resistance of cellulosic fibers. *J. Appl. Polym. Sci.* **2019**, *136*, 47303. [CrossRef]

35. British Standards Institution. *310: 1993 Wood-Based Panels: Determination of Modulus of Elasticity in Bending and of Bending Strength*; BSI: London, UK, 1993.

36. British Standards Institution. *Particleboards and Fibreboards: Determination of Tensile Strength Perpendicular to the Plane of the Board*; BSI: London, UK, 1993.

37. CEN. *317 Particleboards and Fibreboards—Determination of Swelling in Thickness after Immersion in Water*; CEN: Brussels, Belgium, 1993.

38. Ding, R.; Wu, H.; Thunga, M.; Bowler, N.; Kessler, M.R. Processing and characterization of low-cost electrospun carbon fibers from organosolv lignin/polyacrylonitrile blends. *Carbon* **2016**, *100*, 126–136. [CrossRef]

39. Chupin, L.; Motillon, C.; Charrier-El Bouhtoury, F.; Pizzi, A.; Charrier, B. Characterisation of maritime pine (pinus pinaster) bark tannins extracted under different conditions by spectroscopic methods, ftir and hplc. *Ind. Crop. Prod.* **2013**, *49*, 897–903. [CrossRef]

40. Dussan, K.; Girisuta, B.; Lopes, M.; Leahy, J.J.; Hayes, M.H.B. Conversion of hemicellulose sugars catalyzed by formic acid: Kinetics of the dehydration of d-xylose, l-arabinose, and d-glucose. *ChemSusChem* **2015**, *8*, 1411–1428. [CrossRef]

41. Yamada, T.; Ono, H. Characterization of the products resulting from ethylene glycol liquefaction of cellulose. *J. Wood Sci.* **2001**, *47*, 458–464. [CrossRef]

42. Zou, X.; Qin, T.; Huang, L.; Zhang, X.; Yang, Z.; Wang, Y. Mechanisms and main regularities of biomass liquefaction with alcoholic solvents. *Energy Fuels* **2009**, *23*, 5213–5218. [CrossRef]

43. Zhao, Y.; Yan, N.; Feng, M. Polyurethane foams derived from liquefied mountain pine beetle-infested barks. *J. Appl. Polym. Sci.* **2012**, *123*, 2849–2858. [CrossRef]

44. Yang, H.; Yan, R.; Chen, H.; Lee, D.H.; Zheng, C. Characteristics of hemicellulose, cellulose and lignin pyrolysis. *Fuel* **2007**, *86*, 1781–1788. [CrossRef]

45. Pinto, O.; Romero, R.; Carrier, M.; Appelt, J.; Segura, C. Fast pyrolysis of tannins from pine bark as a renewable source of catechols. *J. Anal. Appl. Pyrolysis* **2018**, *136*, 69–76. [CrossRef]

46. Arteaga-Pérez, L.E.; Segura, C.; Bustamante-García, V.; Gómez Cápiro, O.; Jiménez, R. Torrefaction of wood and bark from eucalyptus globulus and eucalyptus nitens: Focus on volatile evolution vs feasible temperatures. *Energy* **2015**, *93*, 1731–1741. [CrossRef]

47. Neiva, D.M.; Araújo, S.; Gominho, J.; Carneiro, A.d.C.; Pereira, H. An integrated characterization of picea abies industrial bark regarding chemical composition, thermal properties and polar extracts activity. *PLoS ONE* **2018**, *13*, e0208270. [CrossRef]

48. Väisänen, S.; Pönni, R.; Hämäläinen, A.; Vuorinen, T. Quantification of accessible hydroxyl groups in cellulosic pulps by dynamic vapor sorption with deuterium exchange. *Cellulose* **2018**, *25*, 6923–6934. [CrossRef]

49. Huntley, C.J.; Crews, K.D.; Abdalla, M.A.; Russell, A.E.; Curry, M.L. Influence of strong acid hydrolysis processing on the thermal stability and crystallinity of cellulose isolated from wheat straw. *Int. J. Chem. Eng.* **2015**, *2015*, 658163. [CrossRef]

50. Kobayashi, M.; Asano, T.; Kajiyama, M.; Tomita, B. Analysis on residue formation during wood liquefaction with polyhydric alcohol. *J. Wood Sci.* **2004**, *50*, 407–414. [CrossRef]

51. Jasiukaitytė-Grojzdek, E.; Kunaver, M.; Crestini, C. Lignin structural changes during liquefaction in acidified ethylene glycol. *J. Wood Chem. Technol.* **2012**, *32*, 342–360. [CrossRef]

Review

The Design Development of the Sliding Table Saw Towards Improving Its Dynamic Properties

Kazimierz A. Orlowski [1,*][iD]**, Przemyslaw Dudek** [2]**, Daniel Chuchala** [1][iD]**, Wojciech Blacharski** [1] **and Tomasz Przybylinski** [3]

1   Department of Manufacturing and Production Engineering, Faculty of Mechanical Engineering, Gdansk University of Technology, 80-233 Gdansk, Poland; daniel.chuchala@pg.edu.pl (D.C.); wblachar@pg.edu.pl (W.B.)
2   Rema S.A., 11-440 Reszel, Poland; przemyslaw.dudek@rema-sa.pl
3   Institute of Fluid-Flow Machinery, Polish Academy of Sciences, 80-233 Gdansk, Poland; tprzybylinski@imp.gda.pl
*   Correspondence: kazimierz.orlowski@pg.edu.pl

Received: 15 September 2020; Accepted: 15 October 2020; Published: 21 October 2020

Abstract: Cutting wood with circular saws is a popular machining operation in the woodworking and furniture industries. In the latter sliding table saws (panel saws) are commonly used for cutting of medium density fiberboards (MDF), high density fiberboards (HDF), laminate veneer lumber (LVL), plywood and chipboards of different structures. The most demanded requirements for machine tools are accuracy and precision, which mainly depend on the static deformation and dynamic behavior of the machine tool under variable cutting forces. The aim of this study is to present a new holistic approach in the process of changing the sliding table saw design solutions in order to obtain a better machine tool that can compete in the contemporary machine tool market. This study presents design variants of saw spindles, the changes that increase the critical speeds of spindles, the measurement results of the dynamic properties of the main drive system, as well as the development of the machine body structure. It was proved that the use of only rational imitation in the spindle design on the basis of the other sliding table saws produced does not lead to the expected effect in the form of correct spindle operation.

Keywords: sliding table saw; spindle; critical rotational speed; static stiffness; dynamic properties; noise; sawing of wood composites

1. Introduction

Cutting wood with circular saws is a popular machining operation in the woodworking and furniture industries. Its popularity is mainly due to the fact that in this method of cutting relatively simple and cheap tools are used, usually saws and disc cutters with small dimensions. In the furniture industry, sliding table saws (panel saws) are commonly used for cutting of medium density fiberboards (MDF) [1], high density fiberboards (HDF), laminate veneer lumber (LVL) [1], plywood [2] and chipboards of different structures [3].

The most demanded requirements for machine tools are accuracy and precision, which mainly depend on the static deformation and dynamic behavior of the machine tool under variable cutting forces [4]. Sliding table saws (panel saws) should guarantee the user straightness of the kerf in the longitudinal direction, perpendicularity of the kerf to the sawn board surfaces, a smooth kerf surface after cutting, as well as lack of washboarding [5] on the sawed surface. In sliding table saws, the blade of the collared saw is rigidly fixed to the driving spindle [6–9]. The qualitative effects of the sawing process depend on the static and dynamic properties of the entire structure of the machine tool, and the cutting system consists of a machine tool, clamping system, a workpiece and a tool. Hu Wan-yi et al. [10] presented the results of empirical works devoted to the noise generated by the sliding table saws during idling. The generated noise has three main sources: air flow around the saw blade and in the suction system (aeromechanic noise), and the noise depending on the structure of the machine tool. Noise caused by mechanical vibrations can result from unbalance of the main saw blade system, eccentricity of the main spindle, incorrect assembly and loosening of the bearing, which are could be caused by wear of bearings race-ways [10]. Vibrations and noise in machine tools are always present simultaneously, hence, if the vibrations are at a lower level, the noise is also lower [11]. For example, self-exciting chatter vibrations have a particularly negative effect on the cutting process effects, a state of the workpiece (waviness, roughness), a machine tool life and tool condition, as well as on the efficiency of machining. In addition, they accelerate spindle-bearing wear and cutting-edge wear (it is not only faster, but even catastrophic). Furthermore, they make it difficult to obtain the required surface quality and cause excessive noise [12]. The manufacturers of sliding table saws tend towards improving sawing accuracy by minimizing the vibration level and noise of their machine tools, and thank to that growing their competitiveness at the market.

Nasir and Cool [13] have done a very thorough review of the literature in which they have shown that the studies of sawing processes are the subject of numerous studies in many scientific centers. Kvietková et al. [14] reported on the results of the effect of number of saw blade teeth on noise level during transverse cutting of beech wood. However, the tests were not conducted on the sawing machine, but with the use of the power tool. For this reason, the expected noise level when using similar circular saw blades on a panel saw is likely to be completely different. The same power tool Kminiak et al. [15] applied in the research on the quality of a machined wood surface while transverse cutting of European beech (*Fagus sylvatica* L.). Very often, experiments of cutting with circular saw blades are conducted on the special laboratory stands, e.g., cross-cutting of green spruce and beech wood [16], or cutting process with feeding in the longitudinal direction of modified beech wood (Bendywood Candidus Prugger Sas, Bressanone, Italy), DMDHEU (1.3-dimethylol-4.5-dihydroxyethyleneurea) (Wood Biology and Wood Products, University of Goettingen, Goettingen, Germany) and Lignamon (a name of ammonia-treated compressed beech wood) [17]. The publication by Mandic et al. [18] was found among the reports from the numerous studies on the process of cutting with circular saws conducted on a sliding table saw. These authors were the few who carried out their empirical research of power consumption and the acoustic emission (AE) on the panel saw the type Minimax CU410K machine tool (SCM Group, Rimini, Italy), and the material to be cut was laminated particle board. Surface roughness produced by rip sawing with circular saw of MDF was evaluated by Aguilera [19] using a stylus technique. In that research while in climb cutting mode the surface roughness was slightly better if the cutting speed was higher. The findings of work by Aguilera and Barros [20] lead to the conclusion that the sound pressure (measured with microphones) generated in the process with circular saw blades is closely related on satisfactory levels of correlation with the surface roughness. The samples were machined in a single-spindle shaper machine. However, in both cases described in works [19,20], the research was carried out on a single-spindle shaper machine, which is much stiffer than a panel saw. Therefore, it would be difficult to expect similar results when cutting MDF boards on a panel saw.

Since, there is the carcinogenic nature of wood dusts [21], the saw dust extraction systems of sawing machines have been also examined in terms of their efficiency [22,23] and acoustic emission [23].

Due to its low inherent stiffness, the circular saw is one of the weakest elements of the machine tool system. Hence, vibrations of the sawblade and any roughness of the cutting surface must be limited. Cutting vibration can be suppressed by uncoupling the two vibration modes of the same nodal diameter number by using outer slots in the saw blade [24,25] and creating high saw body damping using inner slots filled with viscoelastic resin [26]. The behavior of the circular saw blade also depends on its design [27,28]. Nevertheless, the reduction of the transverse vibrations of the saw blade can be obtained if it operates below the value of the critical rotational speed [6,29,30].

Dietrych [31] states that the ability to perform tasks correctly by the new designed machine tool can be evaluated on the basis of quality indices, which include expected life of the machine, reliability, precision and low level of emitted interference (vibration and noise). For a panel saw the objective function should be comparable to the cutting accuracy it achieves. An illustration of the causal relationship between these mentioned quality indicators and vibrations is shown in Figure 1. Too much vibration activity of the machine tool (panel saw) will affect durability (expected life), accuracy and reliability [11,32]. Therefore, at the stage of designing and testing the prototype, it is necessary to find the sources of vibration and noise. Such activities are called vibroacoustic construction (emission) diagnostics [11,32].

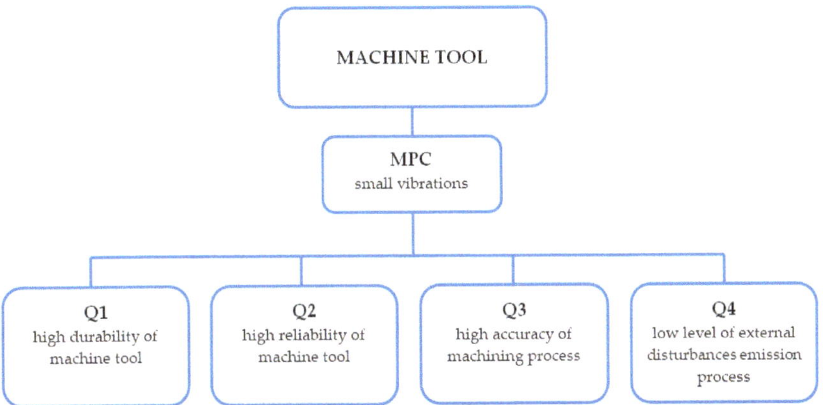

Figure 1. Quality indices of the machine tool and their relation to vibrations, where: MPC—machining process conditions; Q1, Q2, Q3, Q4—quality indices.

Based on the observations, it can be assumed that most manufacturers in the design process use statistical and comparative methods [33], which rely on rational imitation of practically proven sawmill drives of similar design, similar size and similar kinematics, taking into account development trends in a given group of machine tools.

The aim of this study is to present a new holistic approach in the process of changing of the sliding table saw design solutions in order to obtain a better machine tool that can compete in the modern machine tool market. This review presents design variants of saw spindles, changes in increasing the critical speeds of spindles, measurement results of the dynamic properties of the main drive system, as well as the development of the machine body structure.

2. Spindle and Main Driving System

2.1. Design Variants of Main Spindles

In North America, a system with circular saw blades having a spline in the inner hole working with a spindle having an external spline is common to circular sawing machines for primary wood processing, especially [7]. The second way of embedding the saw blades on the spindles of sawing machine tools, common in Europe, is by fixing them with the fastening collars [29,34]. In this case the operation of the saw must be at a rotational speed lower than their critical speed, guaranteeing the stable operation of the tool [29]. However, the clamping of the saw with the help of collars allows for multi-saws to place saws on the spindles with little distances between them. Moreover, the latter solution is commonly used in format saws where the cutting torque is most often transmitted by friction between the mounting collars and the saw blade [8,29,35]. Orlowski and Dudek [35] analyzed the development of the main spindles of the sliding table saws for the last quarter century.

In the last decade of the twentieth century, the long spindles with ratio of the supports spacing L to the inner diameter of the front bearing d of about 12.7, were mounted in the format saws [36]. This type of solution is be still found in the sliding table saw DMMS-40 Classic (REMA S.A., Reszel, Poland, Figure 2a), in which the traditional V-belt has been displaced by the PK belt (v-ribbed belt).

(**a**)   (**b**)

Figure 2. Long main spindle in the sliding table saw DMMS-40 Classic (a, REMA S.A.) and short main spindle (**a**) view from motor side in the sliding table saw Fx550 (**b**), REMA S.A.

In 2006, Altendorf showed at the Drema Fair in Poznan a new generation of sliding table saws F45 Elmo (Altendorf, Minden, Germany) [37], with a short main spindle, with a L/d ratio of about 3 [38]. Since then, the market has been supplying mainly short-wheeled spindles driven mainly with a rear-wheel transmission. This kind of the design was applied in the sliding table saws types as follows: Fx3 (after modernization Fx550 [39], f. Rema SA, Reszel, Poland), UNICA 400 (f. Griggio, Cadoneghe, Italy—company closed in 2018) [40], K 700S (f. Felder Group, Hall in Tirol, Austria) [41] and PF 400S (f. Rojek, Častolovice, Czech Republic) [42]. Each of the aforementioned saws has a stepped main drive, in which the change of rotational speed depends on the position of the PK belt in the belt transmission. There are usually solutions with three pairs of pulleys or less often with four pairs of pulleys. Nevertheless, the most modern solution seems to be the variant with a continuously variable drive, such as in one of the design variants of the F45 saw [37].

2.2. Static and Dynamical Properties of Spindles

The use of only rational imitation in the design on the basis of the other table sliding saws produced does not lead to the expected effect in the form of correct spindle operation, and this mainly concerns the possible exceeding of the tool's lateral run out value. Too much lateral runout of the saw blade can be a source of additional force excitation for the tool, which can cause unwanted machining errors as a result. Errors of this type are not very visible when cutting individual thin wood composite panels, however, very often these materials are cut in packages and then the errors on sawn surfaces can be more observable, especially in case of top boards in the package.

The correctness of the spindle design can be determined based on analytically determined speeds critical, which seems to be a rational approach, especially for circular sawing machines. In the literature [8,43], it can find recommendations to calculate the values of spindle critical rotational speeds n_{cr} from the Equation:

$$n_{cr} = 300 \sqrt{\frac{1}{f_{max}}}, \tag{1}$$

where: f_{max} is a maximum deflection of the spindle in cm (determined on the front or rear end of the spindle). The calculated critical rotational speed should satisfy the inequality:

$$n_{cr} \geq (1.5 \div 2)n_{work},\qquad(2)$$

where: n_{work} is a working rotational speed of the spindle, rpm.

The short spindle of the sliding table saw Fx3 (before modernization) presented at work [34] can develop working speeds of 3500, 4500 and 6000 min^{-1}, depending on the location of the PK belt in belt pulley. This type of solution appeared in this machine tool in 2011 [38]. On the spindle, Ø450, Ø350 or Ø300 mm circular saw blades can be clamped with the Ø125 mm diameter flanges, which defines the rotation speed. On both sides, the spindle was supported on 6206 2RS1 Explorer (SKF) bearings and the ratio L/d = 2.6. Deformation calculations were carried out with the use of the Finite Element Method (FEM) in which the spindle model was loaded on the rear end by a force F_{s-d} from the shaft drive, which was determined using the available software of the company SKF (is an acronym for Svenska Kullagerfabriken, Swedish Ball Bearing Factory, SKF Sweden AB, Göteborg, Sweden) [44]. The circular saw blade at the front end was loaded, with forces which values and location were determined according to the work [8], assuming that full rated engine power of 7.5 kW is available in the cutting zone.

Computations of maximum deformations of the Fx3 (Figure 3) and Fx550 (Figure 4) saw spindles were carried out as static structural linear analyses. In both models linear tetrahedrons type elements with a fully structured mesh of 1 mm size (the mesh of elements was not refined) were applied. The model assumed one degree of freedom in the form of the possibility of rotation in relation to the spindle axis, which in the machine tool's coordinate system it is the Z axis. Moreover, it was assumed that both of the bearing supports are cylindrical and do not allow for translational displacement in X and Y directions, but only allow for rotation in the relation to the axis Z. In computations, the material properties of spindles were as follows: Young's modulus 205 GPa, Poisson's ratio 0.28.

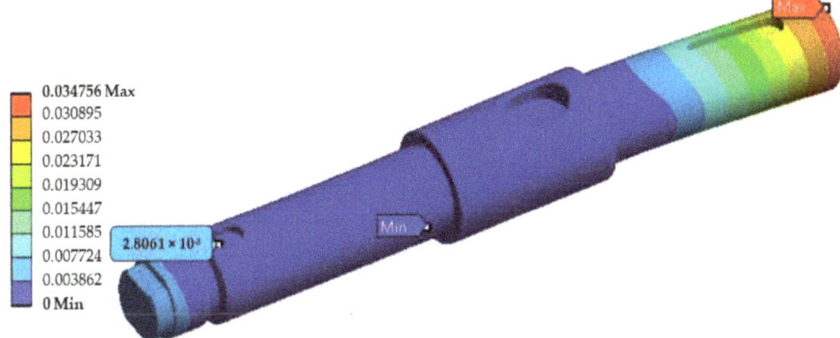

Figure 3. The resultant deformation (in mm) of the spindle in the table sliding saw Fx3 caused by the cutting forces and by the force F_{s-d}, for the operating speed n_{work} = 3500 min^{-1}.

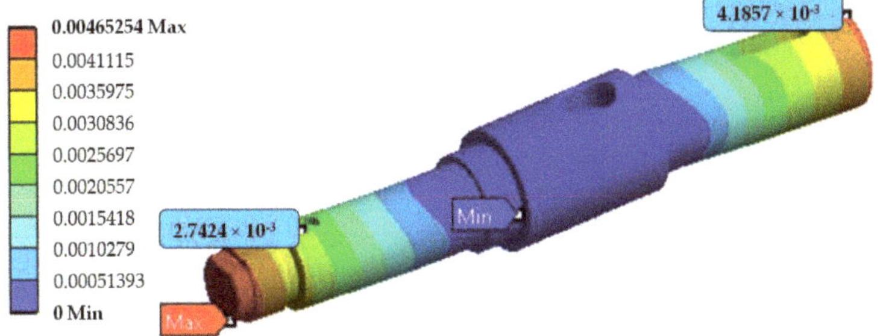

Figure 4. The resultant deformation (in mm) of the spindle in the Fx550 sawing machine caused by the cutting forces and by the force F_{s-d}, for the operating speed n_{work} = 3500 min^{-1}.

In Figure 3, the resultant deformation of the spindle in the Fx3 sawing machine caused by the cutting forces and by the force $F_{s\text{-}d}$, for the operating speed $n_{work} = 3500$ min^{-1} is presented. The maximum deflection of the spindle in this case was equal to 0.034756 mm. Based on the obtained results, it was found that for the two lowest rotational speeds of the saw spindle the condition described by Equation (2) is not satisfied. For that reason, the model of the spindle was redesign.

The new spindle model was calculated with the support diameters equal to Ø35 mm (under 6207 2RS1 Explorer bearing), increasing the L/d ratio to 3.0, according to SKF recommendations for optimum spacing of spindle supports [45]. Due to the predominant influence of the forces from the drive, the diameter of the rear end was also increased. In Figure 4, the resultant deformation of the spindle in the Fx550 sawing machine caused by the cutting forces and by the force $F_{s\text{-}d}$, for the operating speed $n_{work} = 3500$ min^{-1} is presented. The resulting deformation values for the changed spindle turned out to be smaller [34], which resulted in the higher n_{cr}/n_{work} ratios for each case of rotational speed (Figure 5). An improved spindle of the new type (Figure 2b) was implemented in the table sliding saw Fx550. Moreover, in the latter machine tool to increase the sawing aggregate rigidity of the panel saw, the stiffness of the motor plate guiding was increased, and simultaneously rolling guides with higher rigidity were used [38].

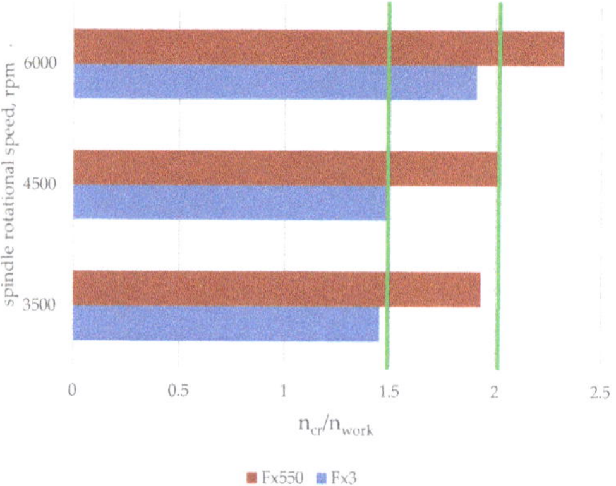

Figure 5. Ratios of critical rotational speeds to working rotational speeds revolutions n_{cr}/n_{work} of spindles of table sliding saws Fx3 (an old applied solution) and Fx550 (a currently applied design) (Rema SA), where vertical bold lines indicate the limits of the recommended values of the ratio n_{cr}/n_{work}.

While the review of the spindle designs, it was observed that the position of the driving wheel on the spindle may be different, therefore, numerical calculations were additionally performed to demonstrate the effect of the pulley position on the spindle with increased rigidity [46] on its critical rotational speeds. The results of the calculations showed that for this type of rigid spindle, in each of the analyzed cases of the wheel position, the critical rotational speed was approximately 14,400 min^{-1} and the ratio of $n_{cr}/n_{work} \approx 2.4$.

3. Dynamic Properties of the Machine Tool

3.1. Dynamic Properties of the Main Driving System

Some selected problems concerning the empirical research of the saw cutting unit of the table sliding saw were described by Orłowski et al. [34]. Cempel [32] expressed that the diagnosis should be limited to one definite industrial case with using repeated methodology. This recommendation was due to the certainty that setting standards of diagnosis is too risky [32]. Before performing of vibroacoustic empirical tests the researcher should consider which parameter should be measured in a given case. The vibration velocity is a common choice, but not always the right one, because the better option is often measuring of displacements or accelerations [32].

Vibration diagnostic experimental analyzes have been carried out on the saw cutting unit of the modernized table sliding saw Fx3 (currently Fx550) (f. REMA S.A., Reszel, Poland). The circular saw blade with main dimensions Ø300 × 3.2 × 30 and number of teeth z = 96 was mounted on the spindle with collars of Ø125 mm. The measured rotational speed of spindle was n_{work} = 5128 rpm. The place of the accelerometer (A in Figure 6) installation was also the measuring point position, and was located on the top of the main body of the spindle system. The Fluke 810 vibration tester (f. Fluke, Everett, Washington, DC, USA) was used to measuring of the accelerations in the X, Y, Z axes, which are axes of the machine tool co-ordination system (Figure 6).

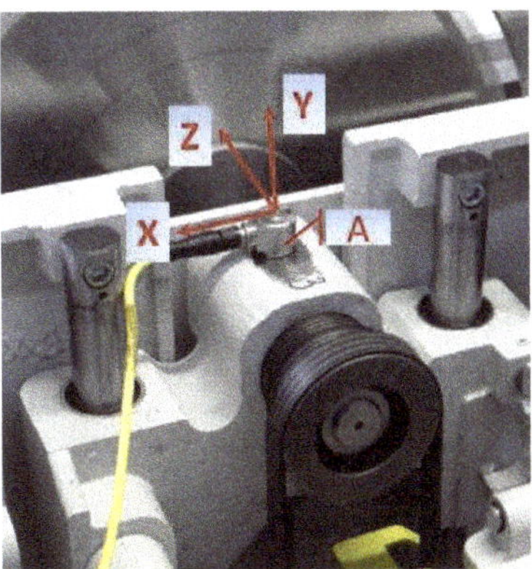

Figure 6. A view of the saw cutting unit of the modernized table sliding saw Fx3 with a position of the accelerometer A, where: *X, Y, Z*—axes of the table sliding saw co-ordination system [47].

The vibration tester Fluke 810 automatically converts the received acceleration signals into a waveform plot (Figure 7a) and creates vibration velocity spectra (Figure 7b). The latter feature is a disadvantage, since, the person conducting the tests has practically no influence on the way the test is performed, and the results are processed.

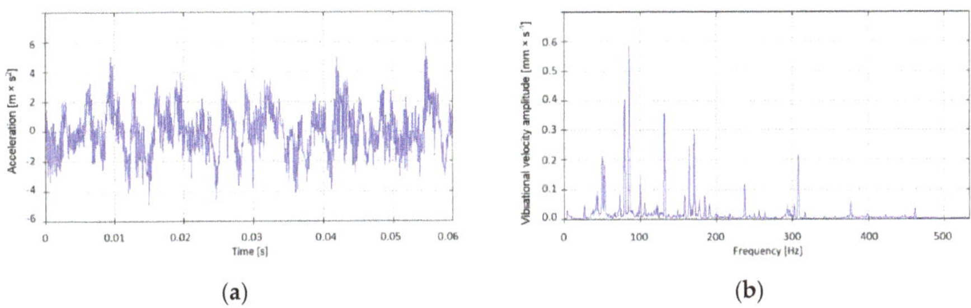

(a)   (b)

Figure 7. Waveform of the acceleration signal (**a**) and vibrational velocity spectrum of the signal (**b**) of the modernized table sliding saw Fx3 in the measurement point (Figure 6) in Z axis [47].

The changes of component velocities v_j in function of time t, can be computed on the basis of vibrational velocity amplitudes (Figure 7b) [47]. The Root-Mean-Square (*RMS*) values of vibrational velocities were calculated for each measurement axis of the modernized table sliding saw Fx3. In the next step, the total vibrational velocity for the measurement point was computed from the equation as follows:

$$v_\Sigma(RMS) = \sqrt{v_X(RMS)^2 + v_Y(RMS)^2 + v_Z(RMS)^2},\tag{3}$$

Figure 8 presents *RMS* values of vibrational velocities in directions of *X*, *Y* and *Z* axes of the coordination system. In Figure 8, the value of the resultant *RMS* of vibrational velocity Sigma was also shown. The analyzed values in three directions *X*, *Y* and *Z* were measured at the measurement point (Figure 6).

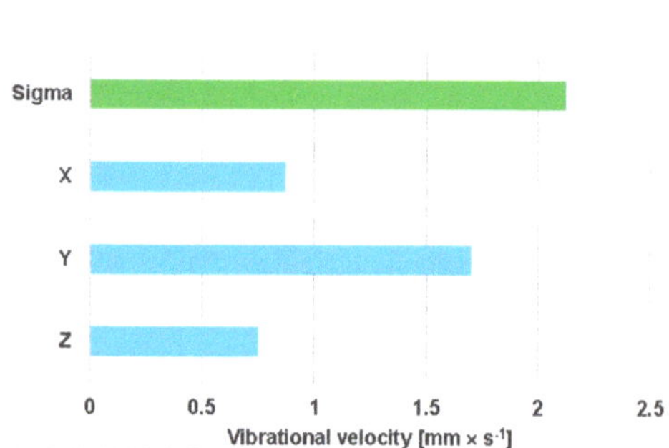

Figure 8. Root-mean-square values of vibrational velocities in directions of *X*, *Y* and *Z* axes of the modernized table sliding saw Fx3 co-ordination system measured at the measurement point (Figure 2a) together with a resultant Root-Mean-Square *RMS* of vibrational velocity Sigma.

Due to the use in literature the different criteria values for making an assessment of the new machine tool [32] in the analyzed investigation were determined two values: *RMS* amplitude of the total vibrational velocity and the peak value (amplitude). Obtained values from experimental test are presented in Table 1. Both Blake and Łączkowski are guided by the values of the peak amplitude [32]. However, for both standards, the experimental values obtained are evaluated differently (Table 1). A similar situation can be observed for the other two machine tool evaluation standards: own standards of the American diagnostic company, (IRD Machanalysis Limited, Maharashtra, India), and the British company VCI Ltd (Strabane, Great Britain). In this case, both standards base their assessment on the value of the *RMS*. Unfortunately, the evaluation of the examined main spindle system of the modernized table sliding saw Fx3 based on the experimental data received is again divergent for both standards (Table 1). The presented analysis shows that there are no unequivocal criteria for assessing the condition of the examined machine tool. Moreover, the choice of assessment criteria can very often be more or less subjective.

Table 1. Experimental values of vibrational velocities with diagnostic evaluation on basis of a few diagnosis standards.

	RMS of Vibrational Velocities, v_Σ mm·s^{-1}	Peak Value of Vibrational Velocities, v_{max} mm·s^{-1}
Experimental Results	2.08	3.27
Standards	**Diagnosis**	**Range *RMS* of vibrational velocities for diagnosis, mm·s^{-1} or *range peak value of vibrational velocities, mm·s^{-1}**
IRD Mechanalysis	Admissible	2–4
VCI Ltd.	Good	1.27–2.54
Blake	Admissible	*2.20–6.00
Łączkowski	Good	*2.50–6.30

3.2. Dynamic Properties of the Machine Tool Body

In the sliding table saw Fx550, which is the follower of the circular sawing machine Fx3, a new machine frame body made of steel sections connected with special lockers and welded was applied (Figure 9). In the previous version of the machine tool body, there was a solution which based on a set of bent body parts connected (mainly welded) with flat steel plates. The new design of the machine tool body has much more stiffness of the machine structure in comparison with the panel saw Fx3, and that kind of the body can be met only in the highest-class panel saws.

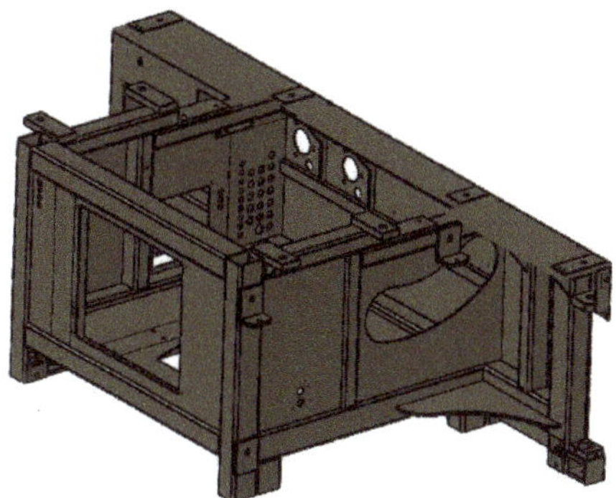

Figure 9. General view of the frame body of the sliding table saw Fx550.

This body solution has given a spectacular reduction of the resultant peak values (amplitudes) v_Σ of the vibrational velocities (Figure 10) measured at the point #4. The measurement point #4 was situated in each case in the middle on the body wall which was parallel to the main spindle axis. The value of the vibration velocity for the new solution is significantly below the permissible minimum value recommended by Łączkowski [32].

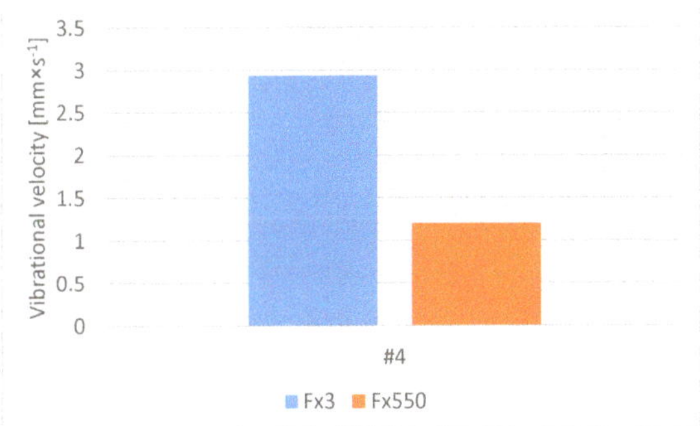

Figure 10. Resultant peak values (amplitudes) of the vibrational velocities v_Σ measured on the frame body at the measurement point #4 of the sliding table saws Fx3 and Fx550.

Testing the unloaded machine tools of both Fx3 and Fx550, maintaining tool setting, machining parameters were carried out in industrial conditions in the frame of the project POIR.01.01.01-00-05888/15 (in Polish Program Operacyjny Inteligentny Rozwój, Smart Growth Operational Programme). Measurements of noise were done with the use of the integrating sound level meter (SON-50) (f. Sonopan, Bialystok, Poland) in the points according to the International Organization for Standardization (ISO) standard (ISO 7960: 1995) [48]. At the location of the microphone in the operator position (about 1.5 m over the hall floor) for the sliding table saw Fx3 the noise on the idling was at the level of

77.5 dB whereas in case of Fx550 it was 73.5 dB. The resultant noise determined as an average from all measurement points was as follows: for Fx3 equalled to 79 dB and for Fx550 was equal 74.6 dB.

Thanks to: a new body of the sliding table saw Fx550, a modernized stiffer spindle, a stiffer system of the electric motor plate guiding [38], and a steady PK belt straining system the described sliding table saw Fx550 has been a source of the general noise on the idling at the level lower about 5 dB in comparison to the sliding table saw Fx3.

4. Conclusions

This review's objective was to present a new holistic approach in the process of changing the sliding table saw design solutions in order to obtain a better machine tool that can compete in the contemporary machine tool market. Based on the review, it can be concluded that:

- In modern design solutions of sliding table saws, the main circular saw blades are clamped by means of collars on spindles with a short support spacing with a ratio of the supports spacing L to the inner diameter of the front bearing d of about 3. It ought to be emphasized, that excessive increase in the diameter of the front bearing d and simultaneous striving for the optimum support spacing is not possible with the sliding table saws, as every manufacturer aims for the smallest possible dimensions of the cutting unit.

- To evaluate the dynamic properties (behavior) of the spindle it is useful to determine their critical values of rotational speeds. The maximum deformations determined in static structural linear analyses showed that in the case of the Fx550 saw spindle they are 10× smaller in comparison with the spindle of the Fx3 saw, which made it possible to estimate critical speeds, which satisfied the inequality presented in Equation (2).

- The use of only rational imitation in the spindle design on the basis of the other sliding table saws produced does not lead to the expected effect in the form of correct spindle operation, and this mainly concerns the possible exceeding of the tool's lateral runout value.

- The errors on sawn surfaces caused by the tool's lateral runout value can be more apparent especially in case of top wood composite boards (MDF, HDF, LVL, plywood or particle boards) in the sawn package.

- An application in a new machine frame body of steel sections connected with special lockers instead of the solution which based on a set of bent body parts between flat steel plates resulted in lower noise values of around 5 dB during idling.

- If the noise of the machine tool is decreased and simultaneously vibrations are at lower level, hence, it could be expected higher accuracy of sawing which is especially important in furniture production.

- Experimental values of vibrational velocities (*RMS* or peak value of vibrational velocities) on basis of a few diagnosis standards allowed us to classify the examined main spindle system of the sliding table saw Fx550 as good or admissible design solution from the point of its dynamics.

Author Contributions: Conceptualization, K.A.O., P.D. and W.B.; methodology, K.A.O., W.B. and D.C.; investigation, K.A.O., W.B., P.D., D.C. and T.P.; writing—original draft preparation, K.A.O., D.C.; writing—review and editing, K.A.O., D.C. and T.P.; supervision, K.A.O.; project administration, K.A.O. All authors have read and agreed to the published version of the manuscript.

Funding: It is kindly acknowledged that this work has been carried out within the framework of the project POIR.01.01.01-00-05888/15, which has been financially supported by the European Regional Development Fund. The authors would also like to acknowledge the company REMA S.A. in Reszel (Poland), which is the beneficiary of the project.

Acknowledgments: It should be acknowledged that the sliding table panel saw Fx550 has been awarded with the MTP Gold Medal of DREMA 2017 in Poznan, Poland. The role of Eng. Karol Duchnicz from Eaton Truck Components Sp. z o.o., ought to be acknowledged for his valuable guidance in spindle modelling.

Conflicts of Interest: The authors declare no conflict of interest.

References

1. Goli, G.; Curti, R.; Marcon, B.; Scippa, A.; Campatelli, G.; Furferi, R.; Denaud, L. Specific cutting forces of isotropic and orthotropic engineered wood products by round shape machining. *Materials* **2018**, *11*, 2575. [CrossRef] [PubMed]
2. Salca, E.-A.; Bekhta, P.; Seblii, Y. The effect of veneer densification temperature and wood species on the plywood properties made from alternate layers of densified and non-densified veneers. *Forests* **2020**, *11*, 700. [CrossRef]
3. Mirski, R.; Derkowski, A.; Dziurka, D.; Dukarska, D.; Czarnecki, R. Effects of a chipboard structure on its physical and mechanical properties. *Materials* **2019**, *12*, 3777. [CrossRef] [PubMed]
4. Hernandez-Vazquez, J.-M.; Garitaonandia, I.; Fernandes, M.H.; Muñoa, J.; López de Lacalle, L.N. A consistent procedure using response surface methodology to identify stiffness properties of connections in machine tools. *Materials* **2018**, *11*, 1220. [CrossRef]
5. Orlowski, K.A.; Wasielewski, R. Study washboarding phenomenon in frame sawing machines. *Holz als-Roh Werkst.* **2006**, *64*, 37–44. [CrossRef]
6. Mohammadpanah, A.; Hutton, S.G. Flutter instability speeds of guided splined disks. An experimental and analytical investigation. *J. Shock Vib.* **2015**, *2015*, 942141. [CrossRef]
7. Mohammadpanah, A.; Hutton, S.G. Maximum operation speed of splined saws. *Wood Mater. Sci. Eng.* **2016**, *11*, 142–146. [CrossRef]
8. Svoreň, J.; Hrčková, M. *Woodworking Machines, Part I*; Technical University in Zvolen: Zvolen, Slovak Republic, 2015.
9. Mohammadpanah, A.; Hutton, S.G. Dynamic response of guided spline circular saws vs. collared circular saws, subjected to external loads. *Wood Mater. Sci. Eng.* **2019**. [CrossRef]
10. Hu, W.-Y.; Qi, Y.-J.; Zhang, Z.-H.; Qi, X.-J. Study on noise of precision panel saw. *J. For. Res.* **2003**, *14*, 335–338.
11. Cempel, C. *Diagnostyka Wibroakustyczna Maszyn*; Politechnika Poznańska: Poznań, Poland, 1985. (In Polish)
12. Szulewski, P.; Śniegulska-Grądzka, D. Systems of Automatic Vibration Monitoring in Machine Tools. *Mechanik* **2017**, *90*, 170–175. [CrossRef]
13. Nasir, V.; Cool, J. A review on wood machining: Characterization, optimization, and monitoring of the sawing process. *Wood Mater. Sci. Eng.* **2020**, *15*, 1–16. [CrossRef]
14. Kvietková, M.; Gaff, M.; Gašparík, M.; Kminiak, R.; Kriš, A. Effect of number of saw blade teeth on noise level and wear of blade edges during cutting of wood. *Bioresources* **2015**, *10*, 1657–1666. [CrossRef]
15. Kminiak, R.; Gašparík, M.; Kvietková, M. The dependence of surface quality on tool wear of circular saw blades during transversal sawing of beech wood. *Bioresources* **2015**, *10*, 7123–7135. [CrossRef]
16. Krilek, J.; Kováč, J.; Kučera, M. Wood crosscutting process analysis for circular saws. *Bioresources* **2014**, *9*, 1417–1429. [CrossRef]
17. Hlásková, L.; Kopecký, Z.; Novák, V. Influence of wood modification on cutting force, specific cutting resistance and fracture parameters during the sawing process using circular sawing machine. *Eur. J. Wood Wood Prod.* **2020**, *78*, 1173–1182. [CrossRef]
18. Mandic, M.; Svrzic, S.; Danon, G. The comparative analysis of two methods for the power consumption measurement in circular saw cutting of laminated particle board. *Wood Res.* **2015**, *60*, 125–136.
19. Aguilera, A. Surface roughness evaluation in medium density fibreboard rip sawing. *Eur. J. Wood Wood Prod.* **2011**, *69*, 489–493. [CrossRef]
20. Aguilera, A.; Barros, J.L. Surface roughness assessment on medium density fibreboard rip sawing using acoustic signals. *Eur. J. Wood Wood Prod.* **2012**, *70*, 369–372. [CrossRef]
21. Barcenas, C.H.; Delclos, G.L.; El-Zein, R.; Tortolero-Luna, G.; Whitehead, L.W.; Spitz, M.R. Wood dust exposure is a potential risk factor for lung cancer. *Am. J. Ind. Med.* **2005**, *47*, 349–357. [CrossRef]
22. Barański, J.; Jewartowski, M.; Wajs, J.; Orlowski, K.; Pikała, T. Experimental examination and modification of chip suction system in circular sawing machine. *Drv. Ind.* **2018**, *69*, 223–230. [CrossRef]
23. Nasir, V.; Cool, J. Characterization, optimization, and acoustic emission monitoring of airborne dust emission during wood sawing. *Int. J. Adv. Manuf. Technol.* **2020**, *109*, 2365–2375. [CrossRef]
24. Yokochi, H.; Nakashima, H.; Kimura, S. Vibration of circular saws during cutting I. Effect of slots on vibration. *Mokuzai Gakkaishi* **1993**, *39*, 1246–1252.

25. Nishio, S.; Marui, E. Effect of slots on the lateral vibration of a circular saw blade during sawing. *Mokuzai Gakkaishi* **1995**, *41*, 722–730.

26. Nishio, S. Stable sawblade. In *Proceedings of the 17th International Wood Machining Seminar*; Scholz, F., Ed.; Retru-Verlag e.K.: Weyarn, Germany, 2005; pp. 418–420.

27. Droba, A.; Javorek, L.; Svoreň, J.; Pauliny, D. New design of circular saw blade body and its influence on critical rotational speed. *Drewno* **2015**, *58*, 147–157.

28. Svoreň, J. The analysis of the effect of the number of teeth of the circular saw blade on the critical rotation speed. *Acta Fac. Tech. (Zvolen–Slovakia)* **2012**, *17*, 109–117.

29. Orlowski, K.A.; Sandak, J.; Tanaka, C. The critical rotational speed of a circular saw: Simple measurement method and its practical implementations. *J. Wood Sci.* **2007**, *53*, 388–393. [CrossRef]

30. Stakhiev, Y.M. Research on circular saws disc problems: Several of results. *Holz als Roh-Werkst* **2003**, *61*, 13–22. [CrossRef]

31. Dietrych, J. *System i konstrukcja, (In Polish, System and Design)*; Wydawnictwa Naukowo-Techniczne: Warszawa, Poland, 1978.

32. Cempel, C. Vibroacoustic condition monitoring. In *Ellis Horwood Series in Mechanical Engineering*; Haddad, S.D., Ed.; Ellis Horwood Ltd.: New York, NY, USA, 1991.

33. Wrotny, L.T. *Projektowanie Obrabiarek. Zagadnienia Ogólne i Przykłady Obliczeń, (In Polish: Designing Machine Tools. General Issues and Examples of Calculations)*; Wydawnictwa Naukowo-Techniczne: Warszawa, Poland, 1986.

34. Orlowski, K.A.; Duchnicz, K.; Dudek, P. Analiza wpływu cech konstrukcyjnych wrzeciona pilarki formatowej na jego obroty krytyczne. *Mechanik* **2016**, *89*, 1112–1113.

35. Orlowski, K.A.; Dudek, P. Analysis of the design development of the sliding table saw spindles. *Mechanik* **2017**, *90*, 849–851. [CrossRef]

36. Wasielewski, R.; Orlowski, K.A. The effect of the power transmission method of the spindle on the position of the circular saw teeth. In Proceedings of the 18th International Wood Machining Seminar, Vancouver, BC, Canada, 7–9 May 2007; pp. 97–103.

37. Altendorf F45. Available online: https://www.altendorf.com/en/f-45.html (accessed on 20 October 2020).

38. Pikała, T.; Rema, S.A., Reszel, Poland. Personal communication, 2017.

39. Rema Fx550. Available online: https://rema-sa.pl/produkty/pilarka-formatowa-rema-fx550 (accessed on 27 August 2020).

40. UNICA 400 and UNICA 500. Available online: https://wtp.hoechsmann.com/en/lexikon/20024/unica_500 (accessed on 20 October 2020).

41. K 700 S. Available online: https://www.felder-group.com/pl-pl/produkty/pily-formatowe-c1947/pila-formatowa-k-700-s-p64674 (accessed on 11 September 2020).

42. PF 400S. Available online: http://rojek.cz/pdf/PF300L_350_400S_New.pdf (accessed on 11 September 2020).

43. Černoch, S. *Strojně Technická Příručka 1*; SNTL-Státní nakladatelství technické literatury: Praha, Czechoslovakia, 1968.

44. SKF Belt Calculator. Available online: https://www.skfptp.com/Publications/Publications# (accessed on 11 September 2020).

45. SKF. Available online: https://www.skf.com/my/products/super-precision-bearings/principles/bearing-selection-process/bearing-execution/system-rigidity (accessed on 2 September 2020).

46. Orlowski, K.; Duchnicz, K. The effect of a belt position in the spindle driving system on critical rotational speeds. *Ann. WULS-SGGW For. Wood Technol.* **2016**, *95*, 303–307.

47. Orlowski, K.A.; Blacharski, W.; Pikała, T. Empirical assessment of the main driving system of the circular sawing machine. *Trieskove Beztrieskove Obrabanie Dreva* **2016**, *10*, 145–151.

48. *ISO-7960: Airborne Noise Emitted by Machine Tools—Operating Conditions for Woodworking Machines*; ISO: Geneva, Switzerland, 1995.

Publisher's Note: MDPI stays neutral with regard to jurisdictional claims in published maps and institutional affiliations.

MDPI

St. Alban-Anlage 66

4052 Basel

Switzerland

Tel. +41 61 683 77 34

Fax +41 61 302 89 18

www.mdpi.com

Applied Sciences Editorial Office

E-mail: applsci@mdpi.com

www.mdpi.com/journal/applsci

9 783036 517728